Complex Electromagnetic Problems and Numerical Simulation Approaches

IEEE Press
445 Hoes Lane
Piscataway, NJ 08854

Complex Electromagnetic Problems and Numerical Simulation Approaches

Levent Sevgi
Electronics and Communication Engineering Department
Dogus University
Istanbul, Turkey

IEEE Antennas & Propagation Society, *Sponsor*

IEEE
IEEE PRESS

WILEY-INTERSCIENCE
A JOHN WILEY & SONS, INC., PUBLICATION

Published simultaneously in Canada.

For general information on our other products and services please contact our Customer Care Department within the U.S. at 877-762-2974, outside the U.S. at 317-572-3993 or fax 317-572-4002.

Wiley also publishes its books in a variety of electronic formats. Some content that appears in print, however, may not be available in electronic format.

Library of Congress Cataloging-in-Publication Data is available.

ISBN 978-0-471-43062-9

Printed in the United States of America.

Let's

do

it !

To my wife Esin Sevgi
and
my son Onurcan Sevgi ......

CONTENTS

PART I
Numerical Simulation Approaches

PART II
Applications

Chapter 1: Ground-wave propagation
(with Funda Akleman)

Chapter 2: Antenna Design and Array Processing

Chapter 3: RCS Prediction and Reduction

Chapter 4: Microwave Network Design

PREFACE

Electromagnetic waves play an increasingly important role in communication, remote sensing, multi-sensor integrated systems, detection and identification (signal processing), microwave hardware design, compatibility, bio-electromagnetics, and many other applications in extremely complex natural and/or man-made environments. Although extremely valuable as reference and physical insights, analytical methods are mostly incapable of handling these challenging real-world problems. Luckily, most of these problems can be attacked by various numerical methods, which become extremely powerful with today's computer technology and intelligent parallel processing approaches. No single method, either analytic or numerical, is best suited for handling all possible cases in these problems; instead, a combination of methods is required to attain greatest flexibility and efficiency. Therefore, the trend is definitely toward "hybridization."

Electromagnetic problems are represented by Maxwell's equations. Mathematically, Maxwell's equations and problem-dependent boundary conditions establish the solution. Postulating the problem, writing it down in terms of integro-differential equations, and showing uniqueness and existence of the solutions may be enough for mathematicians. However, for engineers, observable-based parameterization as well as numerically computable (with rapid convergent properties if in series) forms are essential. Numerical modeling of observable-based numerically computable parameterized complex electromagnetic problems is covered in this book.

The phrases "numerical electromagnetics," "electromagnetic simulation," "computational electromagnetics," etc., have become popular in the last few decades. I have been experiencing difficulty in the conceptualization of *computational electromagnetics* not only for the candidates of internet-oriented, computer-weaned young researchers, but also for experienced engineers and researchers in the electromagnetic community. Some think that computers do everything and they usually neglect to look for the physical explanation behind the numbers. Others believe that computing numbers and drawing a graph from a simple equation with a computer is the substance of computational electromagnetics. Establishing an intelligent balance between these two extremes requires a new perspective in engineering education, which is also one of the aims of this book.

Doing numerical simulation in electromagnetics is as easy (as well as difficult) as doing measurements. It is easy, because you can obtain commercial codes that do almost everything for you, like you can supply computer-controlled devices for measurements. The simulation packages include almost everything, all are user-friendly, all have self-checking routines for control, and all are calibrated, like most of high-tech measurement devices. On the other hand, we are in real trouble if we do not know what to do with the results (the numbers we obtain), or if we have no idea what to expect from the simulation or measurement. Unlike common opinion, we have to understand the physics of the problem, since we may frequently be puzzled even with the simplest mathematical relations.

This book is intended mainly for engineers, software developers, academia, and students. It is an attempt to present challenging electromagnetic problems and numerical simulation approaches. Although I have been working with a broad range of numerical simulation techniques both in time and frequency domains, they form only a subset of all numerical methods. It is intended that the book will be of use to a wide range of readers: to the professional electromagnetic engineers, who are already familiar with many of the problems but require a consolidated overview of the subject; to the students, both graduate and undergraduate levels, as a supplementary text as well as educational and project developing tools; to academia, who wish to use powerful simulators as either educational or research purposes.

A few clarifying notes may be of value, particularly with regard to what this book does not include. First, it does not say that the methods presented here are the only or the most efficient ones. They are solely the methods that the author has been using and has had expertise on. No single method can be the "only one" in electromagnetics. Moreover, efficiency depends not only on the problem at hand, but also on the structure and parameter regimes. Second, the book does not include mathematical details of the problems chosen and assumes that the reader has already some kind of electromagnetic background (the IEEE Electromagnetic Wave Theory Series is an excellent collection of classical books). With this assumption, the majority of the theoretical content is reserved for physical aspects behind electromagnetic problems mentioned here. Finally, the book does not include details of the computation, computer languages (such as Fortran, Matlab, etc), and signal processing techniques (such as discrete and fast Fourier transformations, windowing, zero padding, interpolation, etc.). Therefore, the reader is strongly advised to go over these matters, if he or she is interested in using the simulation packages included, making additions/changes on the source codes supplied, or developing better codes. For example, the reader will be more comfortable if he or she knows the details of the differences between theoretical (mathematical) Fourier transform and discrete (numerical) Fourier transform (DFT) and fast Fourier transform (FFT) (e.g., periodicity, aliasing and spectral leakage effects, etc.).

The Internet broadens our environment and brings people closer. This is an excellent source for everybody as long as we know where to look and how to search. I have many inquiries from all around the world, from India to Brazil, Germany to United States, regarding to assistance on computational electromagnetics. This may be a graduate student who has a term project on RCS modeling, or a young university professor who newly overtakes an undergraduate or graduate course, or a practicing engineer who is hired to design a new antenna system, etc. I try to respond to all of these inquiries step by step, starting from sharing my computer codes (if available) to the end (if possible) until I see them successfully get along with the work by their own. It is astonishing to see how they pass through major steps: self-confident at the beginning, puzzled a lot while making progress, and finally getting experienced at the end. It is these steps that make them good in computational electromagnetics:

- First, they are very excited to have a source code of a powerful simulator, with which they can do "real" engineering job (MODEL AVAILABILITY).
- Then, they soon realize that having a powerful code does not solve their problem. They, themselves, should understand the physics, mathematics, the model, approximations (if any), validity ranges, and parameter optimization (MODEL VALIDTY).
- Finally, they feel strange and confused when they satisfy all pre-requisite, run the simulator and get the results. They all ask similar questions, like:
    - Now, what?
    - What am I going to do with these data?
    - Are they meaningful?
    - What do the numbers mean to me?
    - What should I have expected before the simulation?
    - How do I check my results? (JUDGING and VALIDATION)

It is these relations, interactions, steps, and questions that inspired me a lot in writing this book.

The book is organized into two parts. Introduction, concepts, and approaches and numerical simulation techniques used in the examples are given in Part I. Part II is application-oriented. Characteristic applications, which occupy by far the largest part, are collected in six chapters in Part II. The goal was to make the chapters relatively self-contained, augmented by user-friendly powerful software, enabling the reader to find specific problem solutions in the relevant chapters with minimal interference to other chapters. Each chapter contains sample simulators. The simulators are aimed to include as much different scenarios with maximum flexibility as possible. Although restricted and simulations may be performed with a certain range of parameters, the reader may develop and/or modify the source codes (available on the ftp site at the following address: ftp://ftp.wiley.com/public/sci_tech_med/complex_electromagnetic) to simulate almost any kind of electromagnetic problems.

Chapter 1 describes ground-wave propagation. The problem is to find out propagation characteristics and path loss between two points on or above the lossy surface of the spherical earth including rough surface and non-flat terrain, above which exists radial as well as longitudinal refractive index variations. Observable based, analytical, full-wave (3D) solution has not appeared yet. Instead, analytical approximate reduced formulations have been in use for a couple of decades. Moreover, time- and frequency-domain numerical models have been introduced in the last decade or two and have been applied to 2D problems successfully. This chapter first states the problem in general and outlines spectral solutions in terms of rays and modes. Then, a frequency-domain effective and easy-to-use technique, the split-step parabolic equation (SSPE), is discussed in detail. Finally, the new time domain wave propagator (TDWP) is presented. Sample propagators, namely, GRWAVE, HFMIX, KNIFE, SSPE, and TDWP, are given with characteristic scenarios and canonical examples. The propagators include a variety of test cases, in which the reader may exercise complex electromagnetic interactions and may do simulations to better understand ground-wave propagation phenomena.

Chapter 2 is an exposition of antenna systems. After giving explanations for some of the fundamental antenna parameters that are frequently used, wire antennas and arrays are discussed together with electronic beam steering concepts. MH, HF, and VHF monopoles on and above the ground are analyzed with NEC2 code (with MoM technique). Horizontal and vertical radiation patterns are simulated. Various ground screens are also designed, and their effects in improving antenna gain and array pattern is also demonstrated. Then, an FDTD-based antenna simulator FDANT is presented, where time-domain antenna and array analyses are possible. Canonical comparisons between MoM and FDANT are also included. The simulator FDANT includes linear and planar arrays of wire antennas in free space. The reader may enjoy doing simulations with small-size (maximum two by five) vertical linear arrays with beam steering-capabilities and may obtain radiation patterns at vertical and/or horizontal planes. In order to do that, one only needs to apply off-line discrete DFT by using the output files of FDANT. The routine ANDFT is also enclosed for this purpose.

Chapter 3 deals with radar cross-section (RCS) modeling. First, RCS fundamentals are outlined and RCS reduction techniques are discussed briefly. Then, MoM-, FDTD-, and TLM-based RCS simulators are given as sample simulators, with which the reader may do RCS calculations of complex targets in free space. Two types of sample simulators are designed in a way to handle both mono-static and bi-static RCS problems. Different scenarios are also included to let the reader to do his or her own simulations both in time and frequency domains. The simulators F-SNRCS and T-SNRCS (FDTD-, and TLM-based simulators, respectively) simulate single direction mono-, and bi-static RCS problems. For a chosen target, angle of illumination, and polarization, scattered near fields are simulated inside 3D FDTD or TLM volumes and time-domain far fields are extrapolated along the chosen direction. RCS vs. frequency variations may be obtained off-line via DFT by using the routine SNDFT, which is prepared for this purpose. The simulator F-BIRCS performs the same simulations except for a number of observation directions on a chosen scattering (horizontal or vertical) plane. The DFT routine prepared for this purpose is BIDFT. The reader runs BIRCS, obtains time-domain results and then produces RCS patterns at different planes at different frequencies by applying BIDFT separately at each frequency. The parameters are fixed to yield RCS patterns for maximum $l \approx 6 - 8\ \lambda_{min}$, where l is the longest dimension of the target and λ_{min} is the minimum wavelength.

Chapter 4 is reserved for microstrip network design with FDTD and TLM techniques. After giving fundamental explanations of microstrip lines and networks, FDTD-based sample simulator FDSTRIP is taken into account, where pre-prepared different microstrip networks can be simulated. The reader may choose one of the pre-prepared microstrip networks and calculate short pulse (broadband) *S* parameters. The thickness and relative permittivity of the substrate are fixed to be

h = 1.0 mm and ε_r = 2.2, respectively. The user-selective parameters are W/h ratio, a and b, where a and b correspond to length and/or width, depending on the choice of the scenario. FDSTRIP runs the FDTD simulation twice: first for a simple stripline, then for the network chosen. S_{11} and S_{21} vs. frequency man be obtained off-line via DFT routine MSDFT (supplied for this purpose). The frequency band is fixed from DC to 20 GHz in these simulations. Again, source codes are also available for the reader who wants to investigate his or her own microstrip network. Numerical examples are chosen to include microstrip line impedance matcher, filter, patch antenna, etc.

Chapter 5 discusses electromagnetic compatibility (EMC) and bio-electromagnetics (BEM) modeling. EMC and BEM have become two major areas in electromagnetic engineering. In this chapter, two canonical problems, which are characteristic problems of EMC and BEM, are discussed: the shielding effectiveness (SE) modeling and the specific absorption rate (SAR) calculations. The FDSE and TLMSE packages, which are FDTD-, and TLM-based SE simulators, respectively, are presented. Both of them simulate aperture leakage of a perfectly electrical conductor (PEC) enclosure, which may be assumed, for example, a personal computer (PC) case or a PEC screen of an electronic device, etc. The enclosure size, computation volume, cell sizes, and excitation pulse bandwidth are fixed. The reader may locate different apertures on different planes of the enclosure, and SE simulations may be performed. Electromagnetic leakage is a complex problem and one cannot predict what will happen by looking at the structure, aperture size, polarization, etc. Moreover, high (low)-frequency behaviors can also not be predicted by just calculating or measuring low (high)-frequency variations. SE must be obtained at each case (aperture location, size, and frequency) separately. I think the reader will enjoy to see how electromagnetic leakage varies with frequency for variety of aperture and observation locations. For SAR calculations, FDSAR and TLMSAR packages are presented. Multi-rectangular blocks with different relative permittivity and conductivity are used to eliminate discretization errors, so that FDTD and TLM comparisons may show the difference between their behaviors against numerical dispersion. The reader will also find results and comparisons done via these simulators.

Finally, Chapter 6 presents radar simulation. Radar simulation is a complex problem that includes all aspects of electromagnetics; signal generation and propagation, RCS, antenna effects, noise and clutter elimination, detection, tracking, etc. The answer regarding the need for a radar simulation depends on the mission: For example, it may be to understand radar signal environment; to optimize radar parameters, such as transmitter power, receiver gain, path loss, radar cross section (RCS), noise and clutter, etc.; to find out under what conditions the radar equation is satisfied for a given set of parameters; to evaluate algorithms, such as detection, tracking, sensor fusion, etc.; to use the radar simulator in a multi-sensor surveillance system simulation, to play different scenarios and try to find out the answers to "What if ..."-type questions; and, finally, to use it as a system design tool. In this chapter an emerging radar, high-frequency surface-wave radar (HFSWR) and integrated surveillance concepts are presented.

When the book was planned nearly a year ago, I thought that a collection of computer codes, effective and valid in a broad range of electromagnetic problems including (but not limited to) antennas and propagation, radars and radar cross section, microstrip network design and EMC / BEM, would serve EE students and engineers as educational as well as engineering design and research tools. I also thought that it would be better if the codes were tailored with user friendly graphical user interfaces (GUIs). Since then, I have had many discussions with my colleagues and students on the characteristics and the presentation styles of the codes. During these discussions I have changed my mind frequently about structures of the codes until I thought that I found the most appropriate ones.

The codes are organized as follows:

- Each simulator is prepared for a certain class of electromagnetic problems.
- The simulators are coded in Fortran and compiled with Linux G77.
- Canonical scenarios are prepared for each code, where the reader may test different cases to gain physical insight.
- Some of the parameters are specified and some are left to the reader. The aim here is to present to the reader a simulator that is user-friendly, with as many different scenarios as possible.
- The simulators are included as stand-alone executable (i.e., *.EXE) packages, and Fortran source codes are also included. The philosophy here is to let the reader first use the code to understand and practice and, subsequently, make modifications if he or she is interested in doing so and has a desire to go further.
- The sample simulators have almost identical structures. The reader supplies requested input parameters from an input file (e.g., SSPE.INP or TDWP.INP, etc.). The simulator runs with these parameters and yields output parameters as output files. If, for example, path loss vs. range is simulated for ground-wave propagation, then the output file RANGE.DAT will have two columns; the first column is the range (in meters or kilometers) and the second column is the path loss in (dB). Number of rows depend on the user supplied first range, last range and range increment values. Therefore, the reader may have path loss vs. range graphics with any graphics tool he or she has. For time-domain simulators, frequency-domain results are obtained off-line via discrete Fourier transform (DFT) by using output files of the simulator as input file of the DFT tool.
- Not all sample simulators are prepared in a way to warn or restrict the reader from entering some invalid parameters. For example, the 2D propagation simulator SSPE can handle the effects of an obstacle along the propagation path between transmitter and receiver. The heights and base widths of these obstacles are supplied by the reader. Valid parameter regions are given for each simulator. Although any value may be supplied, SSPE is valid within paraxial region (i.e., for small longitudinal variations), which means that results for sharp and high obstacles are invalid. The reader may observe this invalidity from the output of the SSPE simulator. The contour plots may be used to observe smooth EM wave interaction, cylindrical wave spread, edge and tip diffraction, etc.
- This structure is effective and suitable for GUI preparation. Although supplied one or two samples, preparation of GUIs are left to the interested reader.

Most of the materials used in this book have come from or have been derived from studies carried out under my direction at the Electrical-Electronics Engineering Faculty of Istanbul Technical University and Center for Defense Studies of Istanbul Technical University Foundation. Propagation tools based on ray-mode and parabolic equation approaches were initially prepared by myself. SSPE was written by myself and augmented by Dr. Funda Akleman so that it deals with the impedance boundary condition or varying terrain heights, and TDWP was prepared by Dr. Funda Akleman, during her Ph.D. studies. FDTD-based codes are prepared by myself and my students. Many people have made contributions to them. We follow similar structures in our FDTD-based codes with Dr. K. S. Kuns and Dr. R. J. Luebbers (as they present in their book *The Finite-Difference Time-Domain Method for Electromagnetics*, CRC Press, Boca Raton, FL,1993). The near to far-field transformation algorithm YAKUZ was prepared by myself during undergraduate studies of my student Erdem Başeğmez. All TLM-based codes were developed by Dr. M. Orhan Özyalçın while he was working on his Ph.D. thesis, and they were augmented by him and Dr. Funda Akleman. I express my gratitude to those of my students who had worked in preparation of these computer codes during their undergraduate and graduate theses. Especially, I would like to mention Dr. Funda Akleman and Dr. M. Orhan Özyalçın, who are also co-authors of Chapters 1 and 5, respectively, for their valuable contributions.

Last, but by no means least, I would like to thank four people, who have had great impact on my career. I have the honor of knowing and working with these people. First, many thanks to Ercan Topuz, my Ph.D. advisor, colleague, and friend, from the Electrical-Electronics Engineering Faculty of Istanbul Technical University, from whom I got most of the *inspiration* in my university life. Then, thanks to Leopold Felsen (Leo), my Ph.D. co-advisor, colleague, and friend, from Aerospace and Mechanical Engineering and Electrical and Computer Engineering at Boston University, from whom I learned how to ask the *right* questions. Third, thanks to Nejat İnce, the director of Center for Defense Studies of Istanbul Technical University Foundation (who also was the president of the NATO SHAPE Technical Center in Netherlands during 1961 / 1978), from whom I learned how to deal with *real-life engineering* problems. Finally, thanks to W. Ross Stone, editor-in-chief of the *IEEE Antennas and Propagation Magazine*, my colleague and friend, who has encouraged and guided me a lot for my recent publications.

Levent Sevgi
İstanbul, May 2003

Complex Electromagnetic Problems and Numerical Simulation Approaches

I

Numerical Simulation Approaches

I.1 INTRODUCTION

Many applications in science and technology rely increasingly on electromagnetic (EM) field computations in either man-made or natural environments. Also, today's electronic systems urge system designers to accommodate complex structures close to each other and operate together. Both military and commercial applications require more sensitive electronic equipment be less susceptible. Multi-function, multi-element complex systems have been replacing individual equipment. Moreover, almost all of these system are based on digital technology and supported by intelligent software. Because of these, electromagnetic (EM) problems have become extremely complex and cannot be handled via analytical approaches. Major electrical engineering areas of complex EM problems where numerical simulation approaches are used are listed in Fig. I.1. They include, but are not limited to, propagation, antenna and RCS engineering, biomedical engineering, EMC / EMI analysis, microwave networks, microwave tomography, sensors, and integrated systems.

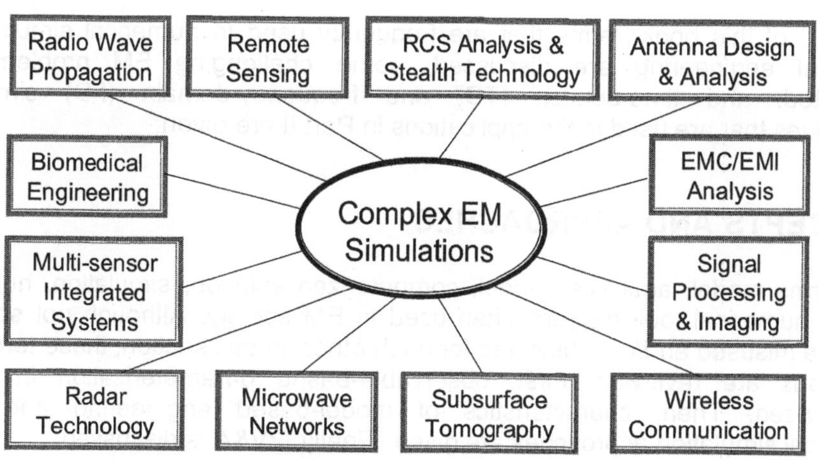

Figure I.1: *Major complex EM problems, in which numerical simulators (based on Maxwell's equations) are used*

On the other hand, today's computer technology makes it possible to simulate more complex real-world systems and processes. Models have become more detailed and reflect more aspects of the reality. Extremely complex systems, such as nationwide telephone networks, nationwide cellular system propagation coverage, multi-sensor integrated surveillance systems, etc. are being simulated as stochastic processes. Parallel to these features, verification, validation and accreditation (VV&A) have become a more and more important issue.

The need to solve realistic problems of arbitrary geometry in a reliable way with high accuracy is one of the major driving factors in developing numerical simulators. Numerical models might be viewed differently from developers' and users' perspectives [1]. Developers deal primarily with the conceptual suitability and implementational steps of the codes (verification), while the users are more involved in computation and application. Both are concerned with validation, although users can always be counted on to apply codes in a way never envisioned by the developers, making validation an especially sensitive topic. From a developer's perspective, the process of developing a numerical model involves conceptualization, formulation, numerical implementation, computation, and validation. On the other hand, user's perspective involves the problem, choosing the most suitable approach and the application steps. A developer needs to know how the code works, but the user needs to know what it can do.

Characterization and comparison of the numerical simulators basically relies on accuracy and reliability, efficiency and productivity, and, finally, applicability. One who deals with numerical simulation is in a position quite similar to that of an experimenter. They both need to understand requirements for their particular problems in terms of basic EM phenomena, but both also need to depend on complex tools that they did not design in order to accomplish their own particular goals.

In Part I of this book, terms that are frequently used in numerical simulation in electrical engineering are discussed, some challenging EM problems are presented, and time-domain (TD) and frequency-domain (FD) simulation techniques that are used in the applications in Part II are given.

I.2 CONCEPTS AND APPROACHES

The terms model, analytical model, computerized solution, simulation, numerical model, numerical tool, etc. are often used in EM society. Although not so often, they are misused and/or substituted for each other. In this section, these terms and definitions are reviewed. First, observable-based parameterization in EM is summarized. Then, characteristics of model-based engineering and direct numerical simulation approaches are given. Finally, VV&A is discussed.

I.2.1 Observable-Based Parameterization

Parameterization in EM should be based on discrete observables, which are either easy to compute or simple to experiment. Some examples of useful observable wave objects in the TD and the FD are listed in Fig. I.2.

EM field equations describe implicit space-time-dependent response of a linear field to a known excitation and pose a problem of determining solutions subject to initial and boundary conditions [2 - 6]. Ray and/or mode approaches are the approximate solutions of the second-order de-coupled wave equation derived from Maxwell's equations. While rays are the local plane waves of the Green's function formulation (i.e., response to point source excitation), modes represent environmental characteristics (i.e., source-free solution). Either in the TD or in the FD, solutions based on beams are also widely used approximations, especially in guided wave propagation (such as optics). Rays, modes, or beams are good approximations in terms of observable-based parameterization, but are valid only in restricted regions and/or under a certain set of parameters (such as frequency, conductivity, permeability, etc.). Since a generally applicable solution has not appeared yet, hybridization extends the range of validity and parameter sets.

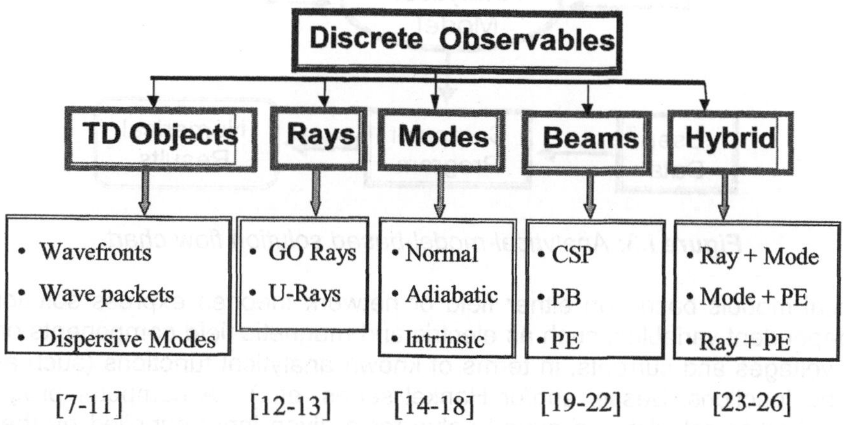

Figure I.2: *Discrete wave observables in EM (TD: time domain, GO: geometric optics, U: uniformized, CSP: complex source point, PB: pulsed beams, PE: parabolic equation)*

I.2.2 A Model-Based Engineering Approach

The solution approaches in EM may be grouped into two: analytical modeling and numerical simulation [27]. A typical flow chart of a model-based engineering approach for analytical modeling is pictured in Fig. I.3. Here, the physical situation is transformed into a mathematical model by projecting or mapping the space-time problem into one or several abstract (sub) domains, where the solution procedure is simplified. For example, ray theory for a source-driven problem inside a resonator, or mode theory in a rectangular cross-section waveguide, etc., are analytical solutions in terms of discrete observables.

Field theory and network (circuit) theory are two major theories that are widely used in EM engineering. Field theory is based on Maxwell equations, and network theory can be obtained from Maxwell equations by applying an approximation valid when dimension of the geometry under investigation is much smaller than the wavelength. However, it should be noted that network theory was not developed by approximating Maxwell's equations, but rather was developed independently from experimentally observed laws. The connection between field and network theories is one of the fundamentals of EM. What is called analytical model in Fig. I.3 is the solution approach to Maxwell's equations in a certain environment (defined by its medium parameters) and under given boundary conditions (the geometry). Analytical model in this flow chart may also be the network theory (transmission line equations) with medium parameters (transmission line constants) under given boundary condition (transmission line termination).

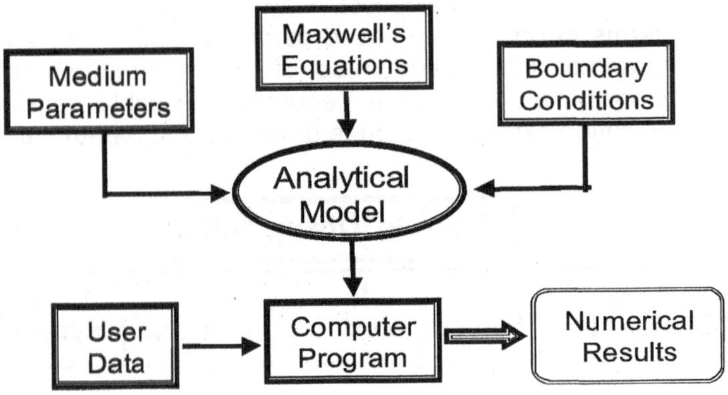

Figure I.3: *Analytical-model-based solution flow chart*

Analytical models based on either field or network theories express solutions for the independent variables, such as electric and magnetic field components or input output voltages and currents, in terms of known analytical functions (such as sine or cosine functions, Bessel and/or Hankel series, etc.). A computer program is only required to calculate an output value for a given input supplied by the user. Although this approach uses a computer program, it is solely used to present an output value for a given input; the solution is already there in the analytical model.

I.2.3 Direct Numerical Simulation Approaches

It is always desirable to obtain solutions in terms of known, easily computable mathematical functions in EM. Unfortunately, these kinds of solutions are only available for a limited class of canonical problems. Practical EM problems can hardly be handled in terms of analytical functions and numerical models are almost mandatory. The flow chart of a numerical model is given in Fig. I.4. The principal algorithm models the intrinsic behavior of EM fields without reference to specific boundary and material configurations. Some well-known and widely used numerical approaches are listed in the figure. For example, FDTD is based on discretization of Maxwell's equation directly in time domain. Similarly, TLM uses Huygen's principle and is based on the circuit theory. As shown, the generic numerical model is applied from the very beginning and augmented by boundary simulators and other peripheral units such as near-, to far-field transformation. Different problems (with respect to geometry and medium parameters) can easily be defined in these numerical models.

FDTD technique [28] depends on the solution of Maxwell's equations directly in time domain, where the physical geometry is divided into small (mostly rectangular or cubical, but non-orthogonal in general) cells. Both time and spatial partial derivatives are handled with finite central difference approximation and the solution is obtained with a marching scheme in iterative form. The characteristics of the medium are defined by three parameters, permittivity, conductivity, and permeability, and three electric and three magnetic field components are calculated at different locations of each cell. Beside the spatial differences in field components, there is also a half-time step difference between electric and magnetic field components, which is called leap-frog computation.

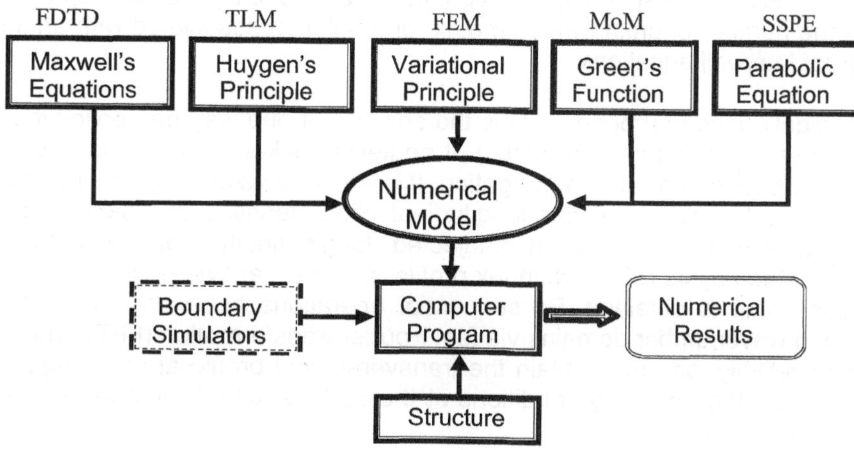

Figure I.4: *Numerical-model-based solution flow chart*

TLM is a numerical technique [29] for solving field problems using circuit equivalent. Unlike others, which are based on mathematical discretization, TLM is a physical discretization. It is based on discretization of fields (the continuous system) by an array of three-dimensional (3D) lumped elements. Thus, TLM technique involves two basic steps: (i) replacing the field problem by the equivalent network and deriving analogy between the field and network quantities, (ii) solving the equivalent network by iterative methods.

FEM analysis [30] of a problem includes (i) discretization of the solution region into finite number of sub-regions or elements, (ii) derivation of governing equations for a typical element, (iii) assembling of all elements in the solution region, and, (iv) solving the system of equations obtained. The step, which makes FEM different from others, is the requirement of derivation of the governing equations for the elements. This means that analytical representation should be obtained for the elements before proceeding the general solution. This is quite similar to the requirement to derive Green's function in MoM technique. One major difficulty encountered in FEM analysis is data preparation, which is basically element generation; therefore special mesh generators have been developed for this purpose.

MoM technique [31] is based on solving (discretizing) complex integral equations by reducing them to a system of linear equations. The equation solved by MoM technique is generally a form of the *electrical field integral equation* (EFIE) or the *magnetic field integral equation* (MFIE). Both of these equations can be derived from Maxwell's equations by considering the problem of a field scattered by a perfect conductor (or a lossless dielectric). The form of integral equation used determines which type of the problems a MoM technique is best suited for. For example, one form of EFIE may be particularly well-suited for modeling thin-wire structures, while another form is better suited for analyzing metal plates. Although it can be applied also in time domain, the majority of these equations are in frequency-domain. Recently, the Fast Multipole Method (FMM), based on MoM, and the Multilevel Fast Multipole Algorithm have been introduced [32], which reduce the number of unknowns in matrix form solution associated with an integral solution of Maxwell equations.

SSPE (widely used in propagation) is the solution of one way parabolic-type wave equation (see [33] and its references). It neglects back-scatter effects and is valid in regions close to near-axial propagation. It is an initial value problem, mostly used in 2D, where transverse and/or longitudinal characteristics can be included. An initial transverse field distribution is injected, longitudinally propagated through a medium defined by its refractive index profile, and the transverse field profile at the next range step is obtained. By sequential operations accessing the transverse spatial and wavenumber domains via fast Fourier transformation (FFT) and inverse FFT, respectively, one may obtain the transverse field profile at any range. SSPE cannot handle the boundary conditions at the surface, and it must be handled via special techniques.

The necessity for hybrid methods has already been recognized in the past [34,35]: for example, in scattering and antenna problems (see, e.g., [36 - 44]), techniques have been devised that combine the Method of Moments (MoM) and the geometrical theory of diffraction (GTD) or physical optics (PO) [45,46]. Similarly, numerical methods such as finite elements (FEM) or finite differences have been considered in conjunction with MOM [47,48], with integral equations [49,50], with boundary integrals [51], with modal techniques [52], with multipole methods [53], etc. Combinations of other methods, e.g. boundary-contour and mode-matching [54] or hybrid electric field integral equations (EFIE) and magnetic field integral equations (MFIE) denoted as HEM [55], have also been proposed. This list of contributions, though necessarily incomplete, indicates that this topic is of considerable interest. The methods enumerated above have typically been applied to solve a specific class of problems efficiently by matching the method to the perceived phenomenology, as in [39 - 41]. Despite these apparent diversities there are certain features common to all hybrid methods; namely, that the overall problem gets partitioned self-consistently, in a problem-matched manner, into interacting subdomain (SD) problems. An architecture, i.e. a structure that addresses complexity systematically and with reasonable generality, can be helpful here. An architecture does not solve a problem, but it can provide a systematic framework for posing proper questions.

Whether analytical or numerical, models need to be coded for calculations by computer. While the model in analytical solutions is constructed according to the geometry of the problem (i.e., boundary conditions and medium parameters), the numerical model is general and the geometry of the problem (together with the input parameters) is supplied after the model is built. That is, the boundary and/or initial conditions are supplied externally to the numerical model together with the medium parameters, operating frequency, signal bandwidth, etc. Once these are specified, simulation is run and a set of observable-based output parameters are computed for the given set of input parameters. As well-known and explained above, physical models are well-established in these numerical models. The algorithms in Fig. I.4 are all model proof; i.e., there is no need to test the validity of Maxwell equations for FDTD or circuit theory for TLM. However, one must establish whether the model is built "*correctly*" and "*under what conditions*" it is valid. This leads to testing for stability (since the solution is iterative) and for numerical dispersion (since the wave physics is traced in the discrete environment).

For very complex multi-function systems the verification strategy is generally not apparent a *priori*. Assume a typical scenario, which involves two microwave radars and one HF (high frequency) radar to cover a coastal region bounding an ocean with surface and air-borne targets. How should the simulator be built and what will the requirements be for model verification? There is no single procedure for constructing this simulator; it essentially depends on the problem requirements. For example, one may need to simulate a target signal as a detailed time sequence that accounts for target fluctuations, noise, clutter, and interference. This requires stochastic modeling of each target and its environment. If a target tracker is to be simulated, a target can only be represented as a single "detected" or "missed" decision. Therefore, verification, validation, and accreditation (VV&A) has become one of the major concerns in numerical simulation procedures, and they are problem-dependent.

I.2.4 Verification, Validation and Accreditation (VV&A)

Simulation, in general, is to pretend that one deals with a real thing while really working with an imitation. In science and engineering, *simulation* is the process of representing the dynamic behavior of one system by the behavior of another [56]. In engineering, numerical simulation refers to the use of computation to implement a *model* of some dynamic system or phenomenon. The purpose is to make experimental calculations or predict behavior, thus moving the laboratory into computer environment. A model is defined as a physical or mathematical abstraction of a real world process, device, or concept, which are all simply called "*the problem.*" A simulator is a program or a series of programs developed to implement and execute simulations with well-defined relations between model objects. These relations are, by definition, mathematical relations, numerical or not. Model objects and their properties in different states are normally specified by numerical variables related to real measurements. Experiments can then check measured states predicted from model relations.

Scientists and engineers deal with the complexity of the real world (the problem) in terms of simplified models. Simplification resides on the fact that to what *extend* and *detail* the model represents the real world. This is the issue of both existence of the system and observability of the input and output. Therefore, observable based parameterization (either measurable or computable) of the real world (the problem) is the initial step.

The fundamental building blocks of a simulation [57] are the real-world problem entity being simulated, a conceptual model representation of that entity, and the computer model implementation of the conceptual model (see Fig. I.5). The suitability of the conceptual model, verification of the software along with the data validity are the technical processes that must be addressed to show that a model is *credible*. Credibility resides on two important checks that must be done in every simulation; validation and verification. As summarized in [58], validation is the process of determining that *the right model is bu*ilt, whereas verification is designed to see if *the model is built right.*

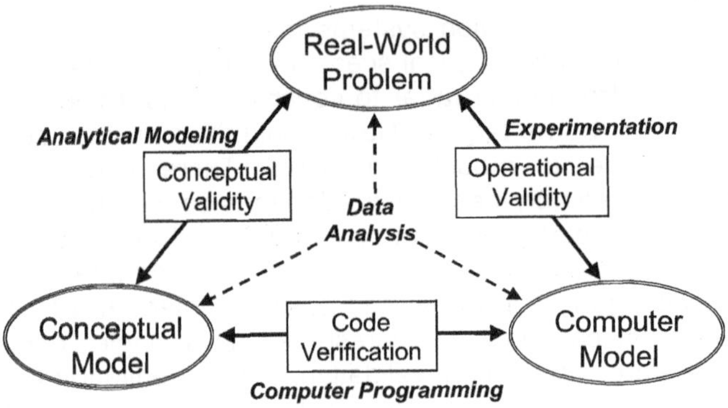

Figure 1.5: *Simulation concepts: Verification, validation, and accreditation*

Verification involves debugging and ensuring consistency between the computer code and the model. *Validation* involves comparing the model's output with the behavior of the real problem, and focuses on three questions [55]:

1. Does the model adequately represent the real-world system?
2. Are model-generated behavioral data consistent with characteristics of the actual system?
3. Does the user have confidence in the model's results?

These three questions determine conceptual validity, operational validity and reliability, respectively.

Credibility (or accreditation) of the simulation depends also on the characteristics of the real-world system (the problem). Three versions of the real world have to be distinguished [55]:

- The system exists and its outputs can be compared with the simulated results
- The system exists but is not available for direct experimentation (e.g., a plane crash, blasting a buried mine, firing a missile, etc.)
- The system is yet to be designed; validation here is in the sense of convincing ourselves and others that observed model behavior represents the proposed reference system if it is implemented

Calibration is the process of *parameter estimation* for a model. It is concerned with tuning of existing parameters and usually does not involve the introduction of new ones, which would change the model structure. Calibration is also referred to as parameter optimization involved in system identification or during experimental design.

Majority of engineering applications, which are very complex in nature, rely on measurement. Simulation is a substitute for real experimentation, and it is usually conducted in the absence of a complete set of *real data*. Where experimentation towards real data is expensive, dangerous, or sometimes impossible because a new system may not yet be available, simulation is almost a *must* and extremely cost-effective. The last but by no means the least, in listing advantages of the simulation is that it provides a prototype system with which questions of a "what if?" nature can be answered by running the simulator with different set of parameters in different scenarios.

I.3 CURRENT CHALLENGING EM PROBLEMS

As discussed above, simulation is required or preferred, since (i) it is too expensive and/or dangerous to do real tests and/or experiments with the existing system, and (ii) the system is not available or almost impossible to prepare a real scenario. For example, testing a missile without doing extensive simulations will sure be too expensive, flying a test aircraft instead of simulating may be too dangerous, or locating hundreds of surface and/or air targets in many different scenarios to test the performance of an integrated electronic surveillance system is almost impossible.

Analysis and design of any electronic system or group of systems that is going to operate

- In the ionosphere
- Above and under the ground
- On ocean surface

is a real challenging problem. Communication and radar systems that use the ionosphere as propagation medium, detection and tracking of objects embedded in land and ocean clutter, or subsurface imaging are among these challenging problems. Non-linear, time-varying, inhomogeneous nature of these environments can only be modeled as stochastic processes via powerful computers. For example, EM effects in the ionosphere varies with sun cycle, and one needs to make observations for at least a few decades in order to gain some physical insight and record data to understand its characteristics. Similarly, land, ocean, and underground clutters are extremely difficult to model. The more we understand these environments the better we model and eliminate their undesired effects.

In this section, some challenging EM problems, where numerical simulations are almost the only mean towards the solution, are outlined.

I.3.1 Integrated Surveillance and Control Systems

Electronic surveillance systems have been widely used for both military and commercial purposes, from intrusion alarms to integrated defense systems. In addition to surveillance, communication, control, command, and fire-control may also be required in some of the surveillance systems. The simulation of these systems requires us to understand sensor physics and to prepare intelligent detection and tracking algorithms. A typical integrated surveillance system is pictured in Fig.I.6, which is a tough, challenging EM simulation problem. Here, a surveillance system to continuously monitor surface activity on the Black Sea (with two surface wave HF radars) and air activity (with an onboard microwave radar) in the eastern part of Turkey is given as an example [59 - 63]. Assume that an integrated system is going to be designed and that reliable sensor simulators are available.

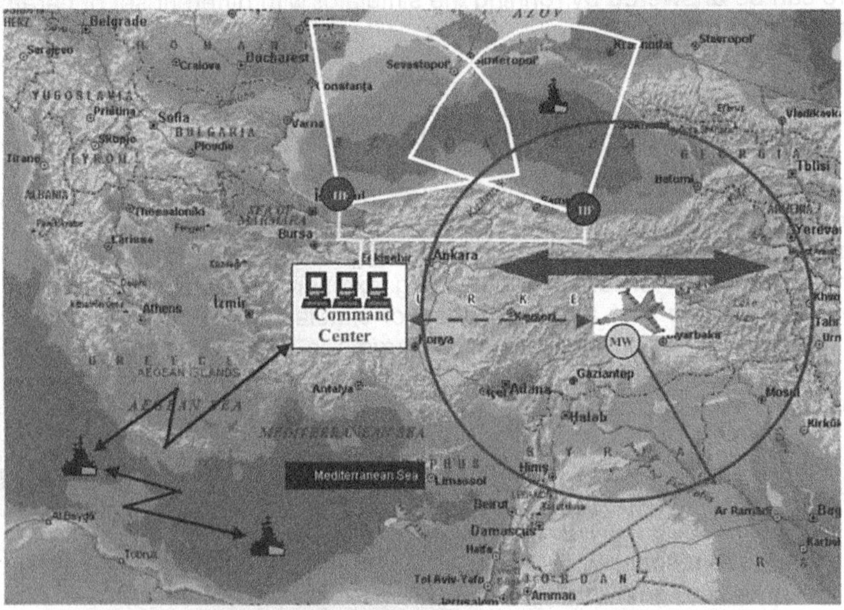

Figure I.6: *A typical integrated surveillance system for Turkey*

Some questions that still need to be answered may be listed as follows:

- Where should HF radars be located for optimum coverage?
- What should the site specifications be?
- At what altitude the aircraft will fly for optimum air coverage?
- What should the speed of the aircraft be for minimum revisit time?
- How many targets can continuously be monitored?
- What speed, capacity, etc., are required for communication links, software and hardware?
- What will the overall performance of the integrated system be for different scenarios?

It is obvious that simulation is the only tool to try to find out answers like the ones given above. Moreover, simulation is the only tool to exercise different scenarios and try to find out the answers of "*what if* ?"-type questions.

Another complex EM problem scenario is given in Fig. I.7, where a rugged mountain border-line surveillance scenario is pictured. Assume that a border line between two neighboring (in conflict) countries is to be controlled. The region has huge mountains and deep valleys. Then, the type of sensors (e.g., HF, VHF, microwave radars, optical, thermal cameras, etc.), their parameters and optimum locations can be determined via testing extensive amount of different scenarios and it can only be done via computer simulations [64,69].

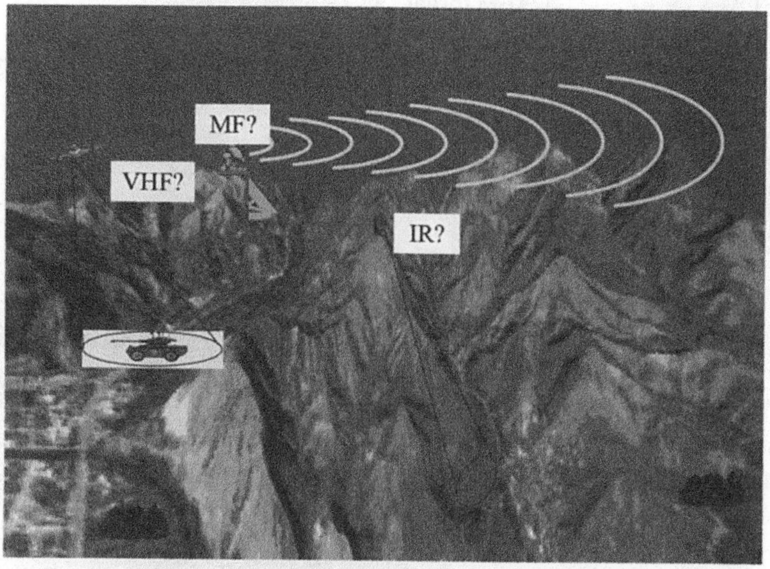

Figure I.7: *Typical border-line multi-sensor surveillance system (MW: microwave radar, VHF: very high frequency radar, IR: infrared sensor)*

The last example for electronic surveillance and control systems is given in Fig. I.8, where tactical air defense scenarios are shown. Today, ballistic (BM)- and cruise (CM)-type missiles that cost million $ US are almost the most dangerous threats for air defense systems, since

- they have very low RCS, which makes them almost invisible to defense radars,
- infrared detection of CM is difficult because they use turbofan engine, which emits little heat.
- CM can fly at very low altitudes (as low as 30 - 40 m), which embeds them into land clutter,
- they are strengthened with intelligent navigation tools as well as decision-making algorithms.

For example, the Tomahawk missile (which was used in the Gulf War, Bosnia, and Afghanistan) is a highly survivable weapon. It includes a Global Positioning System (GPS) receiver, an optical digital scene matching area correlator (DSMAC), a time of arrival control system, as well as inertial and terrain contour matching radar (TERCOM) guidance [70]. The TERCOM radar uses a stored map reference to compare with the actual terrain to determine the missile's position. If necessary, a course correction is then made to place the missile on course to the target. Terminal guidance in the target area is provided by the optical DSMAC system, which compares a stored image of target with the actual target image.

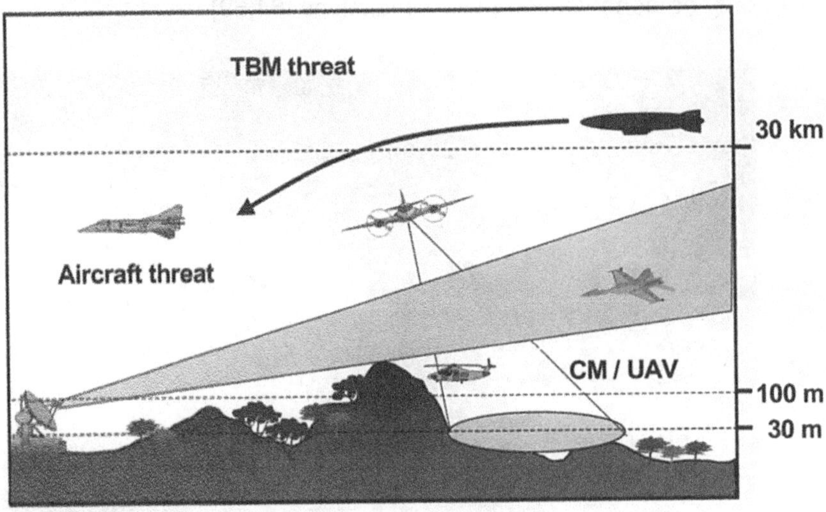

Figure I.8: *Air defense against ballistic- and cruise-type missiles*

Surveillance simulators for all of these purposes are among the top challenging complex EM problems. First of all, they require realistic sensor simulation. Sensor simulation needs to understand almost all aspects of EM, from radio wave propagation to target RCS, from clutter characteristics to noise. Moreover, these aspects are also quite different for HF and microwave radars as well as for day and night cameras (for example, while thermal noise is dominant at microwaves, environmental noise determines the detection threshold at HF band). Then, it requires sensor fusion, intelligent decision-making algorithms, and maybe expert systems [61 - 63].

I.3.2 Stealth Target Design

Another challenging area in numerical simulation is RCS prediction and stealth target design. As shown in Fig. I.9, low RCS targets may be designed by (i) playing with the geometry, (ii) using radar absorbing materials (RAM), and (iii) using active/passive cancellation techniques. The above techniques and/or their hybridized forms are used in RCS modeling and stealth target design. FDTD, TLM, FEM, and MOM (especially FMM) techniques are successfully applied in RCS modeling, especially for resonance RCS region, where the dimension of the target is on the order of radar wavelength. Resonance RCS regime is the region where the structure of the target totally contributes to RCS behavior. These techniques are especially efficient in this RCS regime, since they can handle any geometry.

For example, RCS analysis of air and surface targets can be done with these techniques for HF radars [71 - 73]. As the frequency increases, even with parallel processing techniques, the requirement to model the target in detail exceeds the capacities of the techniques mentioned above. At this point, hybridization of the techniques becomes essential. Hybridization may be done either by combining two numerical techniques or by combining one numerical technique with one of the high-frequency asymptotics, such as geometrical optics (GO), physical optics (PO), geometrical theory of diffraction (GTD), physical theory of diffraction (PTD), or uniform theory of diffraction (UTD). RCS problems of large bodies with small cracks, cavities or apertures are well-suited to these hybrid techniques.

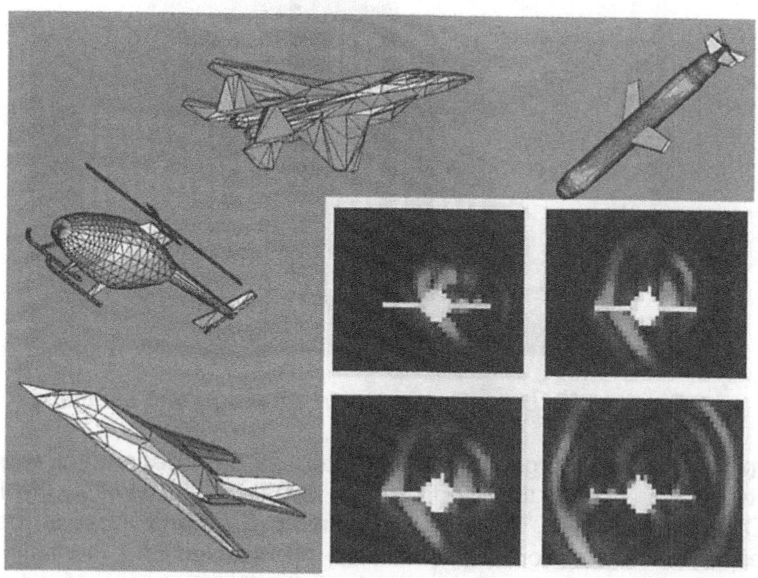

Figure I.9: *Typical low RCS targets and simulations and near field FDTD simulations*

I.3.3 Mobile Communication and Bioelectromagnetics

The widespread use of cellular phones is a recent phenomenon. Their use has escalated over the last decade, and a number of worldwide subscribers is projected to be more than 1.6 billion by the year 2005. The extensive use of mobile phones has been accompanied by (i) EM engineering requirements and (ii) public debate about possible adverse health effects [74,75].

Parallel to coverage requirements in both rural and urban areas studies have been carried out in developing reliable propagation simulators. Since it is a cellular system, the higher the number of subscribers, the higher the number of cells. This results in a challenging problem in area coverage: maximum performance with optimum cell planning and power consumption. A typical example is pictured in Fig. I.10. A propagation simulator should be capable of estimating power levels not only along the streets of a city, but also inside the buildings. Specular reflections, edge and tip diffractions, surface waves, etc., are complex wave phenomena that should be handled with these simulators.

Bioelectromagnetics (BEM) is the other issue related to mobile communication, where mobile phone-human head interaction attracts top interest. Although it covers many aspects (physical, chemical, biological, psychological, etc., effects) thermal effect is still the only parameter that can be measured and/or numerically simulated. Available safety threshold levels are determined via the parameter, specific absorption rate (SAR).

Figure I.10: *Urban base station coverage and simulation*

The non-governmental organization accepted as an independent expert body by the United Nations (UN), the World Health Organization (WHO), the International Labor Organization (ILO), the European Union (EU), and most of the developed countries is the International Commission on Non-Ionizing Radiation Protection (ICNIRP). The Commission is established for the purpose of advancing non-ionizing radiation (NIR) protection for the benefit of people and the environment and in particular to provide guidance and recommendation on protection from NIR exposure. In 1998, the ICNIRP published its latest guidelines for exposure limits, which were based on the evaluation of all available and known analytical/numerical/experimental study results [76] (it should be noted that safety rules are stated by characteristic words such as *shall* for mandatory rules, *should* for recommendations, and *may* for guidelines). In these guidelines as well as in others [76 - 78], in the frequency range of 10 MHz to a few GHz (including 900-MHz and 1800-MHz cellular phone frequencies), SAR is accepted as a main physical parameter. Available experimental evidence indicates that the human body cannot regulate itself under a permanent temperature increase of more than 1°C, which corresponds to a whole-body SAR value between 1 and 4 W/kg under 30 min of EM exposure for resting humans.

The most recognized EM exposure standards, including ICNIRP, adopt the SAR, averaged over the whole body (SAR_{WB}), as the basic parameter to establish the safety of an exposure. The value of 4 W/kg is accepted worldwide as the threshold for the induction of biological thermal effects, and introducing a safety factor of 50 for general public exposure gives rise to a basic limit on SAR_{WB} equal to 0.08 W/kg.

Sometimes, although SAR_{WB} becomes below the basic limit, the local SAR values can be rather high in a confined body region, where the power absorption takes place, such as a human head near an antenna of a mobile phone. For such conditions, limits on local SAR averaged over tissue masses of 1 or 10 g have been introduced in the standards, as 1.6 W/kg over 1 g or 2 W/kg over 10 g.

SAR can only be obtained by simulation, either by using human phantoms that have human electrical parameters, or by computer simulations over discrete human models derived from, for example, nuclear magnetic resonance (NMR) image. Most of the techniques mentioned above (e.g., FDTD and TLM) have almost become reference numerical simulation tools in simulating human head-cellular phone interaction (e.g., visit http://www.fdtd.org, or search internet engines, for example, for computational EM, BEM, human tissue-EM wave interaction, etc., for recent special events and meetings).

In EM engineering, human tissue-mobile phone interaction is twofold. BEM is concerned about the effects of EM energy absorbed by human tissues, but antenna engineers are also concerned about degradation in antenna radiation performance because of the existence of human head and hand in the near-field region of the antenna. Powerful FDTD and TLM simulators can be used as effective SAR simulators and antenna design tools in EM engineering. Two typical examples to illustrate these are given in Fig. I.11 and I.12 [79].

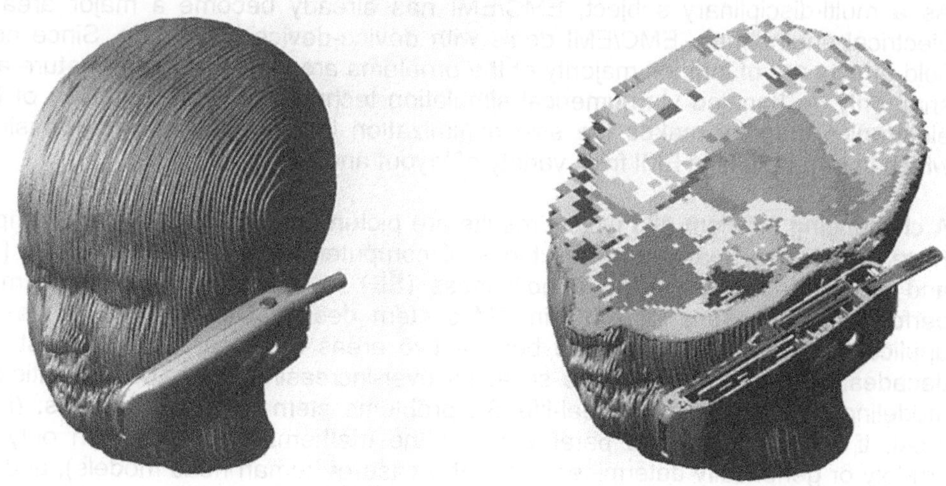

Figure I.11: *A fabricated human head phantom and simulated SAR variations*

In Fig. I.11, a fabricated human head phantom and simulated SAR variations are illustrated. In Fig. I.12, radiation patterns of two different planar inverted F antennas (PIFA) are simulated when human head and hand exist near the antenna.

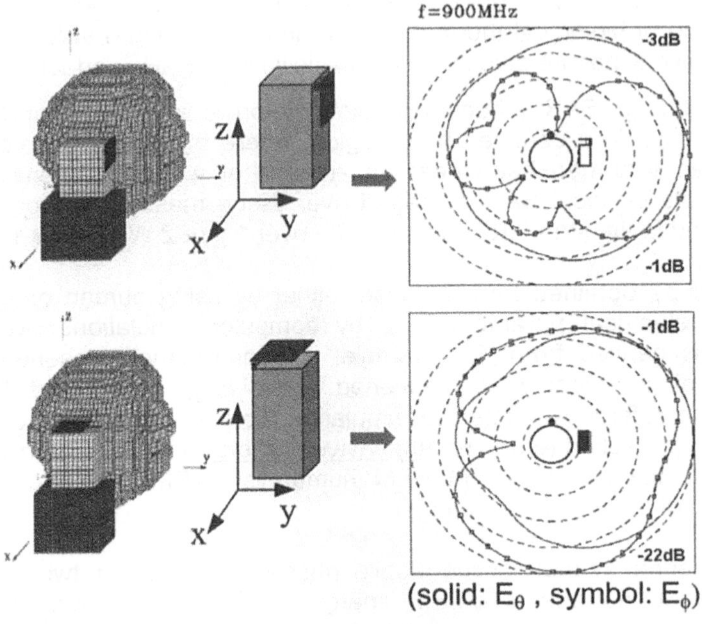

Figure I.12: *Radiation patterns of two PIFA type antennas when human head and hand are present*

I.3.4 EMC / EMI

As a multi-disciplinary subject, EMC/EMI has already become a major area in electrical engineering. EMC/EMI deals with device-device interaction. Since near field effects are of interest majority of the problems are very complex in nature and shall only be handled via numerical simulation techniques. The evaluation of the electronics industry makes the size minimization in emerging designs possible, which in turn results in EMI from variety of layout and design features.

A challenging problem and typical results are pictured in Fig. I.13, as an example. Here, EM energy leakage of two types of computers cases were measured [79] and are plotted as shielding effectiveness (SE) vs. frequency. SE is the main performance measure for EMC in EM system design. Due to wide ranges of application SE calculations have been active areas of research in the past two decades, and it continues to be so at an ever-increasing rate. The difficulties in modeling and simulation of real-life SE problems stem from two sources: (i) In most, if not all cases, the parameters of the mathematical model can only be crudely or generically determined (as in the case of human head models), and (ii) in general accurate solutions cannot be obtained for the resulting mathematical model even with the most advanced computational facilities available today. These difficulties preclude the generation of analytical or measured reference solutions for the class of problems referred to as "complex EM problems." Moreover, it has not been possible to assess the validity of these FDTD results due to lack of reference and/or independently generated alternative solutions. The main objective of this investigation is to fill in the above mentioned void.

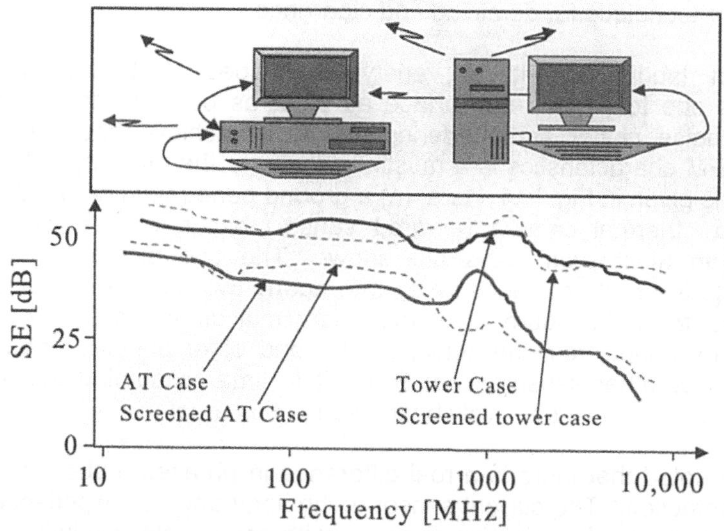

Figure I.13: *Typical EMC / EMI problem environment*

As observed in Fig. I.13, AT and tower case computer EM leakage are quite different. The cases are screened (screws are seemed, doors are hardened, etc.) and measurements are repeated. Results for these cases are also plotted in Fig. I.13 as dashed lines. It is obvious from the figure that there is no general rule with which one can predict EM leakage at lower and/or higher frequencies. The only way is to either do measurements or perform simulations at every single frequency.

I.3.5 Subsurface imaging

EM waves are widely used in subsurface imaging. The applications include, but are not limited to, tumor detection; landmines and unexploded ordnance; archeological discoveries; crack detection inside metals, roads, and bridges; search and rescue in earthquakes; and in avalanche snow; etc. EM waves from HF to IR have the ability to penetrate into the materials. The degree of the penetration depends on electrical parameters (conductivity and permittivity) as well as the frequency. For perpendicular incidence, the higher the conductor and frequency the lower the penetration depth. For example, EM waves can penetrate into average ground (ε_r = 10 -15, σ = 0.01 - 0.05 S/m) tens of meters at 5 MHz, but only a few centimeters at microwaves (e.g., at 1 GHz). Therefore, different sensors have been in use for subsurface imaging.

The crucial application of subsurface imaging is landmine detection. Landmines and unexploded ordnance are among the top worldwide public concern. They are significant barrier to economic and social development in many parts of the world, seriously affecting countries such as Vietnam, Cambodia, Laos, and Afghanistan in this region. The UN lists nearly 70 countries affected by mines, where more than

100,000,000 mines are estimated to be in the ground. Experts say that demining and clearance may take approximately 500 years with today's technologies. Because of all these, worldwide efforts have been put on developing a better device and/or technique for demining and clearance.

Detection of landmines with a variety of shapes, materials, weight, etc., necessitates one to (i) map subsurface as clear as possible and (ii) distinguish targets with noise, clutter, and interfering objects. To achieve these, understanding subsurface EM characteristics is a must. To illustrate the difficulty of the problem, an example is given in Fig. I.14. Here, (a) a ground penetrating radar (GPR) image, and (b) two thermal camera (Infrared sensor) pictures, with fictitious buried objects, taken at different days are shown. The two pictures on the right correspond to the real IR images taken with nearly two months difference [80,81]. As observed from the figure, it is extremely difficult to detect as well as to distinguish the mines from the noise, clutter, and other objects. The situation is quite similar for other sensors, too. Since it is time-consuming and expensive, simulation studies have extensively been used in this area.

It should be noted that there is a real difference in phrases "subsurface imaging" and "mine detection." The current sensor technology and our experience could be enough for most of subsurface applications with acceptable resolution. There are good amount of effective hardware and software for subsurface imaging. On the other hand, one needs to be careful when phrasing "mine detection." The problem is extremely complex, and reliable detection needs intelligent reconstruction, multi-sensor, and data fusion algorithms. Surface and subsurface clutter are major drawbacks in realization these algorithms. Some of the promising sensors are metal detector (inductive coil), GPR, microwave tomography, infrared camera, acoustic sensor, etc.

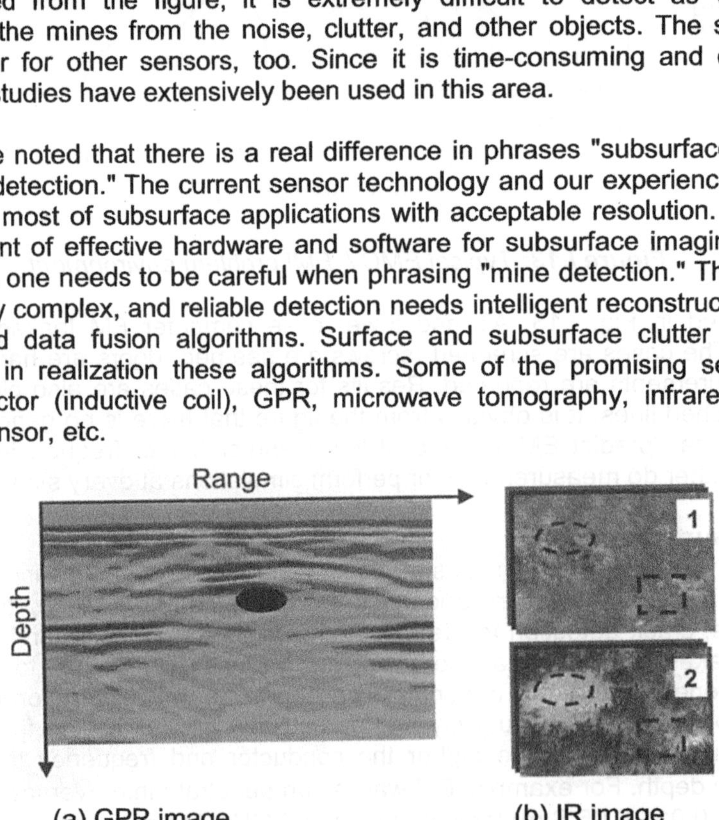

(a) GPR image (b) IR image

Figure I.14: *Typical thermal camera images of various buried mines at different times*

I.4 COMMERCIAL PACKAGES vs. SPECIAL APPROACHES

The aim of computational EM is to develop accurate, reliable, efficient, and user-friendly powerful codes, and it has a wide range of applications. Everybody is aware of non-existence of a generally applicable single code that can handle all kinds of applications. There are many powerful methods, both in time and frequency domains, with their own advantages and drawbacks. One needs to know and compare different methods in order to select optimum code for his/her needs.

For the developers who design codes, combination and generalization of different techniques are the most promising procedure toward better codes. To obtain useful combinations and generalizations (also called hybridization), a detailed analysis of original methods is required. A developer who introduces a new method should compare his/her new code with the existing ones to demonstrate the power of their idea. There are many benchmark problems proposed for reliable comparisons. Unfortunately, these benchmark problems are rarely chosen for the comparisons. Instead, comparisons over only a few specifically selected and tuned problems are used as if the task is to show only the advantages of the code. This is often useless or even misleading for those who want to improve the codes or for potential users. Moreover, most of the codes can only be tuned by the developers and/or experienced users in such a manner that a relatively high performance may be obtained. Because of these, a developer may obtain higher performance than an inexperienced user on the same benchmark problem.

A developer who wants to introduce a new code and/or improve existing ones needs reliable and fair comparisons and a detailed analysis not only of the advantages, but also of the drawbacks of codes. Actually, removing the drawbacks is the first step in improving and requires a correct analysis of disadvantages. When a developer does not correctly detect main problems of a code, all attempts to improve it will fail. These analyses of disadvantages and drawbacks usually show that any two methods may be complementary in the sense that one has the advantages where the other one has drawbacks and vice versa. In this case, hybridization seems to be promising. Although the idea is simple, the procedure for constructing a useful hybrid code is quite difficult. First of all, deep knowledge and careful analysis of the methods to be combined is required; that is, similarities, differences, and corresponding mathematical background and validity ranges should be well understood. This generally poses a problem, since many developers are experts on a single method. Furthermore, it is not guaranteed that the hybrid code removes the drawbacks of individual ones, improves the performances and extends the range of validity. There are different ways of hybridization. A developer may combine two different codes into a single one or the problem may be split into parts and each part is solved with a different code.

A developer's final desire is the generalization of his/her code. Generalization is a powerful engineering concept in physics, which aims to uniformize theories. Generalization in numerical analysis is somewhat different from generalization of theories in physics. Generalization of a code usually leads to relatively complex codes that have a variety of parameters. These parameters allow experienced users to tune the code for obtaining efficient solutions of a specific problem. Without this tuning, the code will be rather inefficient. On the other hand, codes which require (experienced) multi-parameter tuning by the user are not user-friendly. Hybrid codes require more parameters to be tuned as well as to decide which method will be used under what conditions. Obviously, they have more parameters to be tuned and they are not user-friendly either. A developer usually overcomes these problems of generalization by preparing intelligent graphical user interfaces (GUI). The generalized and/or hybridized code supported by GUI is that (i) it gets the user-supplied parameters and (ii) it decides which method shall be used for this set of parameters. An alternative way of designing a generalized code is to introduce automatic tuning. In this case, an external optimization procedure that optimizes the set of tuning parameters within the user-supplied set shall be used.

In order to achieve a maximum of user-friendliness, a developer leaves only a small set of tuning parameters that will be supplied by users. The user does not require any a priori knowledge to run such codes. When there are many tuning parameters, a priori knowledge or extended experience may be required to obtain accurate results or to reduce the computation time. Although most of the time inexperienced users dislike these kinds of codes, these codes are more flexible and powerful than the others.

A priori knowledge and numerical experience are essential to better profit a code. For example, the user can first test the code on a canonical problem, where he/she knows the answer. Similarly, a perturbed structure can be used in these tests. When the solution of a problem is known that is close to the problem at hand, the user can consider the current solution as a superposition of the known (unperturbed) problem and a series expansion that only models the differences between the two problems. Or, for example, when iterative solvers are applied, the number of iterations are drastically reduced when the user starts with a good initial guess of the parameter set. Such a guess may be obtained from a priori knowledge, i.e., from previous computations of similar problems. Some features of using commercial packages or developing one's own code are listed in Table I.1.

Table I.1: Designing a code vs. using a commercial code

EVALUATION PARAMETER	COMMERCIAL PACKAGE	USER CODE
User-friendly	BEST	POOR
Generalization	GOOD	BEST
Open to modifications	POOR	BEST
A priori knowledge	LESS REQUIRED	REQUIRED
Expertise	LESS REQUIRED	REQUIRED
Error analysis	POOR	BEST
Scientific comparisons	GOOD	BEST
Output evaluation	GOOD	BEST

I.5 TIME-DOMAIN (TD) SIMULATION TECHNIQUES

The majority of time domain simulators presented in Part II are based on the finite-difference time-domain (FDTD) and the transmission line matrix (TLM) methods. There are excellent books in the literature, which discuss the theories of the FDTD [82 - 84] and TLM [85 - 86] and their application to variety of EM problems. Therefore, the reader is advised to refer to those references for the details, capabilities, and limitations of these methods. In this section, FDTD and TLM are briefly summarized and are discussed in terms of programming aspects.

I.5.1 FDTD Method

FDTD is widely regarded as one of the most popular full-wave computational EM algorithms. With FDTD, it is possible to analyze EM problems directly in time domain for a broad range of frequencies, only in one simulation. This method was first introduced in 1966 by Yee [28], but because of huge memory size and high speed requirements, initially it had limited application. However, with the rapid development in computer technology during the last two decades the problem areas which can be addressed via FDTD have increased considerably.

The differential time domain Maxwell's equations in a linear, isotropic and non-dispersive medium are

$$\nabla \times \vec{E} = -\frac{\partial \vec{B}}{\partial t} \quad \text{and} \quad \nabla \times \vec{H} = \frac{\partial \vec{D}}{\partial t} + \vec{J} \tag{I.1}$$

$$\nabla . \vec{D} = \rho \quad \text{and} \quad \nabla . \vec{B} = 0 \tag{I.2}$$

where $\vec{D} = \varepsilon \vec{E}$ and $\vec{B} = \mu \vec{H}$. Here,

E [V/m]	= electric field	ρ [q/m^3	= vol. charge density
H [A/m]	= magnetic field	ε [F/m]	= permittivity
D [q/m^2]	= el. displacement vector	μ [H/m]	= permeability
B [Wb/m^2]	= mag. flux density	σ [S/m]	= conductivity.
J [A/m^2]	= the current density		

All the parameters are vector quantities, except ε, σ, ρ, and μ. This is all the information needed for linear isotropic materials to completely specify the field behavior over time so long as the initial field distribution is specified and satisfies the Maxwell's equations. Conveniently, the field and sources are set to zero at the initial time, often taken as time zero. The two divergence equations are in fact redundant as they are contained within the curl equations and the initial boundary conditions. Thus, the starting point for the FDTD formulations is the curl equations.

FDTD method depends on the solution of Maxwell's equations directly in time domain. Using the Taylor's expansion and discretizing partial derivatives directly in time and spatial domain yield the well-known FDTD iterative equation. In Yee formulation, the physical volume of interest is divided into cubic reference cells (Fig. I.15) where characteristics of the medium are defined by three parameters, permittivity, conductivity and permeability.

Three electric and three magnetic field components are calculated at different locations within the reference cell in such a way so as to minimize the computational effort after the discretization of two curl equations by using central-difference approach (or by taking up the second order terms in their Taylor's expansion). Besides, there is also a half time step difference between electric and magnetic field components, which is called as leap-frog computation (i.e., electric and magnetic fields are calculated at time instants $t = \Delta t$, $2\Delta t$, $3\Delta t$, ... and $t = \Delta t/2$, $3\Delta t/2$, $5\Delta t/2$, ..., respectively, where Δt is the time step size).

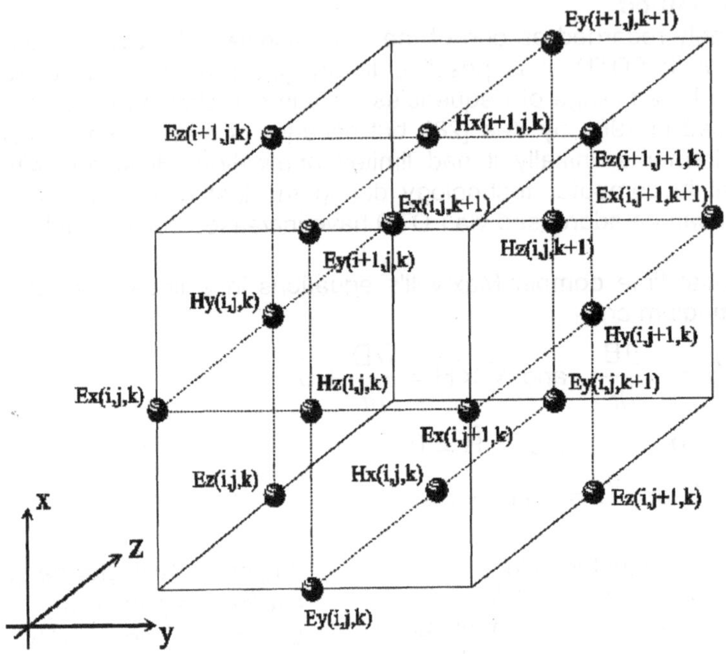

Figure I.15: *Classical YEE Cell*

For a lossy and source-free region, iterative FDTD equations are given as

$$H_x^{\tilde{n}}(i,j,k) = H_x^{\tilde{n}-1}(i,j,k) + \frac{\Delta t}{\mu_0 \Delta z}\left[E_y^n(i,j,k) - E_y^n(i,j,k-1)\right] - \frac{\Delta t}{\mu_0 \Delta z}\left[E_z^n(i,j,k) - E_z^n(i,j-1,k)\right] \tag{I.3}$$

$$H_y^{\tilde{n}}(i,j,k) = H_y^{\tilde{n}-1}(i,j,k) + \frac{\Delta t}{\mu_0 \Delta x}\left[E_z^n(i,j,k) - E_z^n(i-1,j,k)\right] - \frac{\Delta t}{\mu_0 \Delta z}\left[E_x^n(i,j,k) - E_z^n(i,j,k-1)\right] \tag{I.4}$$

$$H_z^{\tilde{n}}(i,j,k) = H_z^{\tilde{n}-1}(i,j,k) + \frac{\Delta t}{\mu_0 \Delta y}\left[E_x^n(i,j,k) - E_x^n(i,j-1,k)\right] - \frac{\Delta t}{\mu_0 \Delta x}\left[E_y^n(i,j,k) - E_y^n(i-1,j,k1)\right] \tag{I.5}$$

$$E_x^n(i,j,k) = \frac{2\varepsilon - \sigma\Delta t}{2\varepsilon + \sigma\Delta t} E_x^{n-1}(i,j,k) - \frac{2\Delta t}{(2\varepsilon + \sigma\Delta t)\Delta z}\left[H_y^{\tilde{n}}(i,j,k) - H_y^{\tilde{n}}(i,j,k-1)\right] + \frac{2\Delta t}{(2\varepsilon + \sigma\Delta t)\Delta y}\left[H_z^{\tilde{n}}(i,j,k) - H_z^{\tilde{n}}(i,j-1,k)\right] \tag{I.6}$$

$$E_y^n(i,j,k) = \frac{2\varepsilon - \sigma\Delta t}{2\varepsilon + \sigma\Delta t} E_y^{n-1}(i,j,k) - \frac{2\Delta t}{(2\varepsilon + \sigma\Delta t)\Delta x}\left[H_z^{\tilde{n}}(i,j,k) - H_z^{\tilde{n}}(i-1,j,k)\right] + \frac{2\Delta t}{(2\varepsilon + \sigma\Delta t)\Delta z}\left[H_x^{\tilde{n}}(i,j,k) - H_x^{\tilde{n}}(i,j,k-1)\right] \tag{I.7}$$

$$E_z^n(i,j,k) = \frac{2\varepsilon - \sigma\Delta t}{2\varepsilon + \sigma\Delta t} E_z^{n-1}(i,j,k) - \frac{2\Delta t}{(2\varepsilon + \sigma\Delta t)\Delta y}\left[H_x^{\tilde{n}}(i,j,k) - H_x^{\tilde{n}}(i,j-1,k)\right] + \frac{2\Delta t}{(2\varepsilon + \sigma\Delta t)\Delta x}\left[H_y^{\tilde{n}}(i,j,k) - H_y^{\tilde{n}}(i-1,j,k)\right] \tag{I.8}$$

Here, Δx, Δy, and Δz are the spatial steps (cell dimensions) in (x,y,z) directions, respectively. One FDTD Yee cell occupies a $\Delta x \times \Delta y \times \Delta z$ volume. The spatial steps Δx, Δy and Δz may be either taken as equal (Yee cube) or different (Yee rectangular prism). Everything inside this cell is assumed to be constant. Calculations are performed at distinct instants t_1, t_2, t_3,..., where $t_1 = \Delta t$, $t_2 = 2\Delta t$, $t_3 = 3\Delta t$,..., with a chosen time step Δt. The integer n is used to denote a number of time steps since the iterations starts. The integers i, j, k are used to mention number of cells from the origin in x, y, and z directions, respectively. A half time-step difference between electric and magnetic fields is denoted by $\tilde{n} = n + 1/2$.

The reason of the locations of electric magnetic field components inside the Yee cell can easily be understood from the iterative equations (I.3) -(I.8). For example, eq.(I.4) tells us that H_y is calculated from its previous time step value and surrounding E_x and E_z values. Therefore, if their locations are chosen as given in Fig. I.16, then for H_y calculations only the components on the left vertical side of the Yee cell will be required and enough. Similar explanations are valid for all other sides and components.

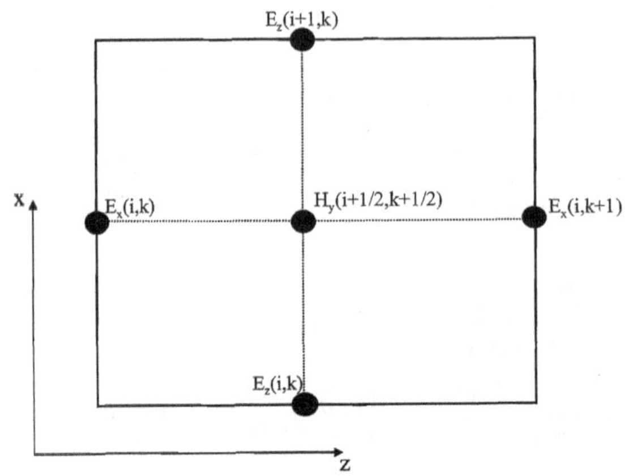

Figure I.16: *Vertical left side of the Yee cell*

Although electric and magnetic field components are updated during the time simulation, voltages and currents in an *(i,j,k)* cell may be obtained directly from Gauss and Faraday laws. Narrow and broadband responses may be readily obtained via FDTD simulations. Since its first introduction, other algorithms, such as free-space simulations (i.e., absorbing boundary simulations to remove the unwanted reflections caused by the boundaries of the computation space) and near- to far-field transformations, have supported FDTD. Now, FDTD can be applied to a variety of complex EM problems, including RCS prediction and antenna simulations, biomedical modeling and planar microwave network analysis, and EMC simulations (visit http://www.fdtd.org for extensive and up to date references).

During the applications of the FDTD method, the parameter values have to be chosen carefully. As it is seen in (I.3)-(I.8), electric (magnetic) fields are calculated using their values at the last iteration time and the magnetic (electric) field values at the neighbor cells, resulting in an iterative (open-form) solution. In order to guarantee the stability of the solutions, there should be a relation between the cell and time step sizes, which is called Courant stability criterion, and for FDTD, it is defined as

$$c\Delta t \le \left[\frac{1}{(\Delta x)^2} + \frac{1}{(\Delta y)^2} + \frac{1}{(\Delta z)^2}\right]^{-\frac{1}{2}} . \qquad \text{(I.9)}$$

The Courant criterion guarantees only the stability, not accuracy, of the solutions.

In FDTD applications, the phase velocities of EM waves could differ from velocity of the light *c*. These differences depend on frequency, cell size and propagation direction and are called numerical dispersion. Numerical dispersion reduces the reliability and accuracy of the method for broadband analysis of EM problems. There is a relation between wavelength of the maximum frequency (λ_{min}) and cell size (similar to the Nyquist criterion between time and frequency). Theoretically, it is required to choose the cell size less than the half of the λ_{min} to trace wave propagation during the time simulation. However, in applications this is not sufficient. In most of the problems, the cell sizes should be less than $\lambda_{min}/10$ (i.e., maximum frequency wave component should be sampled at least with ten cells) and may require as much as $\lambda_{min}/100 - \lambda_{min}/200$. When the Courant stability criterion and requirements for the minimization of numerical dispersion effects are satisfied, the Nyquist criterion between time step size and maximum frequency is also automatically implemented into the algorithm.

The flow chart of the FDTD simulations is given in Fig. I.17.

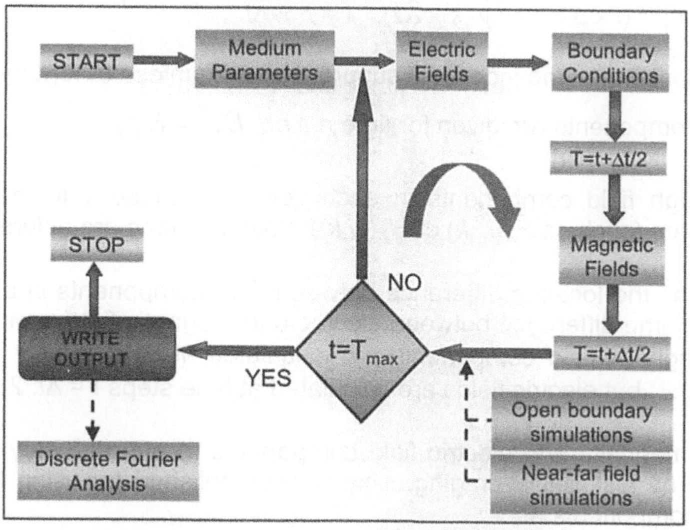

Figure I.17: *Flow chart of FDTD simulation*

The main routine is a time loop, which calculates

(i) Three electric field components at all Yee cells (three loops one inside the other are required)
(ii) The time is increased by another half time step
(iii) Three magnetic field components at all Yee cells (three loops one inside the other are required)
(iv) The time is increased by half time step; and if time reaches the end of simulation time, the loop ends.

The geometry of the problem and all other medium parameters are determined before the main loop.

Observed parameters are written to output files at the end of the simulation and frequency-domain characteristics are obtained by off-line DFT analysis. If the geometry is an open or semi-open region, then absorbing (open) boundary simulators are used inside the main loop at each time step. Similarly, if it is a kind of antenna and/or RCS problem and far fields are required, near-, to far-field transformation routine is used inside the main loop, which extrapolates far-field values directly in TD.

For the numerical implementation of FDTD method, the observations listed below could be important:

- There are three electric and three magnetic field components in each Yee cell which is distinguished by an (i,j,k) label. The discretization steps are Δt and Δx, Δy, Δz for time and position, respectively, which gives the coordinate (i,j,k) and the time n as

$$(i, j, k) = (i \times \Delta x, j \times \Delta y, k \times \Delta z), \quad t = n \times \Delta t \tag{I.10}$$

(for clarity, the time index n is suppressed and unless otherwise is stated the field components are given for time n, i.e., $E_x = E_x^n$).

- Although field components in each cell are labeled with the same (i,j,k) numbers (such as $E_x(i,j,k)$ or $H_z(i,j,k)$), their locations are different.

- Besides the location difference between field components in a cell, there is also a time difference between electric and magnetic field components. That is, magnetic field components are calculated at time steps $t = \Delta t/2$, $3\Delta t/2$, $5\Delta t/2$, ..., but electric fields are calculated at time steps $t = \Delta t$, $2\Delta t$, $3\Delta t$,...

- Both magnetic and electric field components in any cell may be moved to origin by just cell averaging. This is accomplished by using two magnetic field components as

$$H_x(i, j, k) = \frac{1}{2} \times \left[H_x(i, j, k) + H_x(i+1, j, k)\right] \tag{I.11}$$

but four electric field components as

$$\begin{aligned} E_z(i, j, k) = \frac{1}{4} \times [E_z(i, j, k) + E_z(i+1, j, k) \\ + E_z(i, j+1, k) + E_z(i+1, j+1, k)] \end{aligned} \tag{I.12}$$

- As can be seen from the iterative FDTD equations, any object may be simulated by the medium parameters ε, μ, σ. Two of them, ε and σ, appear in electric field components and the third, μ, appears in magnetic field components. Three different ε and σ values may be assigned for three electric field components, so that different objects may be located within the Yee cell. Similarly, different μ values may be given for H_x, H_y and H_z for the same purpose. For example, an infinitely thin wire (starting from (i,j,k) and having length of □= $4\Delta z$) located along z-axis may be simulated by taking;

$$E_z(i,j,k)=0$$
$$E_z(i,j,k+1)=0$$
$$E_z(i,j,k+2)=0$$
$$E_z(i,j,k+3)=0$$

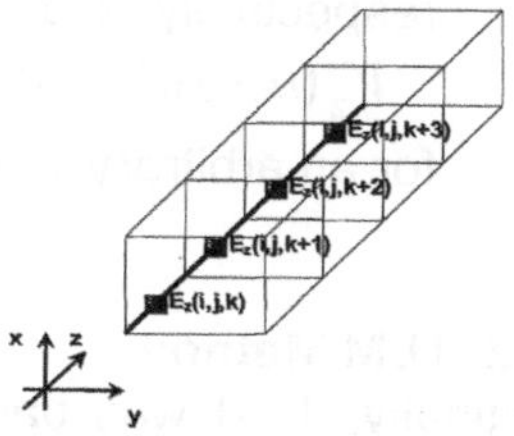

during FDTD iterations. Similarly, an infinitely thin PEC plate may be located at the bottom of (i,j,k) cell along xy-plane by just taking

$$E_x(i,j,k)=E_x(i,j+1,k)=0$$
$$E_y(i,j,k)=E_y(i+1,j,k)=0$$

- Locating an object is nothing but specifying electrical parameters at points where electric field components are calculated. Therefore one should be careful, especially while locating plates. Assume a 2-by-1 cell (with zero thickness) PEC plate will be located as in the figure. The two cells $(L+1,M+1)$ and $(L+2,M+1)$ are of interest. To represent 2-by-1 PEC plate E_y at $(L+3,M+1)$ and E_x at $(L+1,M+2)$ and $(L+2,M+2)$ must also be included.

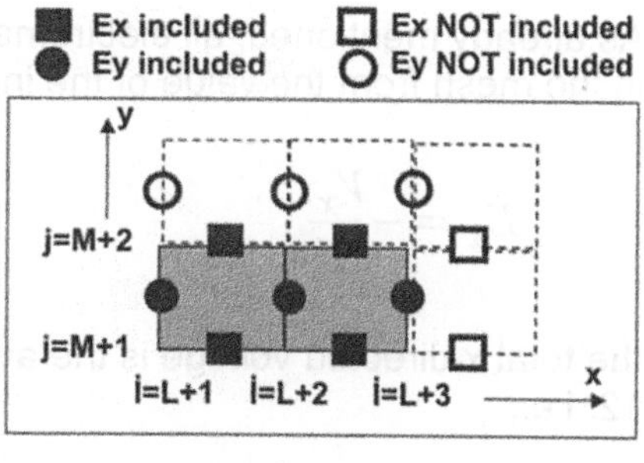

- Although electric and magnetic field components are updated during the time simulation, the voltage and currents in (i,j,k) cell may be obtained from Gauss and Ampere laws directly. For example, V_z and I_z can be obtained as

$$V_z(i,j,k)=-E_z(i,j,k)\times\Delta z \tag{I.13}$$

$$I_z(i,j,k)=\left[H_x(i,j-1,k)-H_x(i,j,k)\right]\times\Delta x + \left[H_y(i,j,k)-H_y(i-1,j,k)\right]\times\Delta y \tag{I.14}$$

- Both sinusoidal and pulse type source simulations can be handled in FDTD. For example, a voltage source $V_z(t)$ at (i,j,k) cell may be injected via

$$E_z^n(i,j,k) = E_z^n(i,j,k) + \frac{V_z(t_n)}{\Delta z} \tag{I.15}$$

where

$V_z(n \times \Delta t) = A_o \times \sin(2\pi f_o t_n)$ and $t_n = n \times \Delta t$,

for sinusoidal source with amplitude and frequency values A_o and f_o, respectively, and

$V_z(n \times \Delta t) = A_o \times g(t_n)$

for an arbitrary source function $g(t)$.

I.5.2 TLM Method

Originally, TLM was based on the analogy between the EM field and mesh of transmission lines [29]. SCN was developed to improve the analysis of electromagnetic waves by the TLM method [87,88]. It is exactly the TLM technique, only where the unit cell is reconstructed to overcome the asymmetry and asynchronous problems. This handy node has the advantage of condensing the field components to one point in space (to the reference point in the TLM cell). The SCN-TLM cell structure for a homogeneous medium is given in Fig.I.18. It is found to be convenient to preserve the idea of the two-wire transmission line, which are shown on each side of the node.

As already mentioned, all electromagnetic quantities may be obtained at any point in the mesh from the value of the incident voltage pulses. Since

$$E_x = -\frac{V_x}{\Delta \ell} \tag{I.16}$$

the total x-directed voltage is the average of the total voltage on ports 1, 2, 9, and 12; i.e.,

$$V_x = \frac{1}{4}\left[\left(V_1^i + V_1^r\right) + \left(V_2^i + V_2^r\right) + \left(V_9^i + V_9^r\right) + \left(V_{12}^i + V_{12}^r\right)\right] \tag{I.17}$$

It has already been shown and it can be confirmed directly using the scattering matrix, that for charge conservation, the sum of incident voltages to these ports is equal to the sum of reflected voltages. Hence,

$$V_x = \frac{1}{2}\left[V_1^i + V_2^i + V_9^i + V_{12}^i\right]$$

and

$$E_x = -\frac{V_1^i + V_2^i + V_9^i + V_{12}^i}{2\Delta \ell} \tag{I.18}$$

where i and r stand for incident and reflected, respectively.

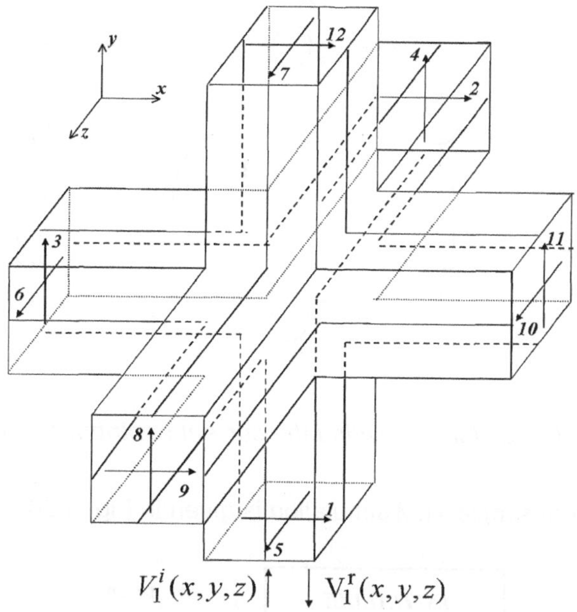

Figure I.18: *Six port SCN - TLM node and related voltages*

Using a similar procedure, the remaining field components may be calculated.

$$E_y = -\frac{V_3^i + V_4^i + V_8^i + V_{11}^i}{2\Delta\ell} \tag{I.19}$$

$$E_z = -\frac{V_5^i + V_6^i + V_7^i + V_{10}^i}{2\Delta\ell} \tag{I.20}$$

The magnetic field component H_x, may be obtained from $H_x = I_x / \Delta\ell$. The current I_x, in turn, is calculated from the circuit shown in Fig. I.19, where each line is replaced by its Thevenin equivalent, to give

$$H_x = \frac{V_4^i - V_5^i + V_7^i - V_8^i}{2Z\Delta\ell} \tag{I.21}$$

where Z is the characteristic impedance of the line.

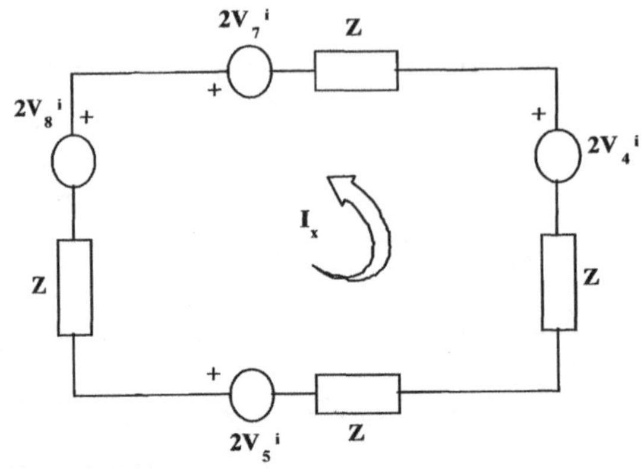

Figure I.19: *Circuit for calculating the current contributing to the x-directed field*

The flow chart of simple TLM simulation is given in Fig. I.20.

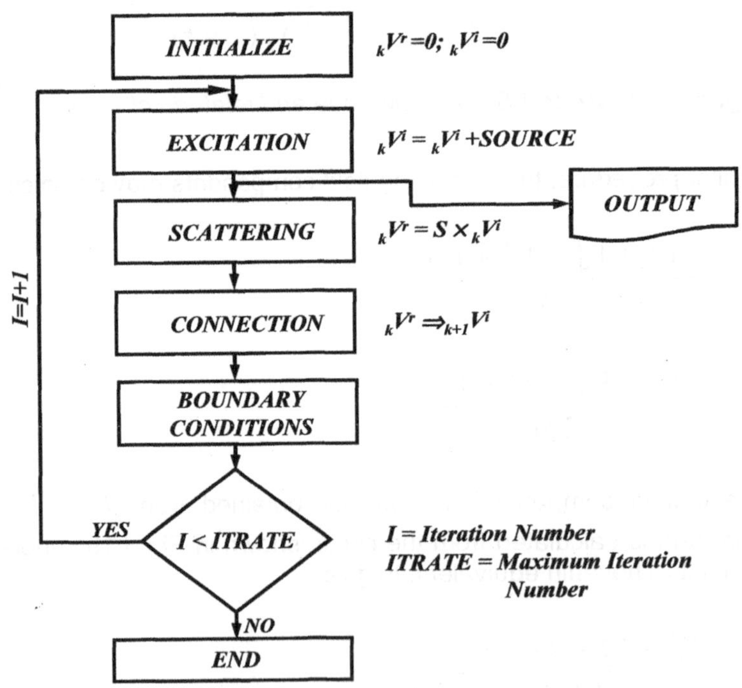

Figure I.20: The flow chart of the TLM computer code

For the SCN - TLM node (or cell) configuration in Fig.I.18, the points listed below should always be remembered during the simulations:

- The TLM method is based on the network theory where voltages are the independent observation parameters.

- The physical environment is sub-divided into small cells as shown in Fig.I.18. Each cell is distinguished with (i,j,k) label. Also, the label n is used to represent the current simulation time. The labels $(i,j,k;n)$ are all integers. As a result, $V(x,y,z;t)$ is replaced with $V(i,j,k;n)$, where $x = i \times \Delta x$, $y = j \times \Delta y$, $z = k \times \Delta z$ and $t = n \times \Delta t$. The total simulation volume extends from (0,0,0) to (X,Y,Z) which is subdivided into $NM = N_x \times N_y \times N_z$ nodes where $\Delta x = X/N_x$, $\Delta y = Y/N_y$ and $\Delta z = Z/N_z$.

- In TLM, EM fields are represented by voltage pulses scattered in the nodes and propagating through transmission lines between neighboring nodes.

- An SCN - TLM node may be assumed as a six port device where two different (i.e., vertical and horizontal) polarizations are represented at each port.

- In any direction, the propagation is simulated via two pairs of transmission lines, which do not couple with each other (these transmission lines are totally separated in space).

- Between the nodes, time-domain propagation is simulated via the connection of incident and reflected pulses by the scattering matrix. The notation ${}_nV_h^i(i,j,k)$ is used for TLM voltages, where the superscript i is used for an incident voltage, and subscript h is used to represent any of the 12 voltages in the node (see Fig.I.18).

- The scattering matrix S, relating the reflected voltages, V^r, to the incident voltages, V^i, is a 12×12 matrix for the node structure given in Fig.I.18, but becomes 18×18 to model any inhomogeneity in the medium of interest [87]. Moreover, lossy structures can be modeled by 21×18 scattering matrix.

- Any medium can be modeled in terms of the parameters; permittivity ε, permeability μ, and conductivity σ.

- The medium parameters, ε and μ, are modeled by open- and short-circuited stubs with the lengths of $\Delta\ell/2$ (where $\Delta\ell = \Delta x = \Delta y = \Delta z$), respectively. But, σ can be modeled by infinitely long (matched terminated) stubs [87]. One extra stub per electric field component is necessary for lossy dielectrics, which causes the total number of ports to be 21.

- Twelve pulses incident upon the node via the link transmission lines produce 12 reflected pulses. The incident and reflected pulses appear on the terminals of the transmission lines at ports which are numbered and directed according to the voltages shown in Fig.I.18.

- The modeling of PEC structures in SCN - TLM is simple. For example, an infinitely thin PEC plate normal to the *x*-axis in the SCN - TLM node may easily be simulated by the voltages on ports 10 and 11. Forcing the incident voltages 10 and 11 to be equal to negative of the corresponding reflected voltages during all the time simulation is enough for this purpose. That is,

$$
\begin{aligned}
{}_{n+1}V_{10}^{i} &= -{}_{n}V_{10}^{r} \\
{}_{n+1}V_{11}^{i} &= -{}_{n}V_{11}^{r}
\end{aligned}
\tag{I.22}
$$

- Similarly, PEC bodies (like dipole antennas etc.) are modeled in SCN-TLM by just short-circuiting the relevant node. To do this, the diagonal elements of the scattering matrix of the node are set to -1.

- Twelve voltage pulses given in Fig. I.18 describe the electric and magnetic field components. For example, the pulses contributing to the E_x field are at ports 1, 2, 9 and 12. So, these relevant voltage pulses define E_x. It is obvious that one should take corresponding stub voltage pulses modeling the medium properties into consideration in defining the field components (e.g., voltage 13 for E_x). One can readily write down the *x*-component of the electric field as

$$E_x = -\frac{2(V_1^i + V_2^i + V_9^i + V_{12}^i + \hat{Y}_x \times V_{13}^i)}{\Delta x(4 + \hat{Y}_x)} \tag{I.23}$$

$$\hat{Y}_x = 2\varepsilon \frac{\Delta y \Delta z}{\Delta x \Delta t Y_0} - 4 \tag{I.24}$$

where $\hat{Y}_x$ is the normalized stub admittance in *x*-direction and Y_0 is the characteristic admittance of the lines.

- To excite an SCN - TLM system with a Gaussian pulse, the following process is made for the relevant voltages related to the desired field component (e.g., voltages 1, 2, 9 and 12 are used for E_x)

$$V_\ell^i(i,j,k;n) = V_\ell^i(i,j,k;n) + 0.5 \times \Delta x \times g(t_n) \tag{I.25}$$

(ℓ =1, 2, 9 and 12) and $g(t_n)$ is the Gauss function

$$g(t_n) = A_0 \exp\left[-\frac{(t_n - t_0)^2}{\tau^2}\right]; \; t_n = n \times \Delta t; \; t_o = M \times \Delta t \tag{I.26}$$

Here A_o , τ and M represent the amplitude, pulse duration, and time shift in terms of Δt, respectively.

In TLM, the cell structure is based on the power or charge conservation in transmission lines. Therefore, there is no equivalent circuit that can be used in analytical discussions such as dispersion relation in TLM. It is solely characterized by the scattering matrix. The user, therefore, has to perform a linear transformation using the scattering matrix (modeling the media) at each nodal point.

The TLM method with SCN introduced by Johns [29] has proven to be a very powerful method in electromagnetic field computation. However, studies have still been carried out to overcome the redundancies in the computational effort. For this reason, alternatives to the current TLM nodes such as alternating (ATLM), rotated (RTLM), and alternating rotated transmission line matrix (ARTLM) schemes have been introduced recently [88]. Studies related to improve the TLM structure and to optimize computational efforts have still continued.

I.5.3 Parameter Selection and Excitation

These kinds of TD simulators are preferred because of

(i) their ability to handle complex EM environments and
(ii) short pulse simulations, which yield broadband behaviors at once.

Running simulations with either FDTD or TLM techniques require parameter optimization. Spatial mesh sizes, Δx, Δy, Δz, time step Δt, simulation time ($T_{obs} = n \times \Delta t$), source bandwidth B, and pulse duration τ are characteristic parameters that should be optimally selected prior to the simulation (it should be noted that pulse duration and bandwidth may be defined in different ways: For example, they may be defined between the points, where amplitude drops to 50% or 10%).

Parameter selection has two initial steps:

1. *The frequency analysis*: The TD simulations are generally used to obtain frequency characteristics of a given EM structure. For example, the radiation characteristics or input impedance of an antenna structure, RCS behavior of a chosen target, transmission and/or reflection characteristics of a microstrip network, propagation characteristics of a waveguide, resonance frequencies of a closed enclosure, shielding effectiveness of an aperture, etc. are among important problems. Whatever the problem at hand is, the characteristic is desired for a given frequency range; from a given minimum frequency f_{min} to a maximum frequency f_{max} with Δf incremental frequency steps.

2. *TD discrete simulation requirements*: The discretization parameters Δx, Δy, Δz, Δt, and observation time $T_{max} = n \times \Delta t$, together with source parameters B_{eff} and T_{eff} are determined according to the desired frequency analysis.

The user first specifies minimum and maximum frequencies, f_{min} and f_{max} together with frequency sensitivity Δf. Suppose the problem is to find the frequency characteristics of an antenna from DC to 1 GHz with 50 MHz frequency steps. As shown in Fig. I.21, the signal spectrum extends from DC to a maximum frequency 1 GHz. An ideal waveform for broadband TD simulations is one with finite duration in both time and frequency domains. However, such a property is not possible in a physical source. For example, a TD pulse with rectangular duration is an ideal duration-limited signal. On the other hand, its frequency duration (which is a sinc function) is infinite (i.e., converging to zero at infinity). A rectangular spectrum corresponds to an infinite-duration pulse in the TD (because of the symmetry property of Fourier transform).

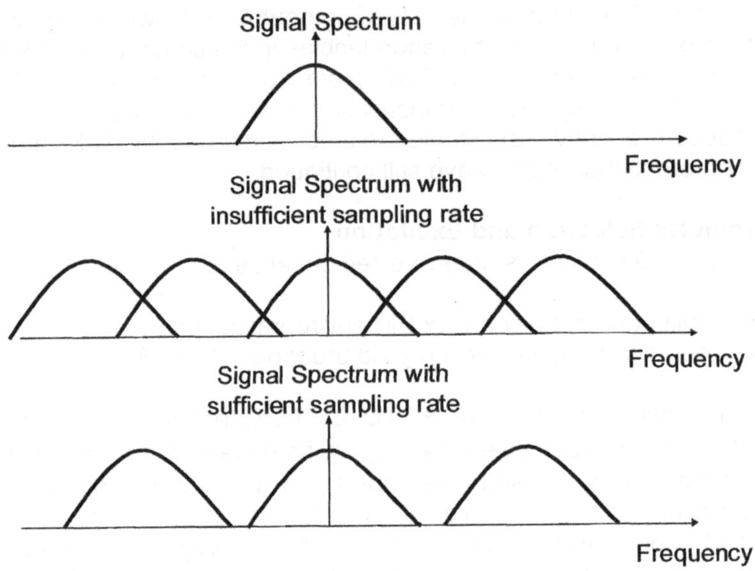

Figure I.21: *Source spectrum and sufficient/insufficient sampling effects*

In general, there is an inverse relationship between source duration and source bandwidth; the shorter the pulse the higher the bandwidth, and vice versa. This is expressed using the time - bandwidth product, which states that, for a source signal the product of Bandwidth and duration is constant. The best source waveform is the one with minimum time-bandwidth product. A Gaussian pulse has this property and it is usually chosen as the source waveform in these kind of simulations. At the end of this section, some source waveforms and their frequency spectra which are preferred in TD simulations are given.

The spectrum of a Gaussian source may be assumed as a band-limited signal as shown in Fig. I.21. As mentioned above, maximum frequency of interest (defined between either 50% or 10% amplitude points) is 1 GHz. According to the Nyquist Sampling criteria, the sampling rate (f_s, sampling frequency) must be at least double of the maximum frequency, i.e., $f_s = 2 \times f_{max} = B$. In Fig. I.21, the spectrum of a waveform sampled (discretized) without and with obeying Nyquist criteria are plotted. Discretization in time domain makes the signal spectrum periodic in frequency domain. Without adequate sampling, these periodic bands coalesce and information is lost. On the other hand, sampling with adequate rate separates these bands in the frequency domain, which makes extracting one of them possible.

Starting from the frequency analysis requirements and sampling criteria, parameter optimization steps can be listed as follows:

- Choose the source waveform with a duration that contains the maximum frequency of interest.

- According to fast Fourier transform (FFT) properties, maximum frequency determines minimum time step, i.e., $\Delta t_{FFT} = 1/(2 \times f_{max})$. This is the hard limit for frequency analysis. Having 1 GHz as the maximum frequency Δt_{FFT} is obtained as 0.5 nanoseconds.

- There are two important points in choosing maximum simulation (observation) time. First, the frequency sensitivity Δf, which is determines the observation time as $\Delta f = 1 / T_{max}$. Second, the simulation should continue until all transient effects finish. Therefore, T_{max} is chosen to satisfy both. Since Δf is given as 10 MHz, T_{max} is determined as 100 nanoseconds. Number of time steps n will then be 200. If all transients decay after 200 time steps then this will be enough for the simulation time. If the structure under investigation is a kind of resonant structure, which corresponds to ringing effects in TD, then much more observation time may be required.

- Up to this point, requirements of discrete Fourier analysis are used in defining time step Δt and maximum simulation time T_{max}. There are two important criteria in TD simulations; Courant stability criteria and numerical dispersion.

- Spatial mesh sizes Δx, Δy, Δz are chosen according to numerical dispersion requirements. It is nothing but satisfying Nyquist sampling criteria in spatial domain. The minimum wavelength λ_{min} component must be sampled with at least two samples, i.e., max of $\{\Delta x, \Delta y, \Delta z\} \leq \lambda_{min}/2$. In practice, at least $\lambda_{min}/10$ is required for acceptable results. Depending on the problem at hand as much as $\lambda_{min}/100$ to $\lambda_{min}/120$ may be required in order to get rid of numerical dispersion effects. In most of the problems presented in Part II, typical values of $\lambda_{min}/25$ to $\lambda_{min}/40$ are chosen, which results in almost perfect agreement in TD simulations. Since λ_{min} is 30 cm, $\Delta x = \Delta y = \Delta z = 1$cm may be chosen if $\lambda_{min}/30$ is good enough for eliminating numerical dispersion effects.

- Once spatial mesh sizes are determined, time step Δt_{FDTD} may directly be chosen from the Courant stability criteria. Since, $\Delta x = \Delta y = \Delta z$ and is equal to 1cm, Δt may be chosen as $1/(\sqrt{3}c)$, where c is the speed of light. This gives $\Delta t \approx 20$ picoseconds. As obtained here, in general, FDTD time step Δt_{FDTD} (20 picoseconds) is much less than the FFT requirement Δt_{FFT}, (0.5 nanoseconds) therefore Δt_{FDTD} is taken into account. With this time step, the simulation time, n will be 5000.

Choosing the *best* source with the desired frequency content is also important in TD simulations. A typical source with minimum time-bandwidth product is the Gauss function $g(t)$ given as

$$g(t) = A_0 \times e^{-\alpha \times (t - t_S)^2} \tag{I.27}$$

where A_0, α and t_s are the amplitude, duration parameter (proportional with the bandwidth, $\alpha \sim B^2$) and time shift, respectively. Gauss function has a DC component (i.e., its frequency bandwidth starts from zero) which usually causes numerical problems, therefore its once-differentiated form, $g'(t)$

$$g'(t) = -2.3 \times A_0 \times \sqrt{\alpha} \times (t - t_S) \times e^{-\alpha \times (t - t_S)^2} \tag{I.28}$$

is a better choice for most of the applications. Although once-differentiated Gauss function has no DC component, its frequency bandwidth has still good amount of low-frequency energy. To get rid of DC component and low frequency energy in the spectrum, a modulated Gauss function

$$g(t) \times Sin(2\pi f_O t) = A_0 \times Sin(2\pi f_O t) \times e^{-\alpha \times (t - t_S)^2} \tag{I.29}$$

may be used as the source, where f_0 is the carrier frequency. It should be noted that the carrier frequency should be well-chosen not to introduce DC and low-frequency energy. Fig. I.22 depicts time and frequency behaviors of these three sources. Here, 1-nanosecond pulse duration (between points where amplitude drops to its 50% value) corresponds to an approximate bandwidth of 0.5 GHz. The modulation frequency is 750 MHz.

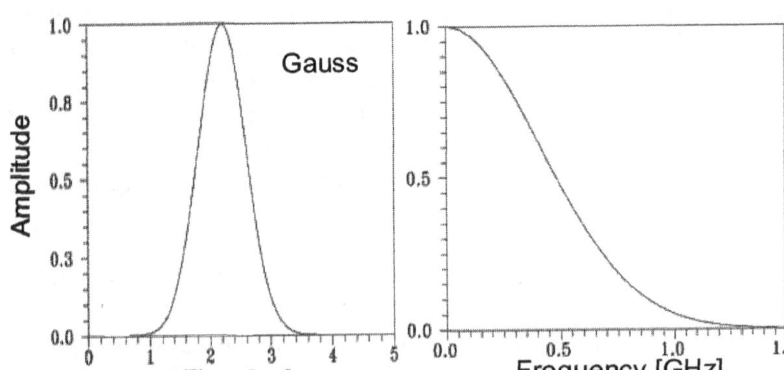

***Figure I.22** : Time and frequency behavior of (a) Gauss, (b) once-differentiated Gauss, and (c) modulated Gauss source functions*

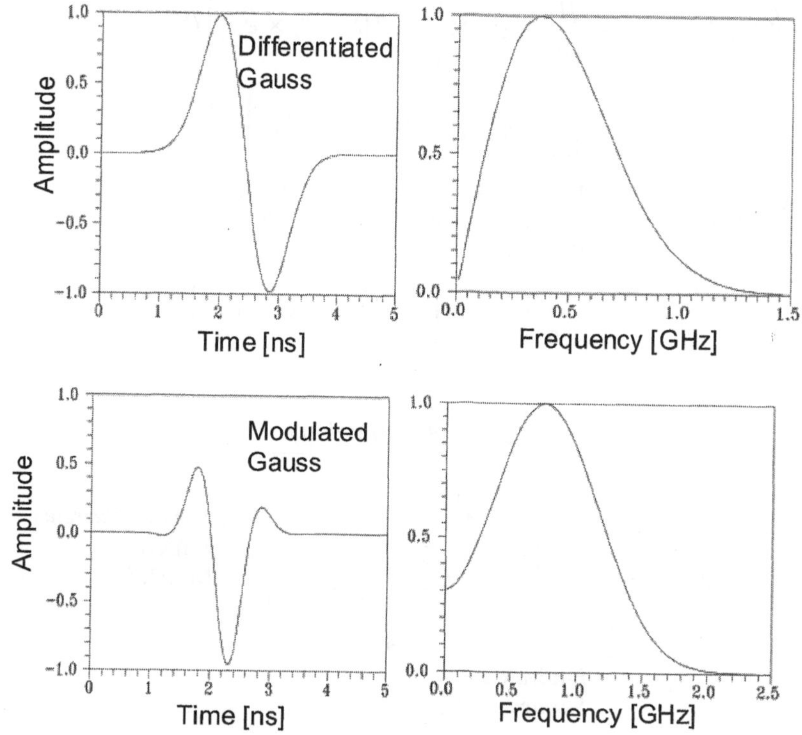

***Figure I.22** : (continued)*

I.5.4 Near- to Far-Field (NTFF) Transformation

Early applications of TD simulations were primarily in areas where EM near-field simulations are of interest such as time-domain pulse response or penetration of EM waves into human tissues, frequency characteristics of an antenna input impedance, etc. Shortly afterwards, TD simulation have begun to be applied to analyze EM scattering from radar targets as well as to calculate antenna radiation characteristics. Since the latter applications need far field calculations, efficient methods for transforming from near zone (inside the TD simulators) to far zone (outside the TD simulators) have been introduced both in FD and TD.

Either in FDTD or TLM, TD simulations are performed in rectangular coordinates. On the other hand, far field values are required in spherical coordinates. Actually, the TD simulation volume, virtual closed surface and the object under investigation are shown in Fig. I.23.

Here, near- to far-field transformation (NTFF), introduced in [89], is used. In NTTF, the TD Stratton - Chu equations [90]

$$\vec{E}^S(\vec{r},t) = -\frac{\mu}{4\pi r}\int_S \vec{e}_r \times \left(\frac{\partial}{\partial t}\vec{J}(t-\tau)\times\vec{e}_r\right)dS' - \frac{1}{4\pi rc}\int_S\left(\frac{\partial}{\partial t}\vec{M}(t-\tau)\times\vec{e}_r\right)dS' \quad (I.30)$$

$$\vec{H}^S(\vec{r},t) = \frac{1}{4\pi rc}\int_S\left(\frac{\partial}{\partial t}\vec{J}(t-\tau)\times\vec{e}_r\right)dS' - \frac{\varepsilon}{4\pi r}\int_S \vec{e}_r \times \left(\frac{\partial}{\partial t}\vec{M}(t-\tau)\times\vec{e}_r\right)dS' \quad (I.31)$$

are used.

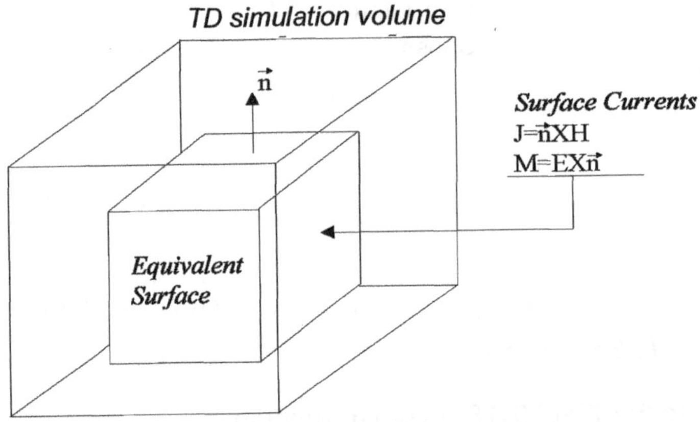

Figure I.23 : *NTFF transformation in TD simulators*

The procedure is as follows:

- The object under investigation is located inside the TD simulators (at the center) and the computation domain is externally terminated by free-space simulators (e.g., PML blocks). Within the external boundary and object under investigation, surface electric and magnetic currents

  $$\vec{J}_S = \vec{n}\times\vec{H}\,, \qquad \vec{M}_S = -\vec{n}\times\vec{E} \quad (I.32)$$

  are calculated on a selected virtual closed surface (which encloses the object). Here the vector *n* represents outward surface normal vector.

- With the help of surface electric and magnetic currents, radiation vectors

$$\vec{w}(t) = \frac{1}{4\pi rc}\frac{\partial}{\partial t}\left\{\int_S \vec{J}_S(t-\tau)dS'\right\} \quad (I.33)$$

$$\vec{u}(t) = \frac{1}{4\pi rc}\frac{\partial}{\partial t}\left\{\int_S \vec{M}_S(t-\tau)dS'\right\} \qquad (I.34)$$

along a chosen direction are accumulated directly in TD, where r is the distance from the origin to the far-field observation point [89].

- The TD simulations continue until all the near fields dissipate out the virtual NTFF transformation surface. Then, far-zone electric fields in spherical coordinates are obtained via

$$E_\theta^S(t) = -\eta w_\theta(t) - u_\phi(t) \text{ and } E_\phi^S(t) = -\eta w_\phi(t) + u_\theta(t) \qquad (I.35)$$

where $\eta = 120\pi\ \Omega$ is the free-space wave impedance.

The NTFF approach summarized here is based on the accumulation of the far-zone vector potentials due to the tangential electric and magnetic fields on a virtual, closed surface surrounding the object under investigation at each time step. Although it is computationally expensive, the outcome is very valuable because broadband frequency responses may be obtained with single TD simulations. Since cell-by-cell TD far fields are calculated during the TD simulations, time and spatial synchronization of the field components must be done by averaging at each time step.

The Fortran code for the NTFF transformation is YAKUZ and is embedded inside all TD simulators that are used in antenna and RCS problems, where far fields are required. It is a modular design and the user can easily embed YAKUZ inside his/her own code. In order to do that, the following steps should be taken into account:

- CALL YAKUZ line must be added inside the main TD simulation routines.
- CALL HESAP line must be added outside the main TD simulation routines, before terminating the program.
- The following declarations must be added into the PARAMETER file, where NSTOP is the number of time steps and NACI is the maximum allowable number of directions for far fields.

```
C********************YAKUZ declarations *********************
      COMMON/YAK/KEH,MAX,DELTA,DT1,CC1,TETA,PHI
      COMMON/EK/UX(0:NACI,-50:NSTOP+50),
     $       UY(0:NACI,-50:NSTOP+50), UZ(0:NACI,-50:NSTOP+50)
      COMMON/MK/WX(0:NACI,-50:NSTOP+50),
     $       WY(0:NACI,-50:NSTOP+50), WZ(0:NACI,-50:NSTOP+50)
```

If, for example, NACI is 360, then a far field pattern with 1 degree of separation can be obtained with this routine.

The subroutine YAKUZ does the accumulation of electric and magnetic vector potentials during the TD simulations. After the TD simulation, the subroutine HESAP does the far field extrapolation along all the directions (one by one), where the number of directions is determined with the parameter MAX. The Fortran code of HESAP is listed in Table I.2.

Table I.2: Fortran Code of HESAP subroutine (KEH=0 for horizontal, KEH=1 for vertical patterns)

```
      SUBROUTINE HESAP
      INCULDE '*.PAR'
C****************************************
C   FAR FIELD CALCULATION     *
C****************************************
        DO L=1,MAX+1
          IF(KEH.EQ.0) THEN
               PHI =(L-1)*DELTA
          ELSE
               TETA=(L-1)*DELTA
          ENDIF
        END DO
C*************************
C  PLANE LOOP     *
C*************************
        DO L=1,MAX+1
          IF(KEH.EQ.0) THEN
            PHI =(L-1)*DELTA
          ELSE
            TETA=(L-1)*DELTA
          ENDIF
          DO N=-MM,NSTOP+MM
            WTETA=WX(L,N)*COS(PHI)*COS(TETA)
     *             +WY(L,N)*SIN(PHI)*COS(TETA)-WZ(L,N)*SIN(TETA)
            WPHI=-WX(L,N)*SIN(PHI)+WY(L,N)*COS(PHI)
            UTETA=UX(L,N)*COS(PHI)*COS(TETA)
     *             +UY(L,N)*SIN(PHI)*COS(TETA)-UZ(L,N)*SIN(TETA)
            UPHI=-UX(L,N)*SIN(PHI)+UY(L,N)*COS(PHI)
            SABIT=12.E8*PI*DT
            FETETA = (-ETA0*WTETA-UPHI)/SABIT
            FEPHI  = (-ETA0*WPHI+UTETA)/SABIT
            WRITE(30,77) (L-1)*DELTA,N*DT*1.e9,FETETA,FEPHI
          END DO
        END DO
   77 FORMAT(2(F18.10,1X),2(F18.8,1X))
      CLOSE(UNIT=30)

       RETURN
       END
```

I.5.5 Injecting Analytical Incident Field

Establishing the governing, iterative FDTD equations via the scattered field formulation (e.g., an RCS problem) has the advantage of specifying the incident field analytically. When an RCS problem is of interest, the incident field is spatially a plane wave but has a Gaussion (or once-differentiated Gaussian)-type time variation. Such a plane wave is required for broadband RCS analysis via single FDTD run.

FDTD runs in Cartesian coordinate system, but incident field is specified in spherical coordinates. Following the usual scattering convention, the direction that the incident field is injected is specified with two angles; θ and φ; the angles are measured from z-and x-axes, respectively. Assuming a unit vector $\vec{r}$ point from origin in the direction of arrival, an incident plane wave from this direction can be specified as

$$\vec{E} = \left[E_\theta \vec{e}_\theta + E_\varphi \vec{e}_\varphi\right] f\left\{t + \frac{(\vec{r} \bullet \vec{r}\,')}{c} + \frac{R}{c}\right\} \tag{I.36}$$

$$\vec{H} = \left[\frac{E_\theta}{\eta}\vec{e}_\theta - \frac{E_\varphi}{\eta}\vec{e}_\varphi\right] f\left\{t + \frac{(\vec{r} \bullet \vec{r}\,')}{c} + \frac{R}{c}\right\} \tag{I.37}$$

where $\vec{e}_\theta$ and $\vec{e}_\varphi$ are the unit vectors along the θ and φ axes, respectively, η is the impedance of free space, *c* is the speed of light, $\vec{r}\,'$ is the vector from origin to the point of interest in the FDTD computation space, and *R* is an arbitrary reference distance. For transient calculations, an incident pulsed plane wave must travel along the direction of incidence throughout the FDTD computation space.

Cartesian components of the incident field can be obtained via

$$E_x = E_\theta \cos\theta \cos\varphi - E_\varphi \sin\varphi \tag{I.38a}$$

$$E_y = E_\theta \cos\theta \sin\varphi - E_\varphi \cos\varphi \tag{I.38b}$$

$$E_z = -E_\theta \sin\theta \tag{I.38c}$$

$$H_x = \frac{E_\theta \sin\theta + E_\varphi \cos\theta \cos\varphi}{\eta} \tag{I.38d}$$

$$H_y = \frac{-E_\theta \cos\varphi + E_\varphi \cos\theta \sin\varphi}{\eta} \tag{I.38e}$$

$$H_z = \frac{-E_\theta \sin\varphi}{\eta} \tag{I.38f}$$

In both FDTD and TLM computer programs, the user supplies the angles of incidence, and an incident short-pulse (as Gaussian or once-differentiated Gaussian functions) plane wave is injected automatically as the wave arrives from the given direction, interacts with the object under investigation (in time) and then disappears.

I.5.6 Absorbing Boundary Simulators

TD numerical computation volumes are bounded (finite) because of the computer capacity limitations. This results in full reflections at six boundary planes for outward scattered or propagating waves, unless preventive measures are taken.

Iterative TD equations, by nature, simulates PEC walls at six boundary planes. Therefore, EM phenomena inside a PEC resonator can directly be simulated with these TD simulators, without taking any additional steps.

For many applications, such as RCS and antenna radiation pattern simulations, numerical calculations must be performed in free space. This requires additional implementations at the boundary planes of the computation volume, which is usually called free-space simulation, absorbing boundary termination, etc. The situation is simplified and pictured in Fig. I.24. Suppose a signal propagation along one coordinate (e.g., along *x*) inside a bounded region (e.g., bounded between X_{min} and X_{max}) is simulated in TD.

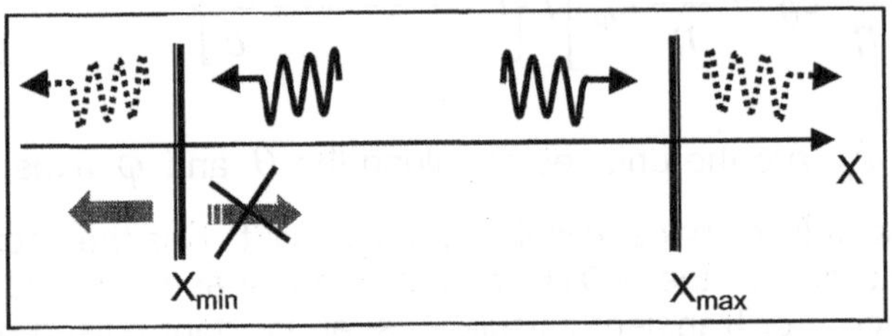

Figure I.24: *One way propagation between X_{min} and X_{max}*

In general, any source inside this region excites waves propagating along both positive and negative directions. The corresponding 2D wave equation is given as

$$\left[\frac{\partial^2}{\partial x^2} - \frac{1}{c^2}\frac{\partial^2}{\partial t^2}\right]U(x,t) = 0 \tag{I.39}$$

where *U* represents a scalar wave function (that may be any component of electric or magnetic fields). This equation represents waves propagating along both positive and negative *x* directions. The problem is well-defined (i.e., with a unique solution) when initial conditions at the starting time (e.g. $U(x,t_s)$ and time derivative $U'(x,t_s)$ at the initial time t_s) and boundary conditions at X_{min} and X_{max} ($U(x_{min},t)$ and $U(x_{max},t)$) are specified. Equation (I.39) can be rewritten as

$$\left[\frac{\partial}{\partial x} + \frac{1}{c}\frac{\partial}{\partial t}\right]\left[\frac{\partial}{\partial x} - \frac{1}{c}\frac{\partial}{\partial t}\right]U(x,t) = 0 \tag{I.40}$$

where the first and second terms on the left represent negative (towards left) and positive (towards rigth) *x*-directed waves, respectively. If a free space is going to be simulated, for example, at $x = X_{min}$ (at the left), then

$$\left[\frac{\partial}{\partial x} - \frac{1}{c}\frac{\partial}{\partial t}\right]U(x,t) = 0 \tag{I.41}$$

must be discretized and used at the left boundary (instead of TD iterative equations), since it represents waves propagating right. Mathematical usage of

(I.41) corresponds to absorption of waves reflected at $x = X_{min}$, physically. The TD discretized iterative form of (I.41) will then be

$$U_{i=1}^{n} = \left[\frac{\Delta x}{\Delta x - c\Delta t}\right] U_{i=1}^{n-1} - \left[\frac{c\Delta t}{\Delta x - c\Delta t}\right] U_{i=2}^{n}. \qquad \text{(I.42)}$$

Here, i = 1 and i = 2 are the first (the left most) and the neighboring cells, respectively, and, n and n - 1 correspond to the current and one previous iteration time steps, respectively. As a summary,

- TD iterative equations are not used at $x = X_{min}$ (at the boundary cell),
- Instead, one time step previous value is used,
- Current and one time step previous values of the neighboring cell are used to simulate free space.

Equation (I.39) depends only one spatial coordinate, which represents two-way plane waves. Using (I.42) results in a perfect absorption (100% performance) at the boundary. In 4D TD simulations (I.39) can be written along y and z axes and used as free-space simulators. Unfortunately, their performance is restricted with the waves propagating perpendicular to the termination boundaries. In practice, waves may propagate and/or be scattered along any direction. Moreover, there may be propagating, specularly reflected, scattered, travelling, creeping, surface, etc., waves. This simple free-space simulation approach can definitely not absorb all these waves coming from all directions. Different free-space simulators have been introduced for the last decade. Some of them, which have found wide applications, are Mur [91], dispersive boundary condition (DBC) [92], modified DBC [93], and perfectly matched layer (PML) [94,95]. Finding effective and generally applicable free-space simulators is still an active research area.

Absorbing boundary termination is also required when infinite length structures are going to be modeled. An infinite waveguide, a microstrip line, an array, etc., are a few examples. A typical problem is given in Fig. I.25. Here, an infinite-length simple microstrip line is simulated. Modified DBC blocks are used at the boundaries of the TD simulation volumes. In the figure, signal vs. microstrip length are plotted at six different time instants. Excitation of the fields under the strip (t_1 and t_2), decomposition of the pulse into left and right propagating waves (t_3 and t_4), and finally, left and right propagating pulses (t_5 and t_6) are clearly observed in the figure. At t_5 and t_6, signal absorption at the left boundary is also observed.

The effectiveness of the absorbing boundary terminations, in general, depend on the number cells and number of previous time steps used for this purpose. For example, PML with 10 cells (PML-10) is definitely more effective than the PML with 5 cells (PML-5). On the other hand, PML-10 requires more memory (therefore longer computation times) than PML-5. The user should find out optimum parameters for the outer boundary simulations. One other consideration is that of determining the distance between the object under investigation and outer boundary. The farther from the object the outer boundary is located, the better the absorption of the outward propagating waves. However, the number of cells that can be placed between the object and outer boundary is limited by computer memory. A common criterion is to use around 10 to 15 cells in between them.

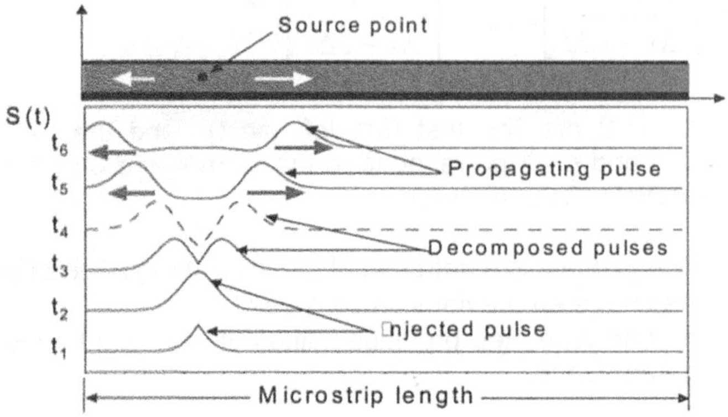

Figure I.25: *Signal vs. length under a microstrip line recorded at different time instants*

I.6 FREQUENCY-DOMAIN SIMULATION TECHNIQUES

I.6.1 Method of Moments

One of the most powerful techniques, which has been in use for more than three decades in frequency domain, is the MoM. The primary formulation of MoM is an integral equation [31] obtained through the use of Green's functions. The technique is based on solving complex integral equations by reducing them to a system of linear equations and on applying *method of weighted residuals*. Actually the terms *method of moments* and *method of weighted residuals* are synonymous. It is Harrington [31] who popularized the term *method of moments* in electrical engineering society. His pioneering efforts first demonstrated the power and flexibility of this numerical technique for solving problems in EM.

All *weighted residuals* techniques begin by establishing a set of trial solutions with one or more variable parameters. The *residuals* are a measure of the difference between the trial solution and the true solution. The variable parameters are determined in a manner that guarantees a *best fit* of the trial functions based on a minimization of the residuals.

The equation solved by MoM technique is generally a form of the *electrical field integral equation* (EFIE) or the *magnetic field integral equation* (MFIE). Both of these equations can be derived from Maxwell's equations by considering the problem of a field scattered by a perfect conductor (or a lossless dielectric). These equations are of the form

$$\text{EFIE:} \quad \mathbf{E} = \mathrm{fe}(\mathbf{J}) \tag{I.43}$$

$$\text{MFIE:} \quad \mathbf{H} = \mathrm{fm}(\mathbf{J}) \tag{I.44}$$

where the terms on the left-hand side of these equations are incident field quantities and **J** is the induced current.

The form of integral equation used determines which type of problem a MoM technique is best suited for. For example, one form of EFIE may be particularly well suited for modeling thin-wire structures, while another form is better-suited analyzing metal plates. Although it can also be applied in time domain the majority of these equations are in frequency domain.

The first step in the MoM solution process is to expand **J** as a finite sum of basis (or expansion) functions,

$$\mathbf{J} = \sum_{i=1}^{M} J_i\,\mathbf{b}_i \qquad (I.45)$$

where $\mathbf{b}_i$ is the ith basis function and J_i is an unknown coefficient. Next, a set of M linearly independent weighting (or testing) functions, w_j, are defined. An inner product of each weighting function is formed with both sides of the equation being solved. In the case of the MFIE, this result s in a set of M independent equations of the form,

$$\langle w_j,\mathbf{H}\rangle = \langle w_j,\mathrm{fm}(\mathbf{J})\rangle \qquad j = 1,2,\ldots,M \qquad (I.46)$$

By expanding **J** using eq. (I.44) a set of M equations in M unknowns are obtained as

$$\langle w_j,\mathbf{H}\rangle = \sum_{i=1}^{M} \langle w_j,\mathrm{fm}(\mathbf{J}_i,\mathbf{b}_i)\rangle \qquad j = 1,2,\ldots,M \qquad (I.47)$$

This can be written in matrix form as

$$[\mathbf{H}] = [Z]\,[\mathbf{J}] \qquad (I.48)$$

where

$$Z_{ij} = \langle w_j, \mathrm{f_m}(\mathrm{b}_i)\rangle$$
$$\mathbf{J}_i = J_i$$
$$\mathbf{H}_j = \langle w_j, \mathbf{H}_{\mathbf{inc}}\rangle$$

The vector **H** contains the known incident field quantities and the terms of the *Z*-matrix are the functions of the geometry. The unknown coefficients of the induced currents are the terms of the **J** vector. These values are obtained by solving the system of equations. Other parameters such as the scattered electric and magnetic fields can be calculated directly from the induced currents.

Depending on the form of the integral equation used, MoM can be applied to configurations of

- conductors only,
- homogeneous dielectrics only, or
- very-specific conductor-dielectric geomeries.

MoM techniques applied to integral equations are not very effective when applied to arbitrary configurations with complex geometries or inhomogeneous dielectrics.

Nevertheless, MoM techniques do an excellent job of analyzing a wide variety of important three-dimensional radiation and scattering problems. General-purpose MoM codes are particularly efficient of modeling wire antennas or wires attached to large conductive surfaces.

Any metallic object is assumed to be a superposition of small segments on which the current distributions are of interest. For example, a PEC wire and a rectangular plate, divided into segments, where currents are assumed constant, are shown in Fig. I.26. Depending on the number of segments, a matrix representation is used and the current distributions are calculated by matrix inversion (see (I.48)). Once the current distributions are calculated, any desired parameter, such as near and far fields, S parameters, impedance, etc., can be calculated.

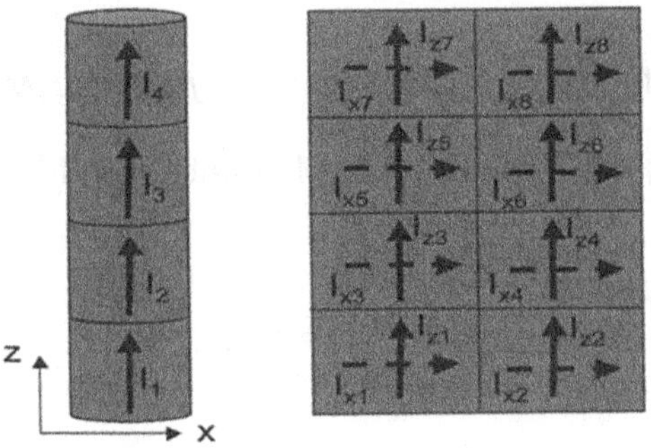

***Figure I.26** : A wire and a rectangular meshed plate.*

In Fig. I.27 and Fig. I.28, two examples that are modeled via MoM technique are given. An aircraft model in Fig. I.27 is modeled by using 255 segments.

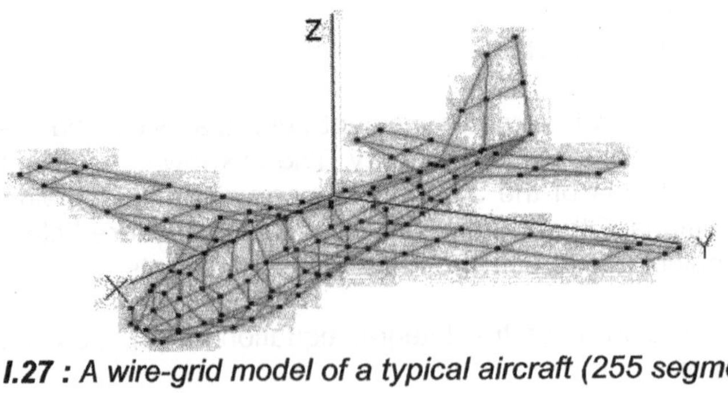

***Figure I.27** : A wire-grid model of a typical aircraft (255 segments)*

Here, a 7.2 m-long aircraft has 7 m wing length with a cockpit that is 1.1 m wide and 1.9 m high. A more complicated structure is shown in Fig. I.28. A three-pole vertically polarized monopoles are erected above a specially designed ground screen, where 120 half wavelength radial wires are used. Nearly 1800 segments are used to prepare a segment model of this antenna system.

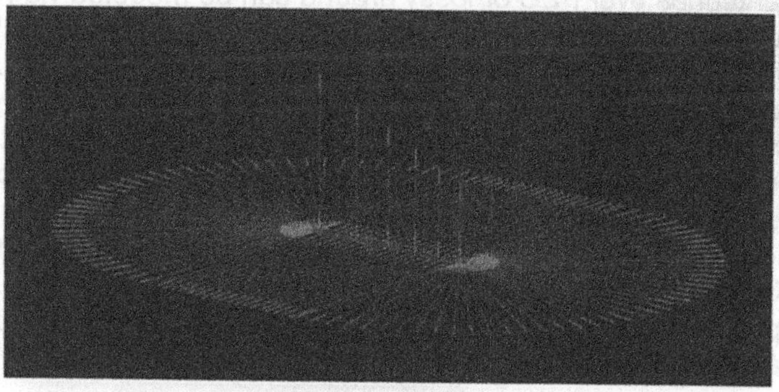

***Figure I.28** : A wire-gird model of an HF transmitter antenna system with a complex ground screen design (nearly 1800 segments)*

These two examples are too complex to handle via either analytical or other numerical solutions. Structures such as shown in this figure can easily be modeled via MoM. The limitations of MoM are

- the number of segments
- and diffraction effects.
- separation of sub-structures

Since MoM is based on the calculation of currents on small segments and since it is formulated as a matrix system, the memory and computation time requirements drastically increase as the number of segment increases. The number of segments depends on the size of the structure and the frequency. Segmentation with one over ten wavelength (i.e., $\lambda/10$) is usually a good choice for most of the EM problems.

Attention should be paid when interpreting MoM results in problems where diffraction effects are important. In MoM, diffraction effects are not taken into account, so edge and/or tip treatments in problem geometries must be handled via some other methods.

Finally, separation of sub elements is important. It should be less than a wavelength (in the near field) since inter-element propagation loss is not taken into account in this form of the MoM code (NEC2).

I.6.1.1 NEC2 Package

The NEC2 package [96] uses a MoM technique that is based on numerical solution of integral equations for the currents induced on metallic structures by sources or by an incident field. Antenna and/or RCS modeling is possible as long as a structure is modeled as a collection of thin wires. The package is especially handy in modeling wire antennas having lengths up to several wavelengths.

The major capabilities and their range of validity may be outlined as follows:

- Wire antennas over PEC or lossy ground can be modeled via NEC.
- Its radial ground screen option can only be used for broadcasting antennas (i.e., for a stand-alone vertical wire antenna over lossy ground).
- Radial and/or rectangular mesh-type ground screens for arrays can only be modeled by the user as a combination of separate wires. Attention should be paid when horizontal wires are located close to ground. Horizontal wires may be as close as $10^{-6}\lambda$ to ground.
- Diffraction effects are not taken into account in NEC, so any difference more than 30 - 40 dB in radiation patterns (or gains) has no practical meaning.
- An option is added into NEC to model two mediums (e.g., poor land and sea) and/or cliffs with a surface impedance approximation.

I.6.2 Split-Step Parabolic Equation Method

The well-known 2D frequency-domain wave propagator SSPE has been effectively used to model ground wave propagation through realistic media for more than a decade [33]. The idea of SSPE is to numerically solve parabolic-type wave equation via discrete Fourier transform, as

$$u(z,x) = \exp\left[i\frac{k}{2}n\Delta z\right]FFT^{-1}\left[\exp\left[-i\frac{k_x\Delta z}{2k}\right]FFT\{u(z_0,x)\}\right] \tag{I.49}$$

where $u(z,x)$ is the scalar wave function, and $\Delta z = z\text{-}z_0$, k_x, FFT and FFT^{-1} correspond to range step size, transverse wave number, Fast Fourier and inverse Fast Fourier transforms, respectively. PE is an initial value problem, therefore, an initial transverse field distribution, $u(z_0,x)$ is injected first. It is longitudinally propagated through a medium defined by its refractive index profile $n(z,x)$ and the transverse field profile $u(z_0+\Delta z,x)$ at the next range step is obtained. The SSPE is a one-way propagator, which neglects back-scatter effects and is valid for near-axial propagation. As long as it is trapped over the earth's surface, SSPE can handle both non-flat PEC terrain and imperfect surface effects (see [33] for details and related references).

A typical 2D SSPE propagation space including a longitudinal smooth terrain profile (colored black) is pictured in Fig.I.29, where signal strength vs. range/height is plotted with color-scale for the standard atmosphere at mid-HF.

For this type of a problem, the implementation of the SSPE algorithm is as follows [97 - 98]:

1. An initial height profile is injected into the SSPE algorithm. A 1D complex array is used to build the initial profile $u(z_0,x_i)$, where x_i (i = 1, .., L) is the sampled height coordinate and L is the number of height points. This array contains amplitude and phase of the fields at each height. Although not a

requirement, a Gaussian beam pattern is often used for the initial vertical field profile.

2. The initial profile is propagated longitudinally from z_0 to $z_0+\Delta z$ via eq. (I.46). The same complex array is used to store the height profile at the next range step.

3. Using the new height profile as the initial profile for the next step and applying the SSPE propagator for the second time yields the height profile at the second range step $z_0+2\Delta z$, etc. The procedure is applied repeatedly and vertical field profiles are stored at each range step until the propagator reaches the desired range z_{last}. A 3D depiction (field strength vs. range/height) is obtained as shown in Fig.I.29.

4. SSPE relies on sequential operations between the x and k_x domains, which are Fourier transform pairs. Since the transforms are implemented via discretization (i.e., via FFT) the domains are truncated at $\pm X_{max}$ and $\pm k_{xmax}$. The transverse and longitudinal step sizes Δx and Δz, respectively, and the maximum height X_{max} are determined from source and/or observation requirements as well as sampling necessities and aliasing effects [33,97]. Once X_{max} is determined, k_{xmax} is found from the Nyquist sampling criteria, $X_{max} \times k_{xmax} = \pi L$, with L being the transform size.

5. Because SSPE cannot handle the boundary conditions (BC) at the surface (PE defines an *initial* value problem), one usually resorts to Dirichlet- or Neumann-type BCs when perfect electrically conducting (PEC) boundaries are of interest; these BCs can be satisfied either by extending the initial vertical profile from $[0,X_{max}]$ to $[-X_{max}, X_{max}]$ (odd or even symmetric) or by applying a sine or cosine FFT, respectively.

6. The problem at hand has a vertically semi-open propagation region, which means an abrupt truncation is required at a certain altitude. This will cause strong artificial reflections, unless measures are taken into account [33,93]. These non-physical reflections can be removed by using (i) windowing techniques, (ii) artificial absorbing layer, (iii) PML termination. Here, (ii) is used. Adding an absorbing layer above the altitude of interest is equivalent to adding a complex part to the refractive index, which makes the propagation in this layer lossy.

7. For a non-flat PEC terrain, the SSPE algorithm is implemented as follows [33]. The 1D vertical field array starts from the bottom (i.e., just above the smooth earth's surface), and the longitudinally changing non-flat PEC terrain profile is injected into the computation sub-space via either a terrain file or a terrain function; various techniques (e.g., staircase approximation or conformal mapping) are used in such applications. To include non-flat terrain effects at each range, (i) array elements below the local terrain (i.e., field values inside the PEC domain) are forced to be zero for staircase approximation, (ii) a new height variable is used with a terrain term added to

refractivity for conformal mapping approximation. Only forward scattering effects are taken into account with this implementation.

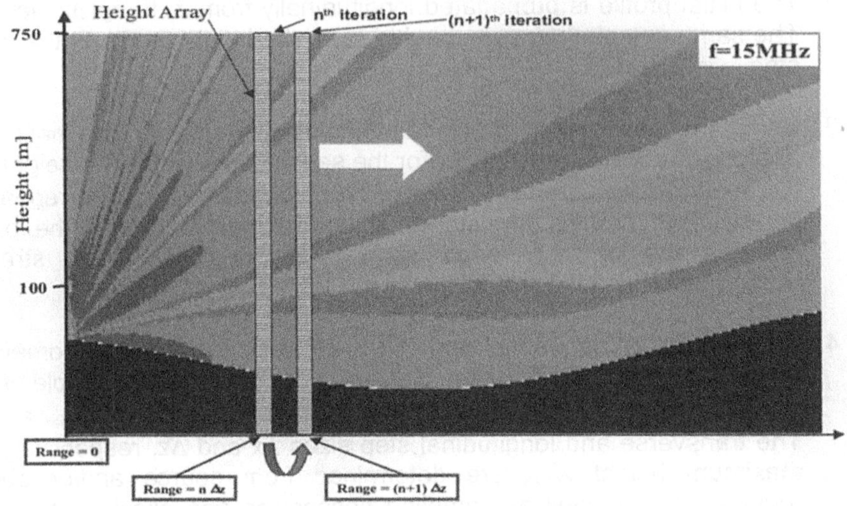

Figure I.29: *A 1D vertical field array propagating towards right*

The discretization parameters are important in SSPE simulations. As explained above, SSPE relies on sequential operations between the x and k_x domains. The user first chooses the altitude of interest X_{max}, which determines the propagation angle (see Fig. I.30). This angle should be inside the paraxial region, since SSPE is reliable in this region. Once X_{max} and α are determined, the transform size L, and height step size, Δx are easily obtained from Nyquist sampling criteria as [98];

$$k_{x\max} = \frac{\pi}{\Delta x}, \qquad \Delta k_x = \frac{\pi}{X_{\max}}, \qquad L = \frac{2k_{x\max}}{\Delta k_x} \tag{I.50}$$

It should be noted that, height resolution should be increased (i.e., step size, Δx should be decreased) in dealing propagation over non-flat terrain (e.g., in hilly terrain). A rough estimation of $\Delta x \approx \lambda/4$ is adequate for a hill with a slope of 5°-10° [33]. Range step size may be 100 m to 500 m which corresponds to a fraction of a wavelength at MF but corresponds to hundreds of wavelengths at UHF and above. [64].

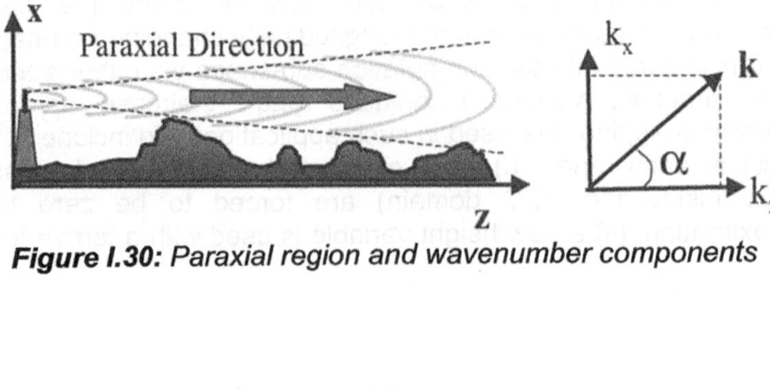

Figure I.30: *Paraxial region and wavenumber components*

A typical example obtained via SSPE is given in Fig. I.31 for a scenario, where non-flat terrain and non-homogeneous refractivity are included.

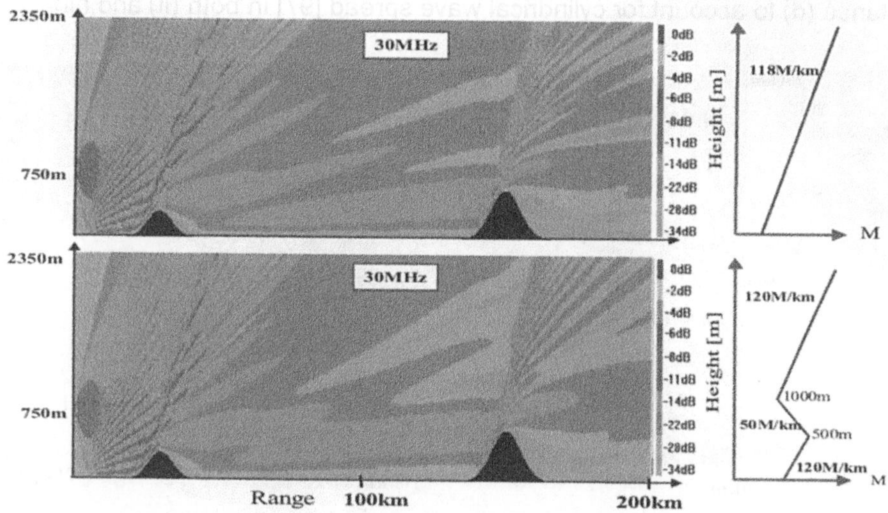

Figure I.31: *Field strength vs. range/height at 30 MHz.*

These non-flat terrain and refractivity profile are chosen because they accommodate various critical wave phenomena (such as diffraction, surface and elevated ducting, guiding-to-antiguiding, etc.). The terrain comprises two separated hills and a 30-MHz transmitter with a Gaussian vertical source distribution is located 750 m above the ground. A Neumann-type boundary condition is applied at the surface. Ground wave propagation through this medium is simulated. As long as the fields are trapped within the paraxial region, propagation over the chosen non-flat terrain profiles in the presence of ducts with arbitrary transverse and/or longitudinal refractivity variation can be modeled via the SSPE propagator.

To illustrate the requirement of the choice of optimum parameters, Fig. I.32 is given. A 300-MHz source is located 100 m above the PEC surface and a standard atmosphere, together with the earth's curvature effect, is also included. Here, the effects of non-physical absorbing boundary termination above the altitude of interest is presented. The top figure shows the physical propagation, where energy detaches the surface and diverges. In the bottom figure, non-physical reflections caused by fictitious absorbing layer termination is shown.

It should be noted that the SSPE solution in (I.49) is the solution of 2D scalar PE-type (longitudinally) one-way equation. Therefore, the scalar wave function $u(z,x)$ must be related to the correct field components of Maxwell's equations in 2D. The PE-type equation may be derived (i) directly in 2D rectangular coordinates, (b) from wave equation in cylindrical coordinates (x,θ,z)) under azimuthal symmetry (θ-independent), and (iii) from wave equation in spherical coordinates (r,θ,φ), under azimuthal symmetry (φ-independent), with approximations in "earth flattened" non-dimensional rectangular coordinates. In (i), *u(z,x)* directly represents the related electric or magnetic field component (depending on TE- or TM-type problems). The scalar wave function *u(z,x)* should be re-normalized with the square root of distance (d) to account for cylindrical wave spread [97] in both (ii) and (iii).

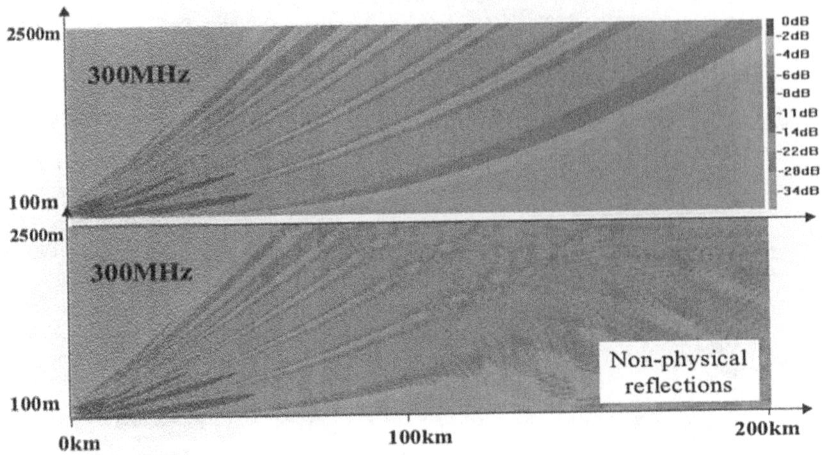

Figure I.32: *Absorbing layer termination. (a) Good (b) Bad*

For example the path loss at a distance d over spherical earth's surface can be calculated from [33]

$$P\ [dB] = -20\log_{10}(u) + 20\log_{10}(4\pi) + 10\log_{10}(d) - 30\log_{10}(\lambda) \quad (I.51)$$

once the function *u* is calculated from SSPE. An example is given in Fig. I.33.

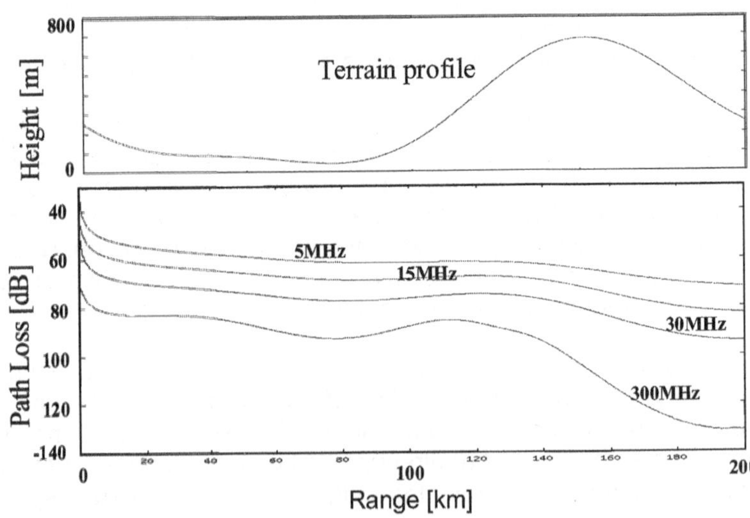

Figure I.33: *Path loss vs. range at HF for propagation over a* PEC hill

As shown in the figure, path loss increases as the frequency increases. Also, path loss decreases (increases) in front of (beyond) the hill, as expected.

I.7 CONCLUSIONS

Numerical simulators have become impressive tools both for scientific researches in academia and for engineers in industry. More comprehensive simulators are becoming available to attack more complex EM problems. Hybridization has also been used in preparation of these simulators, where two or more different numerical approaches are combined to expand the range of validityand accuracy, as well as environment. The need for hybridization arises from practical problems that are usually so large and so complex that no single numerical technique can provide accurate and efficient solution.

The widespread usage of numerical techniques in simulating complex EM problems necessitates us to be very cautious, since the user must

- keep in mind that there is no generally applicable technique, so validity range and accuracy must be controlled
- well-understand the physics behind the mathematics and model of the simulator she uses
- compare the simulation results with reference preferable experimental) data

In Part I of this book, four simulation techniques are outlined from a practical point of view. Major EM problems are taken into account in Part II, where numerical solutions are presented in terms of these simulators.

References

[1] E. K. Miller, *et al.* (Eds.), *Computational Electromagnetics: Finite Difference Method of Moments*, IEEE Press, Piscataway, NJ, 1991

[2] L. B. Felsen, N. Marcuvitz, *Radiation and Scattering of Waves*, Prentice-Hall, Englewood Cliffs, NJ, 1973

[3] L. B. Felsen, M. Mongiardo, P. Russer, "Electromagnetic Field Representations and Computations in Complex Structures: I. Complexity Architecture and Generalized Network Formulation", *Int. J. Numerical Modeling*, Vol.15 No. 1, pp. 93 - 107, 2002

[4] L. B. Felsen, M. Mongiardo, P. Russer, "Electromagnetic Field Representations and Computations in Complex Structures: II. Alternative Green's Functions", *Int. J. Numerical Modeling*, Vol.15 No.1, pp.109 -125, 2002

[5] L. B. Felsen, M. Mongiardo, P. Russer, "Electromagnetic Field Representations and Computations in Complex Structures: III. Network Representations of the Connecting and Subdomain Circuits", *Int. J. Numerical Modeling*, Vol.15 No. 1, pp. 127 -145, 2002

[6] A. Ishimaru, *Electromagnetic Wave Propagation, Radiation and Scattering*, Prentice-Hall, Englewood Cliffs, NJ, 1991

[7] E. Heyman, L. B. Felsen, "A Wave Front Interpretation of the Singularity Expansion Method", *IEEE Trans. Antennas and Propagat.*, Vol. 33, pp. 706 - 718, July 1985

[8] L. B. Felsen, "Progressing and Oscillatory Waves for Source-excited Propagation and Diffraction", *IEEE Trans. Antennas and Propagat.*, Vol. 32, pp. 775 - 796, 1984

[9] L. B. Felsen, F. Niu, "Spectral Analysis and Synthesis Options for Short Pulse Radiation from a Point Dipole in a Grounded Dielectric Layer", *IEEE Trans. Antennas and Propagat.*, Vol. 41, pp. 747 - 754, 1993

[10] F. Niu, L. B. Felsen, "Time Domain Leaky Modes on Layered Media: Dispersion Characteristics and Synthesis of Pulsed Radiation", *IEEE Trans. Antennas and Propagat.*, Vol. 41, pp. 755 - 761, 1993

[11] L. B. Felsen, F. Niu, "Asymptotic Analysis and Numerical Evaluation of Short Pulse Radiation from a Dipole in a Grounded Dielectric layer", *IEEE Trans. Antennas and Propagat.*, Vol. 41, pp. 762 - 769, 1993

[12] L. G. James, *Geometric Theory of Diffraction for Electromagnetic Waves*, IEE Electromagnetic Wave Series-I, Peter Peregrinus, London, 1976

[13] A. Ishimaru, *Electromagnetic Wave Propagation, Radiation and Scattering*, Prentice-Hall, Englewood Cliffs, NJ, 1991

[14] A. D. Pierce, "Extension of Method of Normal Modes to Sound Propagation in an Almost Stratified Media", *J. Acoust. Soc. Am.*, Vol. 37, pp.19 - 27, 1965

[15] A. D. Pierce, "Guided Mode Disappearance During Upslope Propagation in Variable Depth Shallow Water Overlying a Fluid Bottom", *J. Acoust. Soc. Am.*, Vol. 72, pp. 523 - 631, 1982

[16] J. M. Arnold, L. B. Felsen, "Intrinsic Modes in a Wedge-shaped Ocean", *J. Acoust. Soc. Am.*, Vol. 76, pp. 850 - 860, 1984

[17] L. B. Felsen, L. Sevgi, "Adiabatic and Intrinsic Modes for Wave Propagation in Guiding Environments with Longitudinal and Transverse Variations: Formulation and Canonical Test", *IEEE Transactions on Antennas and Propagat.* Vol. 39 No. 8, pp. 1130 - 1136, Aug. 1991

[18] L. B. Felsen, L. Sevgi, "Adiabatic and Intrinsic Modes for Wave Propagation in Guiding Environments with Longitudinal and Transverse Variations: Continuously Refracting Media", *IEEE Transactions on Antennas and Propagat.* Vol. 39 No. 8, pp. 1137 - 1143, Aug. 1991

[19] G. Deschamps, "Gaussian Beams as a Bundle of Complex Rays", *Electronics Letters*, Vol. 7, pp. 684 - 685, 1971

[20] L. B. Felsen, "Complex Rays", *Philips Res. Repts*, Vol. 30, pp.169 - 184, 1975

[21] L. B. Felsen, "Rays, Modes and Beams in Optical Fibre Waveguides", *J. Acoust. Soc. Am.*, Vol. 66 No. 8, pp. 751 - 760, Aug. 1976

[22] E. Heyman, L. B. Felsen, "Gaussian Beam and Pulsed Beam Dynamics: Complex Source and Spectral Formulations within and beyond Paraxial Asymptotics", *J. Opt. Soc. Am.*, Vol.18, pp. 1588 - 1611, 2001

[23] T. Ishihara, L. B. Felsen, "Hybrid (Ray)-(Parabolic Equation) Analysis of Propagation in Ocean Acoustic Guiding Environments", *J. Acoust. Soc. Am.*, Vol.83, pp. 950 - 960, 1988

[24] T. Ishihara, L. B. Felsen, "Hybrid Ray-Mode Parameterization of High Frequency Propagation in an Open Waveguide with Inhomogeneous Transverse Refractive Index: Formulation and Application to a Bilinear

Surface Duct", *IEEE Transactions on Antennas and Propagat.* Vol.39 No.6, pp. 780 -788, 1991

[25] S. H. Marcus, "A Hybrid (Finite Difference-Surface Green's function) method for computing Transmission Losses in an Inhomogeneous Atmosphere ver Irregular Terrain", *IEEE Transactions on Antennas and Propagat.* Vol. 40 pp. 1451 - 1458, 1992

[26] M. F. Levy, "Horizontal Parabolic Equation Solution of Radiowave Propagation Problems on Large Domains", *IEEE Transactions on Antennas and Propagat.* Vol.43, pp. 137 - 144, 1995

[27] W. J. R. Hoefer, "Time Domain EM Simulation for Microwave CAD Applications", *IEEE Trans Microwave Theory Tech*, Vol.40 No.7, pp. 1517 - 1527, July 1992

[28] K. S. Yee, "Numerical Solution of Initial Boundary Value Problems Involving Maxwell' s Equations", *IEEE Transactions on Antennas and Propagat*, Vol. 14 No. 3, pp. 302 - 307, May 1966

[29] P. B. Johns, R. L. Beurle, "Numerical Solution of Two-Dimensional Scattering Problems using TLM", *Proc. IEE*, Vol. 118, pp. 1203 - 1208, 1971

[30] C. S. Cesari, J. F. Abel, *Introduction to the Finite Element Method: A numerical Approach for Engineering Analysis*, Van Nostrand Reinhold, New York, 1972

[31] R. F. Harrington, *Field Computation by Moment Methods*, The Macmillan Co., New York, 1968

[32] W. C. Chew, J. M. Jin, E. Michielssen, J. Song (Eds.), *Fast and Efficient Algorithms in Computational Electromagnetics*, Artech House, Norwood, MA, 2001

[33] M. Levy, *Parabolic Equation Methods for Electromagnetic Wave Propagation*, IEE Institution of Electrical Engineers, 2000

[34] E. K. Miller, "A Selective Survey of Computational Electromagnetics", *IEEE Transactions on Antennas and Propagat.*, Vol. 36, pp. 1288 - 1305, 1988

[35] G. A. Thiele, "Overview of Selected Hybrid Methods in Radiating Systems Analysis", *Proceeding of IEEE*, Vol. 80, pp. 66 - 78, 1992

[36] L.N. Medgyesi-Mitschang, D. S. Wang, "Review of Hybrid Methods on Antenna Theory", *Annals of Telecommunications*, Vol. 44 No. 9, 1989

[37] W. D. Burnside, C. L. Yu, R. J. Marhefka, "A Technique to Combine Geometrical Theory of Diffraction and Moment Methods", *IEEE Transactions on Antennas and Propagat.*, Vol. 23, pp. 551 - 558, 1975

[38] G. A. Thiele, T. H. Newhouse, "A Hybrid Technique for Combining Moment Methods with the Geometrical Theory of Diffraction", *IEEE Transaction on Antennas and Propagat.*, Vol. 23, pp. 551 - 558, 1975

[39] T. J. Kim, G. A. Thiele, "A Hybrid Diffraction Technique - General Theory and Applications", *IEEE Transactions on Antennas and Propagat.*, Vol. 30, pp. 551 - 561, 1982

[40] L. N. Medgyesi-Mitschang, D. S. Wang, "Hybrid Solution for Scattering from Perfectly Conducting Bodies of Revolution", *IEEE Transactions on Antennas and Propagat.*, Vol. 31, pp. 570 - 583, 1983

[41] L. N. Medgyesi-Mitschang, D. S. Wang, "Hybrid Solutions from Large Bodies of Revolution with Material Discontinuities and Coatings", *IEEE Transactions on Antennas and Propagat.*, Vol. 32, pp. 717 - 723, 1984

[42] L.N. Medgyesi-Mitschang, D. S. Wang, "Hybrid Solutions for Large Impedance Coated Bodies of Revolution", *IEEE Transctions on Antennas and Propagat.*, Vol. 34, pp.1319 - 1329, 1986

[43] D. S. Wang, "Current-based Hybrid Analysis for Surface Wave Effects on Large Scatterers", *IEEE Transations on Antennas and Propagat.*, Vol. 39, pp. 839 - 850, 1991

[44] D. P. Bouche, F. A. Molinct, R. Mittra, "Asymptotic and Hybrid Techniques in Electromagnetic Scattering", *Proceedings of IEEE*, Vol. 81, pp. 1658 - 1684, 1993

[45] U. Jakobus, F. M. Landstorfer, "Improved PO-MM hybrid Formulation for Scattering from three-dimensional Perfectly Conducting Bodies of Arbitrary Shape", *IEEE Transactions on Antennas and Propagat.*, Vol. 43, pp. 162 - 169, 1995

[46] U. Jakobus, F. M. Landstorfer, "Improved PO-MM hybrid Formulation by Accounting for Effects Perfectly Conducting Wedges", *IEEE Transactions on Antennas and Propagat.*, Vol. 43, pp. 1123 - 1129, 1995

[47] X. Yuang, D. R. Lynch, J. W. Struhbehn, "Coupling of Finite Element and the Moment Methods for Electromagnetic Scattering from Inhomogeneous Objects", *IEEE Transactions on Antennas and Propagat.*, Vol. 38, pp. 386 - 393, 1990

[48] D. J. Hoppe, L. W. Epp, J. F. Lee, "A Hybrid Symmetric FEM-MOM Formulation Applied to Scattering by Inhomogeneous Bodies of Revolution", *IEEE Transaction on Antennas and Propagat.*, Vol. 42, pp. 798 - 805, 1994

[49] C. Z. T. Cwik, V. Jamnejat, "Modeling three-dimensional Scatterers using a coupled Finite Element Integral Equation Formulation", *IEEE Transactions on Antennas and Propagat.*, Vol. 44, pp. 453 - 459, 1996

[50] C. Z. T. Cwik, V. Jamnejat, "Modeling radiation with an Efficient hybrid Finite Element Integral Equation Waveguide Mode-matching Technique", *IEEE Transactions on Antennas and Propagat.*, Vol. 45, pp. 34 - 39, 1997

[51] K. Ise, M. Koshiba, "Numerical Analysis of H-plane Waveguide Junctions by Combination of Finite and Boundary Elements", *IEEE Trans. Microwave Theory and Techniques.*, Vol.36, pp.1343 - 1351, 1988

[52] M. Mongiardo, R. Sorrentino, "Efficent and Versatile Analysis of Microwave Structures by combined Mode Matching and Finite difference Methods", *IEEE Microwave Guided Wave Letters.*, Vol.3, pp. 241 - 243, 1993

[53] N. Lu, J. M. Jin, "Application of Fast Multipole Method to Finite Element Boundary-Integral Solution of Scattering Problems", *IEEE Transactions on Antennas and Propagat.*, Vol. 44, pp. 781 - 786, 1997

[54] J. M. Reiter, F. Arndt, "Hybrid Boundary-contour mMode-matching Analysis of Arbitrary Shaped Waveguide Structures with Symmetry of Revolution", *IEEE Microwaves and Guided Wave Letters.*, Vol. 6, pp.369-371, 1996

[55] R. E. Hodges, Y. Rahmat-Samii, "An Iterative current-based Hybrid Method for Complex Structures", *IEEE Transactions on Antennas and Propagat.*, Vol. 45, pp. 265 - 276, 1997

[56] P. Roth, "Discrete, Continuous and Combined Simulation ", *Proc. of the Winter Simulation Conference*, CA, 1989

[57] N. İnce (Ed.), *Modeling and simulation Environment for Satellite and Terrestrial Communication Networks*, Kluwer Academic Publishers, Boston, 2001

[58] D. Pedgen, *et al.*, *Introduction to Simulation Using SIMAN*, McGraw-Hill, Nw York, 1995
[59] L. Sevgi, A. S. Sanal "Surface Wave HF Radar Simulator", *Proc. of IEE Radar 97 Conference*, 14-16 October 1997, Edinburgh, UK
[60] L. Sevgi, "Stochastic Modeling of Target Detection and Tracking in Surface Wave High Frequency Radars", *Int. J. of Numerical Modeling*, Vol. 11 No 3, pp. 167 - 181, May 1998
[61] A. N. Ince, E. Topuz, E. Panayirci, C. Isik. *Principles of Integrated Maritime Surveillance Systems*, Kluwer Academic Publishers, Boston, 2000
[62] L. Sevgi, A. M. Ponsford, H.C. Chan, "An Integrated Maritime Surveillance System Based on Surface Wave HF Radars, Part I – Theoretical Background and Numerical Simulations", *IEEE Antennas and Propagation Magazine*, Vol. 43 No. 4, pp. 28 - 43, Aug. 2001
[63] A. M. Ponsford, L. Segi, H.C. Chan, "An Integrated Maritime Surveillance System Based on Surface Wave HF Radars, Part II – Operational Status and System Performance", *IEEE Antennas and Propagation Magazine*, Vol. 43, No. 5, pp. 52-63, Oct. 2001
[64] L. Sevgii L. B. Felsen, "A new Algorithm for Ground Wave Propagation Based on a Hybrid Ray-Mode Approach", *Int. J. of Numerical Modeling*, Vol.11 No. 2, pp. 87 - 103, March 1998
[65] F. Akleman, L. Sevgi, "A Novel Finite Difference Time Domain Wave Propagator", *IEEE Transactions on Antennas and Propagat.*, Vol. 48 No. 5, pp. 839 - 841, May 2000
[66] F. Akleman, L. Sevgi, "Time and Frequency Domain Wave Propagators", *ACES Journal, Special issue on CEM Techniques in Mobile Wireless Communication*, Vol. 15 No.3, pp. 186 - 209, 2000
[67] L. Sevgi, F. Akleman, L. B. Felsen, "Ground Wave Propagation Modeling: Problem-matched Analytical Formulations and Direct Numerical Techniques", *IEEE Antennas and Propagation Magazine*, Vol. 44 No.1, pp. 55 - 75, Feb. 2002
[68] M. O. Özyalçın, F. Akleman, L. Sevgi, "A Novel TLM Based Time Domain Wave Propagator", *IEEE Transactions on Antennas and Propagat.*, (to appear) June 2003
[69] F. Akleman L. Sevgi, "Realistic Surface Modeling in a Time-domain Wave Propagator ", *IEEE Transactions on Antennas and Propagat.*, (to appear) July 2003
[70] Visit http://www.fas.org or http://www.chinfo.navy.mil sites
[71] L. Sevgi, S. Paker, "FDTD Based RCS Calculations and Antenna Simulations", *AEU, International J. of Electronics and Commun.*, Vol. 52, No. 2, pp. 65 - 75, March 1998
[72] F. Akleman, M. O. Özyalçın, L. Sevgi, "Comparison of TLM and FDTD Techniques in RCS Simulation and Antenna Modeling", *Proc. Of AP 2000 (ICAP+JINA) Millenium Conference on Antennas and Propagation*, 9-14 April 2000, Davos, Switzerland
[73] L. Sevgi, "Target Reflectivity and RCS Interaction in Integrated Maritime Surveillance Systems Based on Surface Wave HF Radar Radars", *IEEE Antennas and Propagation Magazine*, Vol. 43 No.1, pp. 36 - 51, Feb. 2001

[74] GAO, "Research and Regulatory Efforts on Mobile Phone Health Issues", US General Accounting Office Report to Congress, GAO-01-545, May 2001 (visit http://www.gao.gov)
[75] M. O. Özyalçın, F. Akleman, L. Sevgi, "SAR Simulations in Wireless Communication and Safety Discussions in the Society", *Special issue of ELEKTRIK, Turkish J. of Electrical Engineering and Computer Sciences*, Vol. 10 No.2, pp. 198 - 216, Feb 2002
[76] ICNIRP "Guidelines for Limiting Exposure to time-varying electric, magnetic, and electromagnetic fields (up to 300 GHz)", *Health Physics*, Vol. 74 No. 4, pp. 494 - 522, 1998
[77] IEEE Standards for Safety Levels with Respect to Human Exposure to Radio Frequency Electromagnetic Fields, 3 kHz to 300 GHz, *IEEE Standard C95.1-1991*, 1992
[78] FCC Guideliness, "Evaluating Compliance with FCC Guidelines for Human Exposure to Radio Frequency Electromagnetic Fields," FCC, Washington D.C., *OET Bull.* 65, August 1997
[79] Visit http://www.speag.com, //www.semcad.com, //www.dasy4.com, //www.indexsar.com, //www3.dogus.edu.tr/lsevgi
[80] TUBITAK- Marmara Research Center, *Land Mine Detection Project*, 1997-2000, Gebze / Kocaeli
[81] F. Akleman, M. O. Özyalçın, L. Sevgi, "Subsurface Imaging and Time Domain simulation Techniques", *Special Session on "Subsurface imaging and reconstruction", PIERS 2002 Progress in Electromagnetics Research Symposium*, MA, USA, 1-5 July 2002
[82] T. Itoh, B. Houshman (Eds.), *Time Domain Methods for Microwave Structures: Analysis and Design, IEEE Press*, Piscataway, NJ, 1998
[83] K. S. Kunz, R. J. Luebbers. *The Finite-Difference Time-Domain Method for Electromagnetics*. CRC Press, Boca Raton, FL, 1993
[84] D. M. Sullivan, Electromagnetic *Simulation Using the Finite-Difference Time-Domain Method, IEEE Press*, New York, 2000
[85] A. Taflove, *Computational Electrodynamics: The Finite-Difference Time-Domain Method*, Artech House, Boston, MA, first edition, 1995, second edition, 2000
[86] A. Taflove, *Advances in Computational Electrodynamics: The Finite-Difference Time Domain Method*, Artech House, Boston, MA, 1998
[87] C. Christopoulos, *The Transmission-Line Modeling Method*, Oxford University Press, Oxford, 1995
[88] N. K. Uzunoglu, *et al.* (Eds.), *Applied Computational Electromagnetics* (Part IV, P. Russer (Ed.), "Transmission Line Method"), NATO ASI Series, Springer Verlag, Berlin, 2000
[89] R. J. Luebbers *et al.*, "A Finite-Difference Time-Domain Near Zone to Far Zone Transformation", *IEEE Transactions on Antennas and Propagat.*, Vol. 39 No. 4, pp. 429 - 433, 1991
[90] S. A. Schelkunoff, "Some Equivelance Theorems of Electromagnetics and Their Applications to Radiation Problems", *Bell Syst. Tech. J.* 92, 1934
[91] G. J. Burke, A. J. Poggio, "Numerical Electromagnetic Code - Method of Moments, Part I: Program Description, Theory", *Technical Document, 116, Naval Electronics System Command* (ELEX 3041), July 1977

[92] G. Mur, "Absorbing Boundary Conditions for the Finite Difference Approximation of the Time Domain Electromagnetic Field Equations", *IEEE Transactions on EMC*, Vol. 23, pp. 377 - 382, 1981

[93] Z. Q. Bi, K. L. Wu, C. Wu, J Litva, " A Dispersive Condition for Microstrip Component Analysis Using the FD-TD Method", *IEEE Transactions on Microwave Theory and Techniques.*, Vol. 40, pp. 774 - 777, 1992

[94] V. Betz, R. Mitra, "Comparison and Evaluation of Boundary Conditions for the Absorption of Guided Waves in an FDTD Simulation", IEEE Microwave Guided Lett., Vol 2, pp. 499 - 501, 1992

[95] J. P. Berenger, "Three Dimensional Perfectly Matched Layer for the Absorption of Electromagnetic Waves", *J. f Computational Physics*, Vol.127, pp. 363 - 379, 1996

[96] F. Akleman, L. Sevgi, "A Novel Implementation of Berenger's PML for FDTD Applications", *IEEE Microwaves and Guided Wave Letters*, Vol. 8 No. 10, pp. 324 - 327, October 1998

[97] L. Sevgi, E. Topuz, "Split Step Parabolic Equation analysis of Coupled Dielectric Waveguides", *ELEKTRIK, Turkish J. of Electronics and Communication*, Vol. 3 No. 2-3, pp. 85 - 92, 1995

[98] L. Sevgi "Split Step Parabolic Equation Solutions in Surface Duct-to-Elevated Duct Transition", *Turkish J. of Physics*, Vol. 19 No. 3, pp. 541 - 551, March 1995

II

Part II

Applications

1

Ground-Wave Propagation

with **FUNDA AKLEMAN**

1.1 INTRODUCTION

In general, propagation through inhomogeneous media, with variable and imperfect boundaries is simulated in this chapter. Although ground waves are of main interest, numerical simulators presented here are quite general and applicable to, for example, propagation through transmission lines, microstriplines, coaxial cables, waveguides, optical films, etc.

The frequency bands of interest are MF (medium frequencies, 300 kHz to 3 MHz), HF (high frequencies, 3 - 30 MHz), VHF (very high frequencies, 30 - 300 MHz), UHF (ultra high frequencies, 300 MHz to 1 GHz), and MW (microwaves, 1 GHz and above).

At frequencies between 3 kHz and 30 MHz, two types of electromagnetic (EM) wave propagation between two points on or above the earth are possible: ground wave propagation and indirect sky wave (ionospheric wave). Ground wave propagation is affected by the tropospheric variations (neutral atmosphere) as well as ground characteristics. Sky wave is reflected from the ionosphere back to the earth at appropriate frequencies and angles of incidence giving rise to long-range propagation. Although the ionosphere acts as a reflector up to frequencies of a few megahertz, it becomes more penetrable for EM waves above 30 MHz as the operating frequency increases; and at a critical frequency, no reflection towards ground occurs. Therefore, propagation along the earth via sky wave is not possible at VHF and upper frequencies. In this chapter, the ground wave propagation is of main interest for the frequency bands given above; and the characteristics of the ionosphere, which is composed of a number of ionized regions, are not considered here.

Ground waves have three components: direct waves, ground-reflected waves, and surface waves. The model environment is a spherical earth with various ground characteristics, above which exists a radially inhomogeneous atmosphere. Having an excitation and observer anywhere on or above the ground, this model has served as a canonical problem. The problem is very complex and neither a full-wave, observable-based, numerically computable analytical solution nor a 3D, generally applicable, numerical solution has appeared yet. Instead, analytical approximate and 2D numerical solutions, valid under different parameter regimes, together with some hybridized solutions, are being used.

The physical characteristics of the propagation depend on many parameters, such as the operating frequency, medium parameters, grazing angle, transmitter and receiver locations, and the geometry (boundary conditions, BC) between them. Some characteristics of the ground wave propagation can be listed as:

- At MF and lower HF, ground-wave propagation is dominated by the surface wave. As long as the transmitter and receiver are close to the surface, direct and ground reflected waves cancel each other and only surface waves can propagate. The earth's surface electrical parameters are important in reaching longer ranges. Sea surface is a good conductor, but ground is a poor conductor at these frequencies. For example, with the same transmitter and receiver characteristics, a 5 MHz signal, which reaches 400-km range over the sea, can only reach up to 40 - 50-km range over the poor ground. Typical electrical parameters (σ, conductivity, ε_r, relative permittivity) are listed in Table 1.1.

- At MF and lower HF, propagation is possible up to the 400 - 500-km range. MF are used for long- and medium-wave radio broadcasting. Lower HF is used for long-range marine communication and HF surface-wave radars. Propagation beyond the line-of sight (LOS, a straight line from transmitter that is tangent to the earth's surface) is possible at these frequencies via surface waves and/or sky waves.

- Surface waves are hardly excited and coupled to the surface, when the transmitter is many wavelengths (e.g., 5λ to 10λ) above the ground. When excited and coupled, they exponentially decay with height; therefore surface waves are rarely used above 10 - 15 MHz.

- At VHF and above (i.e., VHF, UHF and MW), propagation is limited by the LOS, because surface-wave contribution becomes negligible at these frequencies and the ground wave is then known as the space wave, which is the sum of direct and ground reflected waves. For near grazing path, the complex reflection coefficient of the ground for both horizontal and vertical polarization approaches –1, which is appropriate to assume ground as perfectly conducting.

- At VHF, propagation beyond the LOS is still possible by means of diffracted components (typically 5 km to 20-km range beyond obstacles)

Table 1.1: Typical ground parameters

Ground	**σ [S/m]**	ε_r
Sea	5.0	80.0
Medium ground	0.01	15.0
Poor ground	0.001	7.0

Today's communication systems and radars are mostly used within multi-area, multi-sensor, land-based, maritime and/or air-based complex integrated systems (such as an integrated maritime surveillance system or integrated early warning system against tactical ballistic missiles, etc.). Research towards the development and to the performance evaluation of such systems requires powerful computer simulation tools and can only be done via these simulators. Some requirements, related to current integrated system simulations, are as follows:

- Modeling of propagation characteristics and path loss predictions in integrated surveillance system simulations, where optical, microwave, and HF radars are used

- Service planning, in rural and urban locations, for mobile communication systems using 900-MHz and 1.8-GHz systems and for FM radio stations

- Sensor location and service planning for Intelligent Traffic Management Systems (ITMS) in a complicated and densely occupied urban metropolis

- Control and/or detection of very low altitude missiles, camouflaged by the local terrain profile up to the target

- Knowledge of propagation characteristics to overcome problems related to emerging radar technologies (HF and VHF radars), ..., etc.

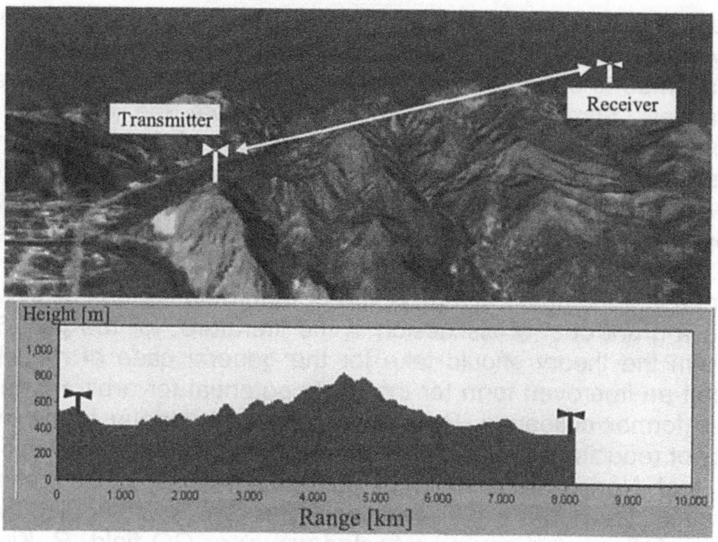

Figure 1.1: *A typical digital map and the height/range function for a chosen propagation path*

In real-world systems, such as those exemplified above, the problem of determining the propagation characteristics between any two selected points (e.g., a communication transmitter and receiver) can best be addressed via an accurate and versatile simulation model. Such a simulator is expected to accept the characteristics of the propagation environment as input (digitized map of the non-flat terrain, as seen in Fig.1.1, ground cover types, parameters of the troposphere, etc.) and provide wave propagation characteristics as output, in a nearly real-time basis. Clearly such a simulation tool would be indispensable to the decision-maker (service planner, site engineer, or the leader of a small military ground contingent) and it has therefore been a continuing challenge to develop a simulator, which satisfies these requirements. The challenge here is to parameterize this complex system in terms of the variety of wave processes that operate there.

1.1.1 Historical Overview

In the review article by J. R. Wait [1] the ancient and modern history of EM ground wave propagation is outlined, and the feature article by N. DeMinco [2] contains a wide-ranging survey of radio wave propagation and antenna modeling. Here we will present a brief history of the investigations related to wave propagation over the earth. For a more detailed account on this subject the reader is referred to the review paper of the J. R. Wait [1].

Beginning with the original investigations of Heinrich Hertz, the influence of the interference caused by the obstacles in the path had been an important subject and then in 1899 Marconi investigated the weakening of the field strength when a hill was located between transmitter and receiver. Early conjectures were that the signal was guided in some fashion by the air - earth boundary, and at that time the possible existence of the ionosphere was also broached. At the turn of the century, the presence of the earth had been a key factor to be reckoned with, and this was the beginning of the subject that is now called "ground-wave propagation."

In 1907, Zenneck analyzed the possibility that the air - earth interface supported a surface wave with low attenuation. Although the guiding-wave mechanism of the air - earth interface seemed to be a leading subject to explain the radio wave transmission over long distances, Sommerfeld took the lead to put the question on a firm mathematical basis. His analysis dealt with the problem of determining the fields of electric and magnetic dipoles located in the insulating half-space over a conducting half-space as well as dipoles located in the planar interface. In 1926, Sommerfeld corrected his original formulation. In 1930s, there was activity on the question of the validity of the Sommerfeld theory, including the modified 1926 version. Pol and Niessen confirmed that the physical form of the Sommerfeld 1926 version was correct. There has been a great deal of discussion in the literature, for the past 50 years or more, on the form the theory should take for the general case of raised terminals. Norton developed an improved form for the Hertz potential for arbitrary heights. J. R. Wait [3] and his former colleague, R. J. King proposed various forms that had the uniform property of reducing exactly to geometrical optics at higher elevation angles. It is pointed out that Norton formulation of surface wave is to be defined as the difference between the total wave field and the easily computed space-wave component (sum of direct and ground-reflected wave) or GO field. R. King has also employed the surface-impedance model, in an independent integral-equation solution.

In order to deal with the actual physical world, a spherical earth model has to be used. In 1918, G. N. Watson began with the classical harmonic series solution for a dipole, in the presence of a homogeneous sphere of radius a with the electrical properties conductivity σ, dielectric permittivity ε, and magnetic permeability μ. The surrounding medium was accepted as free space. Although the solution involving spherical Bessel functions of integer order was mathematically correct, convergence properties were poor. Watson's first step was to represent the series by a contour integral that enclosed the real axis of the complex wavenumber plane, and then Watson wrapped the contour around a manageable number of complex poles, which provided a highly convergent residue series in the shadow region. The improved spherical-earth modal residue series solution is explained in detail in [3,4]. The mixed-path (i.e., land - sea transition) theory [5] as well as irregular terrain concepts [6,7] are also analyzed with analytic approximate approaches by various researches (see [1] for reference list).

1.1.2 Available Approaches

Problem-matched analytic approximate approaches as well as purely numerical techniques have been used and are still being developed to better understand and model wave propagation over a non-flat terrain covered with various ground cover types. Analytic models listed below assume a smooth spherical earth (and/or its earth-flattened approximate equivalent) with various smooth ground characteristics, a radially inhomogeneous atmosphere above and excitation by a vertical or a horizontal electric dipole on or above the earth's surface.

1. The Norton formulation, which extracts from a wavenumber spectral integral representation a ray-optical asymptotic approximation [8].

2. The Wait formulation, which restructures the spectral integral as a series of normal modes propagating along the earth's surface [3].

3. The Ishihara - Felsen formulation, which combines ray fields and mode fields in a self-consistent hybrid scheme [9,10], which however, is valid under less general conditions than those of options 1 and 2 individually.

4. The hybrid algorithm, WAVEPROB, which combines Norton and Wait solutions in an efficient manner to extend the range of validity of options 1 and 2 [11].

5. The adiabatic and intrinsic mode formulations, which extend analytic modal solutions to media with transverse as well as longitudinal variations [12,13].

6. The numerical solution of the parabolic equation (PE) that is derived in an approximate sense neglecting the back-scattered field (one-way propagation) and is most directly applicable within a narrow range of angles between the propagation direction and range (i.e., paraxial approximation) (see [14] for historical review and references for various PE approaches).

7. A split-step Padé PE method for problems involving large angles between propagation direction and range [14].

8. An improved PE method for impedance (Cauchy-type) boundary conditions [14,15].

9. The hybrid ray - PE method, which addresses transitional effects in option 3 [16].

10. A hybrid (finite difference PE) – (surface Green's function) method to account for propagation loss over mixed paths [14,17].

11. A horizontal PE method for the rigorous extension of standard PE solutions to large transverse spatial domains [14].

12. The FDTD-based wave propagator (TDWP), which simulates and traces broadband pulse propagation over long distances [18,19].

13. The TLM-based wave propagator (TLM-WP), which simulates and traces broadband pulse propagation over long distances [20].

14. Simplified Deygout solution for multiple knife-edge diffraction [21] (more sophisticated diffraction models representing the terrain profile as a succession of knife-edges, wedges, cylinders or flat strips are also available [22,23]. Another important family of models uses an integral equation formulation [6,7], which can be simplified by paraxial approximation [24]. All of these techniques assume propagation in a homogeneous or linear atmosphere).

In dealing with the propagation requirements listed above, it is essential to understand propagation characteristics over spherical earth's surface such as (i) energy spread because of curvature effects, (ii) wave scattering because of irregular terrain, (iii) guiding as well as anti-guiding effects because of refractivity variations, etc. Although it is a source-driven (i.e., a Green's function) problem in 3D spherical co-ordinates, because of the azimuthal symmetry met in almost all practical applications this problem can be reduced to 2D and using appropriate transformations the natural co-ordinates of the problem can be converted into normalized Cartesian co-ordinates (see Fig. 1.2) [11,19]. Detailed reviews of these radiowave propagation problems can be found in [1,2,25 - 27].

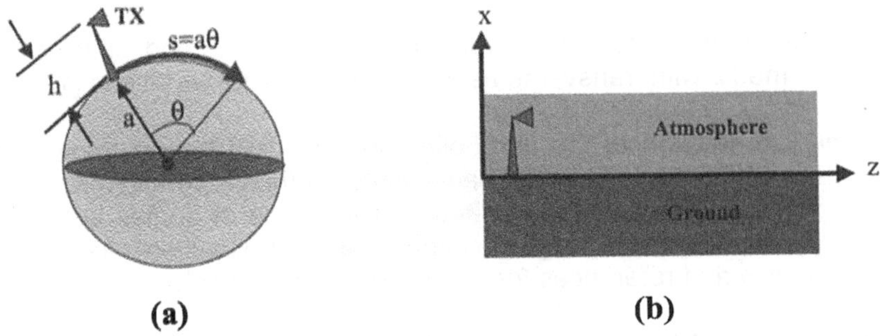

Figure 1.2: *The geometry of the ground-wave propagation problem in (a) spherical coordinates and (b) normalized Cartesian coordinates.*

1.1.3 The Lower Atmosphere

Ground wave propagation through atmosphere (up to nearly 100 GHz) is affected by oxygen and water-vapor molecules. The air can be considered as a non-dispersive medium and can be represented by its refractive index ($n = \sqrt{\varepsilon_r}$). The refractivity of the propagation medium should be well understood, since non-flat terrain and/or earth's curvature may also be implemented via refractivity in most of the analytical as well as numerical approaches. Refractive index of the air is very close to unity (typically about 1.0003); therefore, it is customary to use the refractivity N, defined as

$$N = (n-1) \times 10^6 \tag{1.1}$$

were N is dimensionless, but is measured in "N units" for convenience. N depends on the pressure P (mbar), the absolute temperature T (°K) and the partial pressure of water vapor e (mbar) as [18],

$$N = 77.6\frac{P}{T} + 3.73\times10^5 \frac{e}{T^2} \tag{1.2}$$

which is valid in earth - troposphere waveguides and can be used in ground wave propagation modeling. If the refractive index were constant, radio waves would propagate in straight lines. Since n decreases with height, radio waves are bent downward toward the earth, so that the radio horizon lies further away than the optical horizon. It should be noted that the radio horizon effect is taken into account either by using N with the effective earth radius a_e, or by introducing a fictitious medium where N is replaced by the modified refractivity M,

$$M = N + \frac{x}{a}\times10^6 = N + 157x \tag{1.3}$$

with the height x given in kilometers. In (1.3)

$$a = 6378 \text{ km}, \quad \frac{10^6}{a} = 157 \text{ and } \frac{\partial M}{\partial x} = \frac{\partial N}{\partial x} + 157 \tag{1.4}$$

For the standard atmosphere (i.e., for a vertical linearly decreasing refractive index), N decreases by about 40 N unit/km while M increases by about 117 N unit/km. Sub-refraction (super-refraction) occurs when the rate of change in N with respect to height (i.e., $\partial N / \partial x$) is less (more) than 40 N unit/km.

A linearly decreasing (increasing) vertical refractive index variation forces a wave trap near (diverge from) the earth's surface while propagating. Similar effects are also caused by concave and convex surfaces. Therefore, there is an analogy between refractive index and surface geometry in terms of propagation effects. By using this analogy, te earth's curvature effect is included into the refractive index of the air. The earth's curvature effect is equivalent to a vertical refractivity gradient of 157 N unit/km (i.e., linear vertical increasing refractivity profile). As stated in the previous paragraph, standard atmosphere, together with te earth's curvature, is represented by a vertical refractivity gradient of 117 N unit/km.

1.1.4 Summary

Norton and Wait formulations parameterize the propagation process in terms of different phenomenological models; their ranges of validity, accuracy, rate of convergence, etc., depending on such problem parameters as operational frequencies, source/observer locations, and the physical propagation environment, differ as well, with particular impact on computations. This has led to a new hybrid WAVEPROB algorithm [11], which seeks to combine the best features of the Norton and Wait algorithms in an efficient, adaptive format, with dynamic prediction of the critical height, range, frequency, and number of necessary roots in the algorithm that switches from one method to the other.

Either by using ray-mode theories separately or by using WAVEPROB, one may deal with smooth-boundary problems, such as [11,25]

- Surface wave path loss or field strength variation with respect to range (especially beyond the horizon and when both transmitter and receiver are on the surface).

- Range and/or height propagation variations in interference regions (i.e., when transmitter and receiver are above the surface and within the line-of-sight (LOS)).

Ray-mode and their hybridized techniques cannot handle problems, such as

- Propagation over rough surface terrain (note that roughness capability can be added to WAVEPROB, in an approximate sense, via modification of the surface reflection coefficient; the Rayleigh roughness criterion, which is based on the phase difference between two nearby reflections of the same incident wave due to surface height changes, may be used for this purpose [11]).

- Propagation through surface and/or elevated ducts formed by inhomogeneous vertical as well as horizontal atmospheric conditions (note that analytic formulations are based on a *coordinate separable* model over a smooth spherical earth with standard atmosphere).

Although height gain functions in mode theory [3] can be used to account for transmitter/receiver heights, it is difficult to deal with receiver heights in diffraction regions (beyond LOS) because of numerical problems in calculating higher-order terms in the series representation of Airy functions [26]. Nevertheless, analytic approximate solutions are important because they parameterize a problem in terms of wave phenomenologies, such as rays and modes, provided that the actual problem does not deviate substantially from the analytic prototype.

The Split-Step Parabolic Equation propagator (SSPE) and recently introduced time-domain wave propagators TDWP and TLM-WP are very promising, since they all can handle propagation over a realistic spherical earth's surface (including non-flat terrain profile as well as vegetation modeled as a surface impedance) above which lies an atmosphere with a transverse as well as longitudinally inhomogeneous refractive index.

1.2 ANALYTICAL MODELS

An exact analytical model requires formulation of the real-life problem; calculation of a field component at a given frequency, at any point on or above the earth's surface, caused by a point source on or above the lossy ground, which is characterized by surface impedance boundary condition, in spherical coordinates. The model shall include surface irregularities (rough surface, non-flat terrain, etc.) as well as longitudinal and/or transverse refractivity variations. Unfortunately, this model has not appeared yet, but a model for smooth spherical earth with homogeneous or standard atmosphere (defined in the next sections) has been in use for several decades.

1.2.1 Statement of the Problem

Consider the problem given in Fig. 1.3 and assume wave propagation over the smooth spherical earth with impedance boundary conditions. Further assume homogeneous atmosphere and consider a vertical electric dipole located near the earth's surface as the source. In spherical (r,θ,φ) coordinates the earth's radius is $r = a$, finite conductivity of the earth's surface at $r = a$ is described by the surface impedance Z_S.

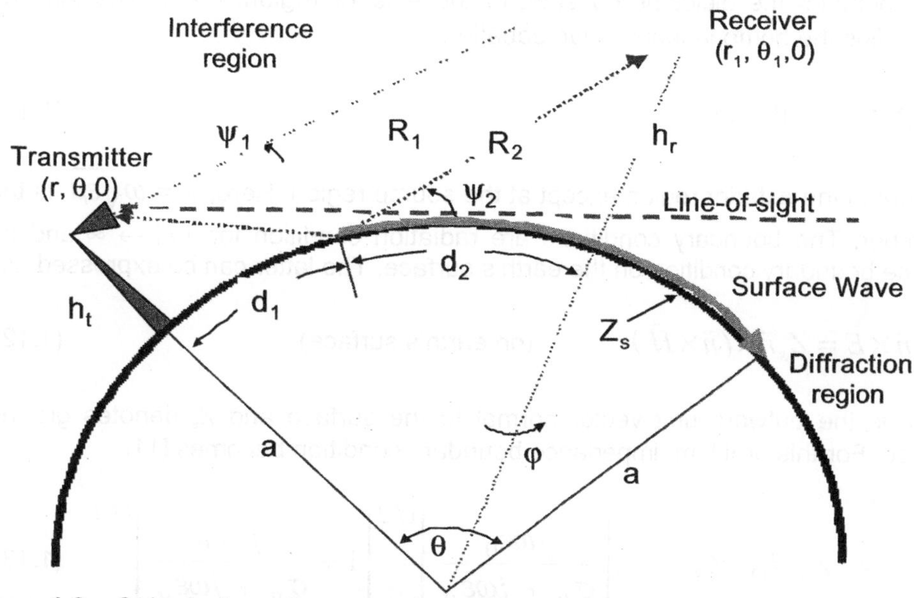

Figure 1.3: *Spherical earth geometry with smooth terrain, in spherical polar coordinates. Also shown are transmitter and receiver heights, as well as path lengths associated with wave processes discussed in 1.2.5. Here,* ψ_1 *denotes the angle between the direct ray and the tangent to the earth's surface at the ray reflection point.*

Under the assumption of azimuthal symmetry for the proposed problem and exp(-jωt) time dependence the fields of a dipole source located on the polar axis θ = 0 at r' ≥ a can then be derived from the radial component $rU(r,\theta)$ of the electric Hertz vector as [3],

$$\vec{E} = \nabla \times \nabla \times (Ur\vec{e}_r) \tag{1.5}$$

$$\vec{H} = -j\omega\varepsilon\nabla \times (Ur\vec{e}_r) \tag{1.6}$$

or in component form [3,4]

$$E_r = -\frac{1}{r\sin\theta}\frac{\partial}{\partial\theta}\sin\theta\frac{\partial U}{\partial\theta} \tag{1.7}$$

$$E_\theta = \frac{1}{r}\frac{\partial^2}{\partial r\partial\theta}(rU) \tag{1.8}$$

$$H_\varphi = j\omega\varepsilon \frac{\partial U}{\partial \theta} \tag{1.9}$$

and

$$E_\varphi = H_\theta = H_r = 0 \tag{1.10}$$

where ε denotes the dielectric constant of the exterior region, $r > a$. The function $U(r,\theta)$ satisfies the homogeneous wave equation

$$(\nabla^2 + k^2)U = 0 \tag{1.11}$$

everywhere in the exterior region except at the source region. Here, $k = \omega\sqrt{\varepsilon\mu}$ is the wavenumber. The boundary conditions are radiation condition for $|r| \to \infty$ and an impedance boundary condition on the earth's surface. The latter can be expressed as

$$\vec{n} \times \vec{E} = Z_s \vec{n} \times (\vec{n} \times \vec{H}) \qquad \text{(on earth's surface)} \tag{1.12}$$

where $\vec{n}$ is the outward unit vector normal to the surface and Z_s denotes ground impedance. For this problem, impedance boundary condition becomes [11]

$$E_\theta = -Z_s H_\varphi,\ Z_s = Z_0 \left[\frac{j\omega\varepsilon_0}{\sigma_g + j\omega\varepsilon_g}\right]^{1/2} \left[1 + \frac{j\omega\varepsilon_0}{\sigma_g + j\omega\varepsilon_g}\right]^{1/2} \tag{1.13}$$

where $Z_0 = 120\pi$ is the free-space wave impedance, ε_g and σ_g are the permittivity and conductivity of the ground, respectively. Substituting from (1.8), (1.9), and (1.10) one obtains

$$\frac{1}{r}\frac{\partial}{\partial r}[rU] = -j\omega\varepsilon Z_s U \qquad \text{at } r = a \tag{1.14}$$

1.2.2 Equivalent Earth's Radius Concept

The atmosphere is assumed to be homogeneous in the above formulations. However, in reality, the dielectric constant decreases smoothly and monotonically with increasing height, $h = r\text{-}a$. When this weak radial inhomogeneity is taken into account, (1.7)-(1.9) become [4]

$$E_r = -\frac{1}{r\sin\theta}\frac{\partial}{\partial\theta}\sin\theta\frac{\partial U}{\partial\theta} \tag{1.15}$$

$$E_\theta = \frac{1}{r\varepsilon_r(r)}\frac{\partial}{\partial r}\left(\varepsilon_r(r) r \frac{\partial U}{\partial\theta}\right) \tag{1.16}$$

$$H_\varphi = \frac{jk_0\varepsilon_r(r)}{Z_0}\frac{\partial U}{\partial \theta} \tag{1.17}$$

$$\frac{\partial}{\partial r}\frac{1}{\varepsilon_r(h)}\frac{\partial}{\partial r}\left[\varepsilon_r(h)rU\right]+\frac{1}{r\sin\theta}\frac{\partial}{\partial\theta}\sin\theta\frac{\partial}{\partial\theta}U+k_0{}^2n^2(h)rU=0 \tag{1.18}$$

$$\frac{1}{r\varepsilon_r(r)}\frac{\partial}{\partial r}\left[\varepsilon_r(r)rU\right]=j\omega\varepsilon_r(r)Z_sU \qquad \text{on } r=a \tag{1.19}$$

where $\varepsilon_r(h)=\varepsilon(h)/\varepsilon_0$ is the dielectric constant relative to permittivity ε_0 of the free space and k_0 is the free space wavenumber. It now proves to be convenient to factor out a rapidly varying phase term by introducing an auxiliary quantity $\widetilde{U}$ [4] defined as

$$\widetilde{U} = e^{-jk_r a\theta}\varepsilon_r(h)r\sqrt{\sin\theta}U \tag{1.20}$$

Substituting from (1.20) into (1.18), it can be shown that the wave equation is satisfied and $\widetilde{U}$ can approximately be reduced to the following parabolic form

$$\frac{\partial^2\widetilde{U}}{\partial r^2}+2j\frac{k_0}{a}\frac{\partial\widetilde{U}}{\partial\theta}+k_0^2\left(\frac{\varepsilon_r(h)}{\varepsilon_r(0)}-\frac{a^2}{r^2}\right)\widetilde{U}=0 \tag{1.21}$$

We will introduce a change of variables via

$$\begin{aligned} h &= r-a \\ s &= a\theta \end{aligned} \tag{1.22}$$

Then using the approximate equality

$$1-\frac{a^2}{r^2}\cong\frac{2h}{a} \tag{1.23}$$

(1.21) can be written as

$$\frac{\partial^2\widetilde{U}}{\partial h^2}+2jk_0\frac{\partial\widetilde{U}}{\partial s}+k_0^2\left(\frac{\varepsilon_r(h)-\varepsilon_r(0)}{\varepsilon_r(0)}+\frac{2h}{a}\right)\widetilde{U}=0 \tag{1.24}$$

In order to combine the effect of both earth radius and inhomogeneous atmosphere in one parameter, the equivalent earth radius a_e is defined as

$$\frac{1}{a_e}=\frac{1}{a}+\frac{\varepsilon_r'(0)}{2\varepsilon_r(0)} \qquad \text{where} \quad \varepsilon_r'(0)=\left(\frac{\partial\varepsilon_r}{\partial h}\right)_{h=0} \tag{1.25}$$

and (1.24) becomes

$$\frac{\partial^2 \widetilde{U}}{\partial h^2} + 2jk_0 \frac{\partial \widetilde{U}}{\partial s} + \frac{2k_0^2}{a_e} h(1+g)\widetilde{U} = 0 \tag{1.26}$$

where

$$g = \frac{a_e}{2\varepsilon_r(0)} \left(\frac{\varepsilon_r(h) - \varepsilon_r(0)}{h} - \varepsilon_r'(0) \right) \tag{1.27}$$

If the refractive index of the atmosphere varies linearly with *h*, then *g=0* and (1.26) reduces to

$$\frac{\partial^2 \widetilde{U}}{\partial h^2} + 2jk_0 \frac{\partial \widetilde{U}}{\partial s} + \frac{2k_0^2}{a_e} h\widetilde{U} = 0 \tag{1.28}$$

On the other hand, in a homogeneous atmosphere, $\varepsilon_r(h) = \varepsilon_r(0)$ for all *h* and using (1.24) and (1.26), one then obtains

$$\frac{\partial^2 \widetilde{U}}{\partial h^2} + 2jk_0 \frac{\partial \widetilde{U}}{\partial s} + \frac{2k_0^2}{a} h\widetilde{U} = 0. \tag{1.29}$$

From (1.28) and (1.29) it is clear that propagation in a "linear atmosphere" can equivalently be considered as the propagation over an earth with a homogeneous atmosphere, provided that the earth radius *a* is modified to be the effective earth radius a_e. Since $\varepsilon_r(0) \approx 1$, (1.24) can be rewritten as

$$\frac{\partial^2 \widetilde{U}}{\partial h^2} + 2jk_0 \frac{\partial \widetilde{U}}{\partial s} + k_0^2 \left(\varepsilon_r(h) - 1 + \frac{2h}{a} \right) \widetilde{U} = 0. \tag{1.30}$$

1.2.3 Transformation to Non-Dimensional Quantities

The normalized coordinates for the relevant source height (*h*) and observation height (h'), which satisfy the inequalities $h'/a \ll 1$ and $h/a \ll 1$ are introduced as [3,4]

$$\zeta = \frac{k_0 s}{2m^2}, \qquad \xi = \frac{k_0 h}{m}, \qquad \xi' = \frac{k_0 h'}{m} \qquad q = jm\varepsilon_r(0)\frac{Z_s}{Z_0} \tag{1.31}$$

where

$$m = \left(\frac{k_0 a}{2} \right)^{1/3} \tag{1.32}$$

If an attenuation function is defined as

$$V(\xi, \xi', \zeta, q) = 2\sqrt{\pi\zeta}\, e^{j\pi/4} F(\xi, \zeta) \tag{1.33}$$

the quantity $\widetilde{U}$ can be expressed in the form of V as

$$\widetilde{U} = \sqrt{\frac{a}{s}}V \quad \Rightarrow \quad \widetilde{U} = \sqrt{\frac{m}{\zeta}}V \tag{1.34}$$

and eq. (1.30) can be approximated in "earth-flattening" non-dimensional rectangular (ξ,ζ) coordinates as

$$\left[\frac{\partial^2}{\partial \xi^2} + j\frac{\partial}{\partial \zeta} + p(\xi)\right]F = 0 \tag{1.35}$$

where the equivalent permittivity function $p(\xi)$ is given by

$$p(\xi) = \xi + m^2\left[\varepsilon_r(\xi) - 1 + \frac{2h}{a}\right] \tag{1.36}$$

The boundary condition at (1.19) is expressed as

$$\frac{\partial F(\xi,\zeta)}{\partial \xi} + qF(\xi,\zeta) = 0 \quad \text{for } (\xi = 0) \tag{1.37}$$

for the transformed rectangular co-ordinates.

1.2.4 Exact Spectral Solution

Consider the Green's function problem [11] corresponding to (1.35) - (1.36)

$$\left[\frac{\partial^2}{\partial \xi^2} + j\frac{\partial}{\partial \zeta} + p(\xi)\right]F(\xi;z) = -\delta(\zeta)\delta(\xi - \xi') \tag{1.38}$$

where the source location is $\zeta = 0$ (s, θ = 0) and $\xi' = k_0 h'/m$, with h' being the height of the source above ground and it is assumed that $h' << a$. This equation is Fourier transformed in ζ-domain using the transform pair

$$F(\xi,\xi'\zeta) = \frac{-1}{j2\pi}\int_{\beta=-\infty}^{\infty} e^{j\beta\zeta} g_\xi(\xi,\xi';\beta)d\beta \,,$$
$$g_\xi(\xi,\xi';\beta) = \int_{\zeta=-\infty}^{\infty} F(\xi,\xi';\zeta)e^{-j\beta\zeta}d\zeta \tag{1.39}$$

and one obtains a one-dimensional spectral Green's function problem in the ξ-domain as

$$\left[\frac{d^2}{d\xi^2} + (p(\xi) - \beta)\right]g_\xi(\xi,\xi';\beta) = -\delta(\xi - \xi') \tag{1.40}$$

The corresponding impedance boundary condition can be obtained from (1.37) by replacing F with g_ξ and a radiation condition should be imposed for $\xi\to\infty$. The solution of (1.40) can be written as

$$g_\xi(\xi,\xi';\beta)=\frac{[f_2(\xi_<,\beta)+R_0 f_1(\xi_<,\beta)]f_1(\xi_>,\beta)}{W[f_1,f_2]} \tag{1.41}$$

where $\xi_<$ and $\xi_>$ denote the smaller or larger values of ξ and ξ', respectively. The Wronskian is given by

$$W[f_1,f_2]=f_1f_2'-f_1'f_2, \qquad f'\equiv\frac{d}{d\beta}f \tag{1.42}$$

and the reflection coefficient R_0 by

$$R_0=-\frac{f_2'(0,\beta)+qf_2(0,\beta)}{f_1'(0,\beta)+qf_1'(0,\beta)} \tag{1.43}$$

In these equations f_1 and f_2 denote two linearly independent progressing (traveling) wave solutions of the source-free equation corresponding to (1.40) which are employed in proper combination in order to ensure that all boundary conditions are satisfied. Clearly, closed-form solutions of the source free equation can only be obtained for a few special permittivity profiles. Then the spectral integral for $F(\xi,\zeta)$ in (1.38) can be evaluated using residue calculations and/or asymptotic saddle point techniques. Expanding the resonant denominator W in (1.41) by power series and evaluating the terms of these series asymptotically at the saddle points yields the ray representation, while evaluating the spectral integral in terms of the residues yields the modal representations.

1.2.5 Ray Asymptotics: Norton Formulation

In the standard atmosphere, three terms are obtained via ray-asymptotic evaluation of the spectral integral in eq. (1.39): the direct wave from the source to the observer, the ground-reflected wave, and the surface wave, with the sum of the first two contributions also referred to as the space wave. In his classic paper [8], Norton gives expressions for the vertical and tangential electric field components in the earth-flattening approximation. The problem has been addressed in [29], where the three components of the ground wave are expressed in cylindrical co-ordinates [11]. The exact solution to the problem of an antenna above a spherical finitely conducting earth can be set up in terms of an infinite series of spherical harmonics with coefficients containing 12 Bessel functions [11,29]. The convergence of these series is extremely slow, and for commonly used radio frequencies, as many as 10^3 to 10^8 terms are necessary for the calculation of the dominant contribution [11,29]. On the other hand the space-wave component (sum of direct and ground-reflected wave) can easily be constructed in spherical co-ordinates directly from the geometrical relations in Fig. 1.3 [11]. The space-wave components can be expressed as

$$E_{ver} = E_0\left[\cos^2\psi_1 + D\Gamma_v \frac{R_1}{R_2}\cos^2\psi_2\, e^{jk_0\Delta R}\right] \tag{1.44}$$

$$E_{\tan} = E_0\left[\cos\psi_1 \sin\psi_1 + D\Gamma_v \frac{R_1}{R_2}\cos\psi_2 \sin\psi_2\, e^{jk_0\Delta R}\right] \tag{1.45}$$

where

$$E_0 = ik_0 Z_0 M_0 \frac{e^{jk_0 R_1}}{4\pi R_1} = -j30k_0 M_0 \frac{e^{jk_0 R_1}}{R_1} \tag{1.46}$$

is the direct (free space) electric field strength at an arc-distance d generated by an electric dipole with moment $M_0 = I\,dl$. The distances R_1 and $R_2 \equiv R_1 + \Delta R$ (incident plus reflected path length) are given by

$$R_1 = \left[(a_e + h_1)^2 + (a_e + h_2)^2 - 2(a_e + h_1)(a_e + h_2)\cos\frac{d}{a_e}\right]^{1/2} \tag{1.47}$$

$$\Delta R = R_2 - R_1 = \frac{2h_1 h_2}{d}(1 - S_1{}^2)(1 - S_2{}^2) \tag{1.48}$$

where

$$S_1 = \frac{d_1}{\sqrt{2a_e h_1}}, \qquad S_2 = \frac{d_2}{\sqrt{2a_e h_2}} \tag{1.49}$$

$$d_1 = \frac{d}{2} + \rho\cos\left(\frac{\psi + \pi}{3}\right), \quad d_2 = d - d_1, \quad \psi = \cos^{-1}\left[\frac{2a_e}{\rho^3}(h_2 - h_1)d\right] \tag{1.50}$$

$$\rho = \frac{2}{\sqrt{3}}\left[a_e(h_1 + h_2) + \frac{d^2}{4}\right]^{1/2} \tag{1.51}$$

$$h_1 = h_t,\ h_2 = h_r \text{ for } h_t < h_r;\ h_1 = h_r,\ h_2 = h_t \text{ for } h_t > h_r \tag{1.52}$$

In the above, d is the arc distance between the transmitter and receiver, d_1 and d_2 are the arc distances from the transmitter and receiver to the reflection point on the ground, a and a_e are the earth's radius and effective radius, and h_t and h_r are the transmitter and receiver heights, respectively. In (1.44) and (1.45), the ray field reflection coefficient at the specular reflection point on the surface is

$$\Gamma_v = -\frac{\bar{\varepsilon}_g \sin\psi_2 - \sqrt{(\bar{\varepsilon}_g - \cos^2\psi_2)}}{\bar{\varepsilon}_g \sin\psi_2 + \sqrt{(\bar{\varepsilon}_g - \cos^2\psi_2)}} \tag{1.53}$$

with the angular coordinates given by

$$\psi_2 = \tan^{-1}\left\{\left(\frac{h_1+h_2}{d}\right)\left(\frac{(1-S_2{}^2)+T^2(1-S_1{}^2)}{1+T^2}\right)\right\} \tag{1.54}$$

$$\psi_1 = \begin{cases} \cos^{-1}\left[\dfrac{-(h_2+a_e)^2+(h_1+a_e)^2+R_1{}^2}{2(h_1+a_e)R_1}\right]-\dfrac{\pi}{2}; & h_t < h_r \\ \cos^{-1}\left[\dfrac{-(h_1+a_e)^2+(h_2+a_e)^2+R_1{}^2}{2(h_2+a_e)R_1}\right]-\dfrac{\pi}{2}; & h_t > h_r \end{cases} \tag{1.55}$$

Finally, in (1.44), D is the ray divergence factor at the reflection point, which accounts for the curvature effects of the spherical earth and is given as

$$D = \left[1+\frac{4S_1S_2{}^2T}{S_1(1-S_2{}^2)(1+T)}\right]^{-1/2}, \qquad T=\sqrt{\frac{h_1}{h_2}}. \tag{1.56}$$

1.2.6 Modal Residue Series: Wait Formulation

Calculating the residues of $g_\xi(\xi,\xi';\beta)$ in the spectral integral (1.39), the function F may be expressed in terms of height gain functions. In this formulation, the vertical (radial) component of electric field for the configuration in Fig. 1.3 is obtained as

$$E_{ver} = E_0 V(\xi,\xi',\zeta,q) \tag{1.57}$$

where E_0 is given in (1.48) and attenuation function, V, in (1.46) is expressed as

$$V(\xi,\xi',\zeta,q) = e^{j\pi/4}2\sqrt{\pi\zeta}\sum_{s=1}^{\infty}e^{j\beta_s\zeta}\frac{d\beta_s}{dq}\frac{W(\beta_s-\xi)}{W(\beta_s)}\frac{W(\beta_s-\xi')}{W(\beta_s)} \tag{1.58}$$

Here β_s are the spectral poles of $g_x(x,x';\beta)$, and

$$\frac{W(\beta_s-\xi)}{W(\beta_s)} \tag{1.59}$$

is the "height-gain" function. All parameters are defined as in (1.31) and (1.32). For the standard atmosphere model including the effects of the earth's curvature (i.e., linearly increasing equivalent vertical profile), the transverse mode functions are the well-known solutions of the Airy equation [30],

$$W(\beta) = \sqrt{\pi}\ \left[Bi(\beta) - jAi(\beta)\right] \tag{1.60}$$

which satisfies the impedance boundary condition on $x = 0$,

$$\left[\frac{d}{d\beta}W(\beta)-qW(\beta)\right]_{\beta=\beta_s} = 0 \qquad (1.61)$$

and the radiation condition at $x \to \infty$. Now, the wave phenomenology is modeled via transverse (*x*-domain) oscillatory modal fields, which progress with fixed wavenumber β_s along *z*. Only the radial component of the electric field excited by a vertically polarized short dipole is given because the tangential component of the field is negligible. The height gain functions are convenient to parameterize the effects of both transmit and receive antenna heights with respect to path loss variations.

1.3 NUMERICAL MODELS

EM waves excited from a vertical electric dipole (VED) located over the ground in 2D Cartesian co-ordinates requires us to study a TM_z problem with the field components E_x, H_y and E_z, where *x* and *z* are chosen as the transverse (height) and longitudinal (range) coordinates, respectively (see Fig. 1.4). For a horizontal electric dipole (HED) located over the ground, a TE_z problem with the field components H_x, E_y and H_z, has to be considered. If the operating frequency is at HF region, horizontally polarized waves are heavily attenuated; therefore vertically polarized antennas are required to generate efficient ground wave propagation. For both VED and HED excitations, the propagation region is a semi-infinite region that extends from $x = 0$ (bottom) to $x \to \infty$ (top), and from $z \to -\infty$ (left) to $z \to \infty$ (right).

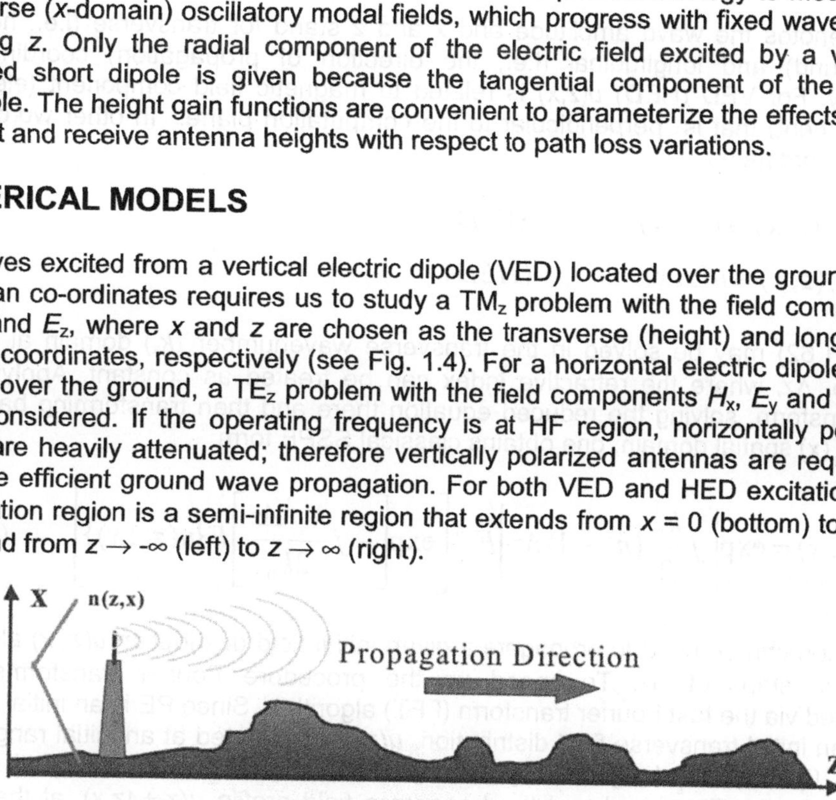

Figure 1.4: *The 2D propagation space for numerical simulators*

Numerical ground wave propagation models included here are SSPE, TDWP and TLM-WP and are basically the solutions of parabolic type 2D wave equation and Maxwell's equations in 2D, respectively.

1.3.1 Split-Step Parabolic Equation (SSPE) Propagator

The Fourier Split Step algorithm used to solve parabolic type equations is common and popular for the modeling of electromagnetic wave propagation in the troposphere. Although there are other propagation models capable of accounting for horizontal refractive gradients, they are restricted to simplistic refractive conditions, lower frequencies, and/or certain regions of space. Leontovich and Fock [4, Chapter 11] are pioneers who described the use of parabolic equation (PE) for electromagnetic wave propagation in a vertically inhomogeneous medium. However, this approach has become famous after the introduction of the Fourier Split-Step algorithm by Tappert [31,32]. Tappert's aim was to describe acoustic wave propagation in the ocean, but his approach then applied to electromagnetic wave propagation in troposphere.

The standard parabolic wave equation is given as

$$\frac{\partial^2 u}{\partial x^2} + 2jk_0 \frac{\partial u}{\partial z} + k_0^2 (n^2 - 1)u = 0 \tag{1.62}$$

where u denotes the wave amplitude and x and z stand for transverse (i.e., height above ground) and longitudinal (i.e., the direction of propagation) coordinates, respectively. For VED (HED) $u(z,x)$ is related to magnetic field component (electric field component) that is perpendicular to the computation plane. In other words, in Cartesian coordinates

$$H_y(z,x) \equiv u(z,x) \qquad (VED)$$

$$E_y(z,x) \equiv u(z,x) \qquad (HED)$$

Equation (1.62) may be solved in the transverse wavenumber (k_x) domain at each range step, Δz, where the refractive index can be treated as constant. Applying a Fourier transform, solving the reduced equation there and then transforming back to transverse (x) spatial domain, one obtains classical SSPE form

$$u(z,x) = \exp\left[j\frac{k_0}{2}(n^2 - 1)\Delta z \right] F^{-1}\left[\exp\left[-j\frac{{k_x}^2 \Delta z}{2k_0} \right] F\{u(z_0, x)\} \right] \tag{1.63}$$

This equation can be used to propagate a given initial field distribution $u(z_0,x)$ along z to $u(z,x)$ in steps of Δz. To speed up the procedure Fourier transforms are implemented via the fast Fourier transform (FFT) algorithm. Since PE is an initial value problem, an initial transverse field distribution, $u(z_0,x)$ is injected at an initial range, z_0, (e.g., $z_0 = 0$) then it is longitudinally propagated through a medium defined by its refractive index profile, $n(z,x)$ and the transverse field profile $u(z_0+\Delta z,x)$, at the next range step, is obtained.

The classical SSPE flow chart is given in Fig. 1.5 and implementation steps are listed in Section 1.6.2 of Part I. The program starts with the user supplied initial vertical scalar field profile, some parameters (such as frequency, maximum height, first range, last range, range increment, etc.) and vertical refractivity profile that may also be range-dependent. By sequential operations accessing the x and k_x domains via FFT and inverse FFT, respectively, one may obtain the transverse field profile at any range. In 2D rectangular coordinates, the earth's curvature is included by modifying the refractivity profile. Extra terms may also be added to model various super or sub-refraction propagation cases. Each run contains one forward and one backward FFTs. After each run, the program checks whether the last range supplied by the user has been reached or not. If not, the loop in Fig. 1.5 continues by replacing calculated vertical field profile as the new input field profile at each range, and when last range is reached, output is written and program terminates.

Boundary conditions must be implemented in SSPE algorithm since it is the solution of an initial value problem. If the ground is assumed to be a perfect electrical conductor,

Dirichlet and Neumann BCs are needed for VED and HED, respectively. Sine (cosine) Fourier transformation automatically satisfies Dirichlet (Neumann) BC. Another way of handling these BCs is as shown in Fig. 1.6. Here, the 2D propagation space that extends from $x = 0$ to $x = X_{max}$ vertically and from $z = z_0$ to $z \rightarrow \infty$ horizontally, and initial vertical field profile of SSPE wave propagator at range $z = z_0$ is shown on the left. The Dirichlet (Neumann) BC at the surface is satisfied by extending vertical profile odd (even) symmetric with respect to $x = 0$ axis (in the middle). After one SSPE run the next vertical field profile $u(z_0+\Delta z,x)$ at the next range step $z = z_0+\Delta z$ is obtained (on the right). On the other hand, Cauchy (impedance) BC is needed mostly at HF and lower VHF regions where the ground cannot be accepted as PEC because of the dominant surface wave propagation. However it is difficult to model impedance boundary condition and needs a lot of computation. Recently, discrete mixed Fourier transformation (DMFT) has been introduced [15] to handle impedance BC in SSPE.

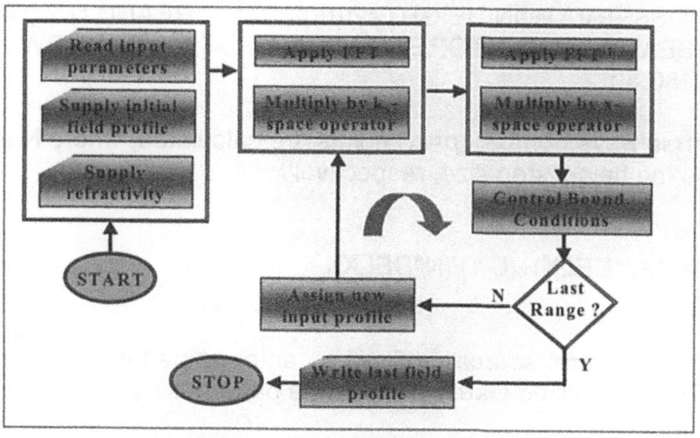

Figure 1.5: *Flow chart of SSPE routine*

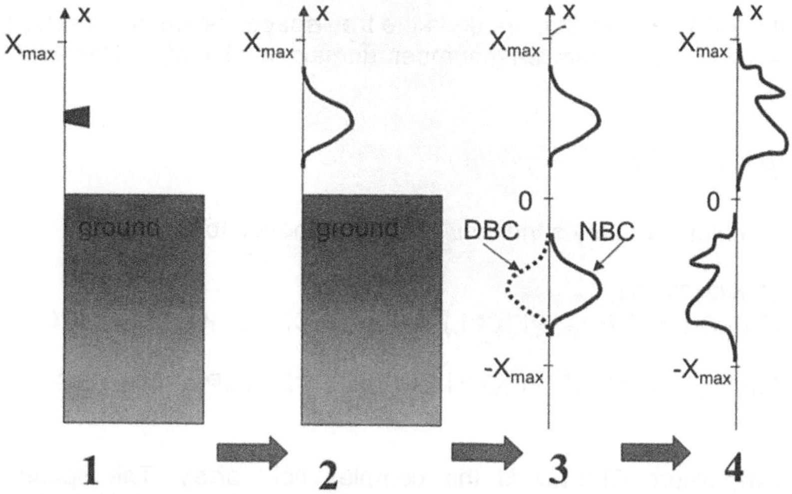

Figure 1.6: *Satisfaction of Dirichlet or Neumann BC in SSPE*

1.3.1.1 Classical SSPE Algorithm

The classical SSPE algorithm is a few-line code as a main routine. The initial vertical field profile array is fed into an FFT routine, multiplied by an exponent term in transverse wavenumber domain, then a inverse FFT, followed by another complex exponent multiplication, is applied. Here, this classical SSPE code, written in FORTRAN programming language, is listed together with explanations. Only the necessary code parts are listed. The variable declaration, array and file definitions, FFT sub-routine, etc., are not included.

The algorithm first requires the user to input initial parameters:

```
WRITE(*,*) 'LAST RANGE, Z2=? (km)'                READ(*,*) Z2
WRITE(*,*) "RANGE STEP SIZE , DELZ=? (km)'        READ(*,*) DELZ
WRITE(*,*) 'SOURCE HEIGHT=? (m)'                  READ(*,*) X_SOURCE
WRITE(*,*) "OBSERVATION HEIGHT=? (m)'             READ(*,*) Y_OBS
WRITE(*,*) 'BEAMWIDTH=[DEGREE]?'                  READ(*,*) BW
BW=BW*PI/180.
```

Then discrete transverse wave number (p=k_x) points are calculated, where N is the FFT size and DELX is the height step size, respectively.

```
DO L=1,N
   KX(L)=-1/(2*DELX)+(L-1)/(N*DELX)
END DO
```

In this sample algorithm, vertical source profile (i.e., antenna pattern) is chosen as a Gaussian function. The normalized Gaussian antenna pattern is

$$f(p)=e^{-p^2w^2/4}, \qquad w=\frac{\sqrt{2\ln 2}}{k_0\sin(BW/2)}$$

where BW is the 3 dB beam-width, and k_0 is the free-space wavenumber. The antenna pattern at z = 0, in transverse wavenumber domain is directly obtained from the function $f(p)$ as

$$\varphi(0,p)=f(p)e^{-jpx_0}-f(-p)e^{jpx_0}$$

to satisfy the Dirichlet boundary condition. This corresponds to

```
IF(L.GT.N/2) THEN
    CFIELD(L)=EXP(-(2*PI*KX(L)*W/2)**2)*EXP(-J* KX(L)*SOURCE)
ELSE
   CFIELD(L)=-EXP(-(2*PI*KX(N-L+1)*W/2)**2)*EXP(J*KX(L)*SOURCE)
ENDIF
```

in the algorithm, where CFIELD is the complex field array. The spatial source distribution at the starting range is obtained by applying an inverse FFT to the function $\varphi(0,p)$. Any type of transverse and/or longitudinal refractive index profile can be implemented in the algorithm. For example, a linearly decreasing refractive index height profile is

```
DO L=1,N
  IF(X(L).GE.0.0) THEN
     REFR(L)=1.00060009+A0*X(L)
  ELSE
     REFR(L)=1.00060009-A0*X(L)
  ENDIF
END DO
```

where X(L) and REFR(L) are height and refractive index arrays, respectively. Here, A0 denotes the height gradient of the refractive index. If the refractive index has range dependence, its definition should be renewed at each range step. In this sample algorithm, the pre-defined source function is propagated at each range step based on the flowchart given in Fig. 1.5. The transverse wavenumber (*p-space*) complex operator is computed as

```
DO L=1,N
     CPSPACE(L)=CEXP(-J*DELZ*(KX(L)*2*PI)**2/(2*K0))
END DO
```

while complex *z-space* operator is computed as

```
DO L=1,N
     CZSPACE(L)= CEXP(J*DELZ*K0*(REFR(L)-1)/2)
END DO
```

for all height points. The same sub-routine is used for both FFT (Flag 1) and inverse FFT (Flag 0) calculations. The main routine is then written as

```
DO Z=Z1,Z2,DELZ
   CALL FFT(CFIELD, N, 1)                 ! Apply FFT
   DO L=1,N
      CFIELD(L)=CFIELD(L)*CPSPACE(L)
   END DO
   CALL FFT(CFIELD, N, 0)                 ! Apply inverse FFT
   DO L=1,N
      CFIELD(L)=CFIELD(L)* CZSPACE(L)
   END DO
END DO
```

SSPE is a reliable technique for the electromagnetic propagation problems, where back-scattered fields are not required and parabolic (paraxial) approximation is acceptable. Parabolic approximation is used for nearly horizontal propagation directions [31,33 - 35]. In other words, it is used in problems where *u(z,x)* varies slowly with range. That also suggests that *n(z,x)* should vary smoothly with *z* as well as with *x*. The parabolic equation approximation affects the accuracy of the calculations, so it is required to analyze the order of error. The parabolic wave equation obtained for tropospheric propagation is exactly analogous to that obtained for acoustic propagation in the ocean; therefore the conclusions of acoustic parabolic equation investigations are instructive.

In certain classes of problems the parabolic approximation results in significant modal phase velocity errors for propagation in directions with angles more than ≈15° from the horizontal axis (range). These errors become evident in problems where an oceanic duct or shallow region gives rise to a coherent sum of several trapped modes propagating at relatively large angles. It is possible to correct these errors using the wide-angle approximation, which is explained in [36,37]. On the other hand, atmospheric refractive index gradients that trap angles larger than ≈1° are rare and a relatively large gradient in troposphere corresponding to a strong duct is at about $|\partial n/\partial x| \approx 10^{-6}\ \mathrm{m}^{-1}$ (i.e. smaller than ≈ $3\times10^{-4}\ \mathrm{m}^{-1}$, which is given in [36] as a typical example case). Therefore, the investigation of most electromagnetic problems with parabolic approximation is adequate and there is no need for the wide-angle propagation approximation. The wide-angle algorithm is required only for the problem geometries (i.e., high antennas at short ranges or the non-flat surfaces, which consist large slopes resulting in wide-spread scattering of the propagating wave) where angles greater than 15° are of interest.

1.3.1.2 Calibration Against Analytical Exact Solution

In order to show the accuracy of the SSPE algorithm, propagation over flat earth with a linearly decreasing vertical refractivity profile is considered. This problem has analytical exact solutions in terms of Airy functions, when the refractive index is defined as

$$n^2(x) = 1 - a_0 x \tag{1.64}$$

Here a_0 is a positive constant. In such a non-homogeneous environment scalar function $u(x,z)$ satisfies the wave equation

$$\frac{\partial^2 u}{\partial x^2} + \frac{\partial^2 u}{\partial z^2} + k_0^2 n^2(x) u = 0 \tag{1.65}$$

subject to homogeneous boundary condition at $x = 0$

$$\alpha_1 \left.\frac{\partial u}{\partial x}\right|_{x=0} + \alpha_2 u(x=0) = 0, \tag{1.66}$$

where α_1, α_2 are constants, and $\alpha_{1,2} = 0$ results in Dirichlet and Neumann conditions respectively. The radiation boundary condition is

$$\begin{aligned} u(x,z)\big|_{x\to\infty} &\to 0 \\ u(x,z)\big|_{z\to\mp\infty} &\to 0 \end{aligned} \tag{1.67}$$

Recalling that the time dependence is taken as $e^{-j\omega t}$, the modal series solution can be expressed as

$$u(x,z)=\sum_{q=0}^{\infty} C_{0q}\psi_q(x)e^{j\beta_q z} \tag{1.68}$$

where β_q is the longitudinal propagation constant k_z. for the related mode represented by index q. The new function $\psi_q(x)$ satisfies the one-dimensional wave equation

$$\left[\frac{d^2}{dx^2}+k_0^2 n^2(x)-\beta_q^2\right]\psi_q(x)=0 \tag{1.69}$$

In order to solve this equation new variables are defined as

$$k_0^2 n^2(x)-\beta_q^2 = K_1 x + K_2, \qquad \rho=-K_1^{-2/3}(K_1 x+K_2) \tag{1.70}$$

and the wave equation becomes an Airy equation

$$\left[\frac{d^2}{d\rho^2}-\rho\right]\psi_q(\rho)=0 \tag{1.71}$$

The solution of this equation with BCs given in (1.66) and (1.67) is constructed in terms of modes that are represented by the first kind of Airy function $A_i(\rho)$, with ρ being real.

$$\psi_q(\rho)=C_{0q}A_i(\rho) \tag{1.72}$$

where C_{0q} is the normalization constant defined as

$$C_{0q}=\left[\int_0^{\infty} A_i^2(\rho)d\rho\right]^{-1/2} \tag{1.73}$$

If $n^2(x)=1-a_0 x$ is used in the one-dimensional wave equation

$$K_1=-a_0 k_0^2 \quad \text{and} \quad K_2=k_0^2-\beta_q^2 \tag{1.74}$$

then

$$\rho=(a_0 k_0^2)^{1/3}x-(a_0 k_0^2)^{-2/3}(k_0^2-\beta_q^2) \tag{1.75}$$

This equation at $x = 0$ is denoted as σ_q, q representing the mode number. The condition at the surface is then reduced to

$$A_i'(-\sigma_q)+\alpha A_i(-\sigma_q)=0 \tag{1.76a}$$

$$A_i(-\sigma_q)=0 \tag{1.76b}$$

$$A_i'(-\sigma_q)=0 \tag{1.76c}$$

for Cauchy-, Dirichlet-, and Neuman-type BCs, respectively. Here, the prime denotes the derivative with respect to vertical coordinate. The longitudinal propagation constant is found as for each mode

$$\beta_q=\pm\left[k_0^2-\sigma_q(a_0k_0^2)^{2/3}\right]^{1/2} \tag{1.77a}$$

and modes are trapped between the surface and their modal caustics determined from

$$X_{cq}=-\sigma_q\left[a_0k_0^2\right]^{-1/3} \tag{1.78b}$$

Finally, the exact modal solution of Airytype wave equation is then found to be

$$u(x,z)=\sum_{q=1}^{N}C_{0q}A_i((a_0k_0^2)^{1/3}x-\sigma_q)e^{j\beta_q z} \tag{1.79}$$

The first and second kind of Airy functions Ai(.) and Bi(.) are real for real arguments. Their variations are pictured in Fig. 1.7. They both have standing wave character when the arguments are negative. While Ai(.) decays exponentially, Bi(.) blows up for positive arguments.

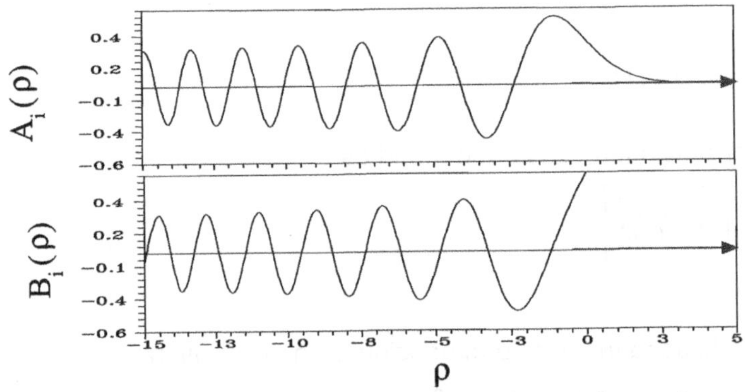

Figure 1.7: *First and second kind of Airy functions for real arguments*

In order to validate the SSPE algorithm, the data computed with (1.79) for the chosen parameters are compared with SSPE solutions, as is seen in Fig. 1.8. Here, $n^2(x)=1-a_0x$ and $a_0=4\times10^{-8}$, which corresponds to the refractive index of standard atmosphere (excluding the effect of earth's radius), causing a surface duct that traps the wave. The ground is assumed to be PEC, satisfying the Dirichlet boundary condition and $k_0=2\pi$. First, (1.79) is computed with the first ten modes at z

= 0 km between x = 0 m and x = 2000 m at 512 different height points, and the result is used in SSPE algorithm as the initial height variation of the scalar function, $u(x,z)$. Then height variation of the function at four different ranges, (z = 25 km, z = 50 km, z = 75 km and z = 100 km) is calculated via SSPE and plotted in Fig. 1.8. The results of exact modal solutions computed via (1.79) at the same ranges and heights are also plotted. Perfect agreement between the exact analytical solution and SSPE, which is a semi-analytic-numerical method, is observed in the figure.

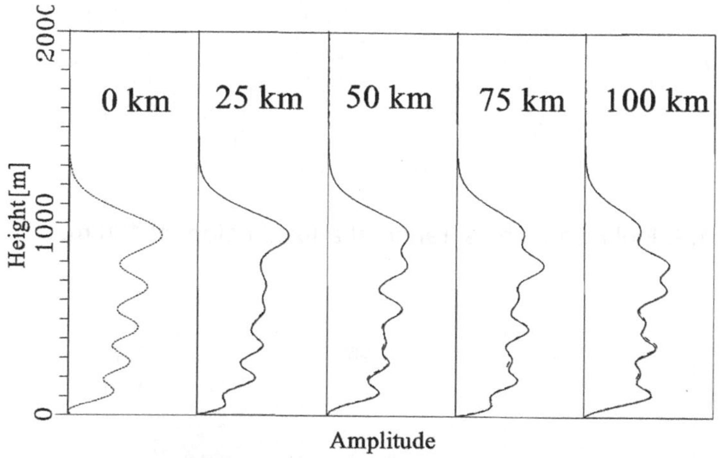

Figure 1.8: *Field vs. height computed with exact analytical solution and SSPE algorithm at different ranges. The height variation at z = 0 is calculated numerically with exact analytical solution and used in SSPE as the initial height variation of the scalar function.*

Modes given in (1.79) are the building blocks (i.e., eigenfunctions) of the geometry (they are orthonormal to each other and have unit energy); therefore, any initial transverse source profile can be represented by superposing them via the proper modal excitation coefficients. Modal excitation coefficients are determined from the completeness relation. Any vertical profile (e.g., a Gaussian field distribution) can be chosen and represented by a modal series. The number of modes can be determined numerically by trial and error. Then, this profile may be used as an input profile of the SSPE program, and results at any range can be obtained both analytically and numerically. This is illustrated in the following example.

Suppose a strongly surface trapping duct, which is specified by (1.64) with a_0 equal to -4×10^{-5}. In this case refractive index at the surface is "1" but reduces to, for example, "0.2" at 20 km height. Frequency is specified as 300 MHz. The source is located 250 m above the ground and has a vertical extend from 100 m to 400 m. This vertical profile is built with the first 100 normal modes, NM (eigenfunctions). Excitation coefficients are between 10^{-4} and 10^{-2} for the first 11 NMs, between 0.1 and 2.23 for NMs 12 - 60, and between 10^{-4} and 10^{-6} for the rest of 40 NMs. Typical comparisons are given in Fig. 1.9 and Fig. 1.10.

In Fig. 1.9, horizontal field profile at 100 m constant height is plotted. Solid and dashed lines correspond to exact analytical and numerical SSPE results, respectively. In Fig. 1.10, vertical field profiles at three ranges (at z = 0 km, 12 km, and 15 km), obtained with both methods are plotted. Very good agreement is clearly observed, even for the highly oscillatory nature of the field profiles, in the figure.

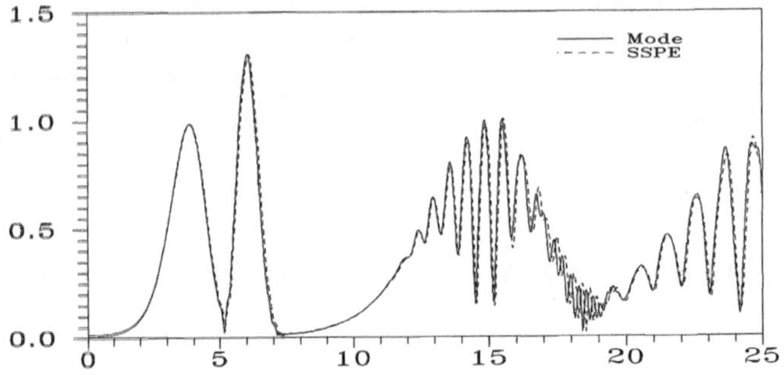

Figure 1.9: Field strength vs. range at a fixed height of 100 m.

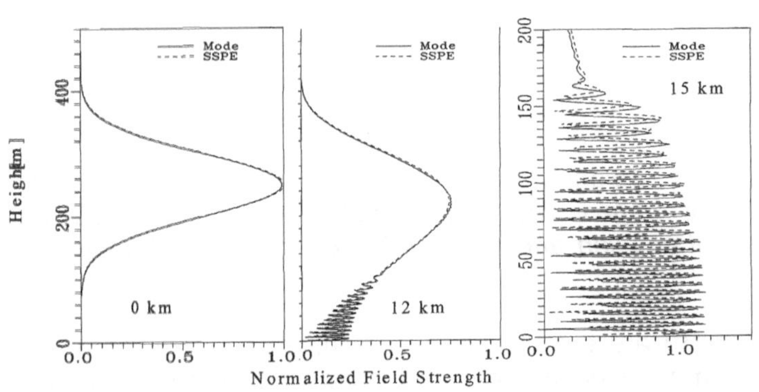

Figure 1.10: Field strength vs. height at z = 0 km, 12 km, and 15 km

Again, perfect agreement obtained between the two methods is clearly observed in both figures.

1.3.1.3 Staircase Terrain Modeling

Longitudinally non-flat terrain can be incorporated into the SSPE algorithm via several different mathematical approaches and it is possible for the user to choose the appropriate one for his/her problem. Here, only the staircase approximation is taken into account. Staircase terrain modeling is simpler, in which slope values are not required (as in the other approximations); only the terrain height for each range is needed. On each segment of constant height, the field is propagated in the usual way for staircase terrain, applying necessary boundary condition at the ground. When terrain height changes, corner diffraction is ignored and the field is simply set to zero on vertical terrain facets. Since the computation height is not changed due to the terrain, there is no need to modify refractive index; therefore it is also easy to implement the staircase terrain modeling into the SSPE algorithm. One just needs to take into account that ground does not support propagation under the constant height for each segment [14]. Although staircase terrain cannot model the terrain as smooth as it must be, comparisons show that the results are quite satisfactory [38]. A typical non-flat terrain profile and its discrete staircase representation is pictured in Fig. 1.11. The smaller the range and height step sizes (Δz and Δx, respectively) the better the approximation in discrete terrain profile.

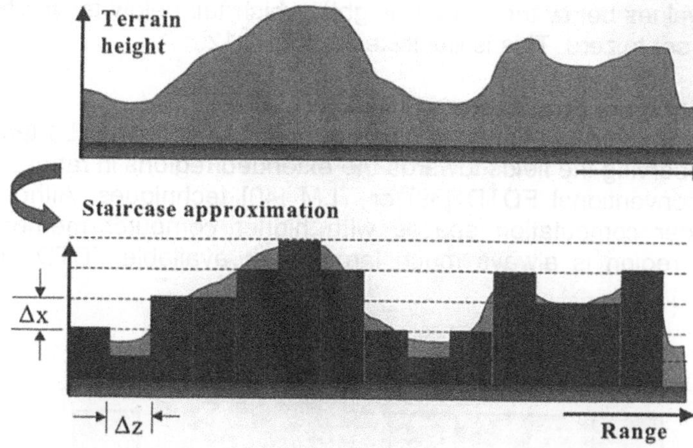

Figure 1.11: *The difference between terrain and its staircase approximation*

In SSPE, terrain effects are included as follows:

- A terrain file contains height vs. range that represents non-flat terrain in the chosen scenario. Zero height corresponds to flat earth's surface.

- A complex array contains field values at discrete height points from the surface (zero height) to maximum height of interest.

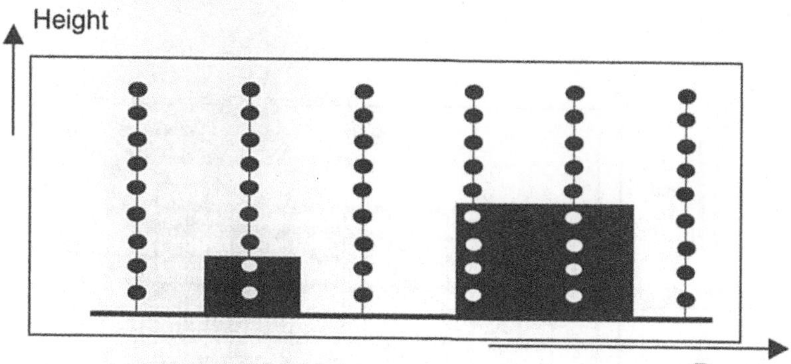

Figure 1.12: *Non-flat terrain implementation in SSPE (black circles are the field values at different height points.; white circles are the ones that are forced to be zero since they are inside the non-flat terrain)*

- Computed vertical field profile at the current range is used as the initial field profile of the next range for next SSPE iteration.
- While propagating step by step the terrain file is checked continuously and height of the terrain at the range of computation is read and array elements (vertical field values below the terrain height), which fall below terrain height at that range are set to zero. This is illustrated in Fig. 1.12.

1.3.2 Time-Domain Wave Propagators (TDWP, TLMWP)

The difficulty of analyzing ground-wave propagation via time-domain (TD) techniques is the necessity of observing the fields towards the extended regions in range, which is impossible for the conventional FDTD [39] or TLM [40] techniques. Although it is possible to use larger computation spaces with higher computer memories, the desired propagation region is always much larger than available FDTD and TLM spaces (see Fig. 1.13).

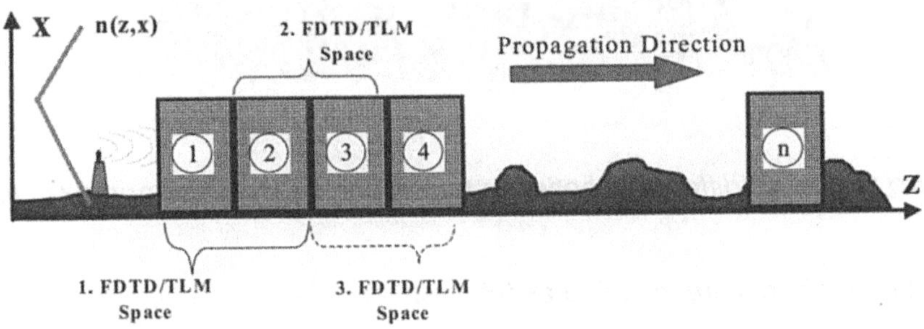

Figure 1.13: *The 2D propagation space, sliding FDTD/TLM windows*

This is overcome by tracing the propagation region with a dynamic sliding window (the FDTD/TLM space). In both TDWP and TLM-WP, the computation space has to be terminated by absorbing boundary blocks (here perfectly matched layer, PML is used) at the left, right, and top of the window. This is pictured in Fig. 1.14 with both TLM and FDTD unit cells. A one-cell transition is used to match FDTD field components to the TLM voltage pulses in order to use the same FDTD-based PML routine in both propagators.

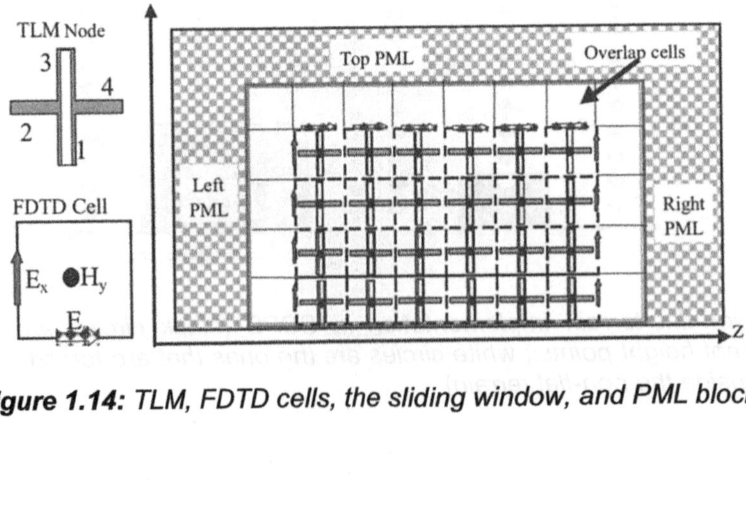

Figure 1.14: *TLM, FDTD cells, the sliding window, and PML blocks*

The related FDTD iterative equations in TDWP are

$$\begin{aligned} E_x^n(i,k) &= \frac{\varepsilon}{\varepsilon+\sigma\Delta t} E_x^{n-1}(i,k) \\ &- \frac{\Delta t}{(\varepsilon+\sigma\Delta t)\Delta z}\left[H_y^{n-1/2}(i,k) - H_y^{n-1/2}(i,k-1)\right] \end{aligned} \tag{1.80}$$

$$\begin{aligned} E_z^n(i,k) &= \frac{\varepsilon}{\varepsilon+\sigma\Delta t} E_z^{n-1}(i,k) \\ &+ \frac{\Delta t}{(\varepsilon+\sigma\Delta t)\Delta x}\left[H_y^{n-1/2}(i,k) - H_y^{n-1/2}(i-1,k)\right] \end{aligned} \tag{1.81}$$

$$\begin{aligned} H_y^{n+1/2}(i,k) &= H_y^{n-1/2}(i,k) \\ &- \frac{\Delta t}{\mu_0 \Delta z}\left[E_x^{n-1}(i,k+1) - E_x^{n-1}(i,k)\right] \\ &+ \frac{\Delta t}{\mu_0 \Delta x}\left[E_z^{n-1}(i+1,k) - E_z^{n-1}(i,k)\right] \end{aligned} \tag{1.82}$$

where ε [F/m], σ [S/m] and μ_0 [H/m] are the permittivity, conductivity and permeability of the propagation medium, respectively. Here, *i* and *k* are the cell numbers in the *x* and z coordinates, respectively, and n tags the time step. FDTD electric field components are calculated at discrete time instants Δ*t*, 2Δ*t*, 3Δ*t*, .., *n*Δ*t*, and the magnetic field components are calculated at Δ*t*/2, 3Δ*t*/2, 5Δ*t*/2, .., (*n*+1/2)Δ*t*, in cubical Yee cells [22] with cell sizes Δ*x*, Δ*y*, Δ*z*.

The corresponding TLM components I_y, V_x and V_z are determined from Thevenin equivalence of the series node [23]

$$H_y = \frac{I_y}{\Delta\ell};\ E_x = -\frac{V_x}{\Delta\ell};\ E_z = -\frac{V_z}{\Delta\ell} \tag{1.83}$$

as

$$I_y = 2\frac{V_1^i - V_2^i - V_3^i + V_4^i + V_5^i}{\hat{Z} Z_{TL}\Delta\ell};\ V_x = V_2^i + V_4^i;\quad V_z = V_1^i + V_3^i \tag{1.84}$$

where $\hat{Z} = 4 + Z_s$ *and* $Z_s = 4(\mu_r - 1)$. The characteristic impedance of each individual transmission line, Z_{TL} becomes $Z_o/\sqrt{2}$, where Z_0 is the characteristic impedance of the free space. The propagation medium has the relative permeability of 1 but the relative permittivity of ε_r. So, by duality, the normalized impedance of the short-circuited stub, Z_s becomes $Z_s = 4(\varepsilon_r - 1)$, meaning that the short circuited stub models the permittivity of the medium.

The implementation of TD simulations is as follows:

- A source with a chosen spatial altitude distribution, having a pulse character in time (that yields broad band analysis with a single simulation), is injected via the necessary field components.
- Earth's surface effects are included either as a terrain profile or a smooth impedance boundary.
- In order to implement a non-homogeneous atmosphere, a refractivity profile over the earth's surface (including earth's curvature) is introduced via relative permittivity of the air, $\varepsilon_r=n^2(x,z)$.
- One-way propagation is traced via a 2D rectangular sliding window. The content of this propagation window is the pulse that carries information related to terrain scattering, refractivity effects, and surface losses.
- The sliding propagation window moves from left to right in the computation space and circulates back to the left when reaches the right most-end, which is the initial profile of the next computation space.
- The process and TD simulations repeat until the wave longitudinally propagates to a desired range. The algorithm stores necessary information (such as range, terrain profile, transverse and/or longitudinal propagation characteristics at chosen observation points).
- Frequency-domain characteristics may be obtained from the TD simulations by off-line discrete Fourier transformation (DFT) analysis.

It should be noted that preparation of TDWP algorithm, based on FDTD plus PML termination, is quite straightforward. On the other hand, one needs to introduce modifications in TLM-WP algorithm [20]. First of all, 2D TLM computation space is constructed via a series node representation (see Fig. 1.14) to represent a TM_z problem. Although four-arm node is shown in the figure, the fifth arm (with a short-circuit termination) is also used to model permeability of the node, which in turn represents permittivity because of the duality principle. Secondly, the FDTD-PML algorithm, prepared for TDWP, is used in TLM-WP to absorb scattered fields [20,41]. In order to do that, a transition cell is located between TLM computation space and FDTD-PML blocks, and field components of TDWP are matched to the voltage pulses of TLM-WP in this cell, unlike [41], where overlap cell is directly taken as the first cell in FDTD-PML region. Also, attention should be paid in matching FDTD fields to TLM voltage pulses at the left FDTD-PML-TLM interface, since FDTD field components in a cell are not symmetrically located as the voltages are in the TLM node. If for example, $k = m$ (in the z-direction) is the overlap cell, then vertical electric field component (E_x) of the next cell ($k = m+1$) should also be used to write down TLM voltages in terms of FDTD fields (or vice versa).

The implementations of non-flat PEC terrain and surface impedance in TD propagators are as shown in Fig. 1.15. Here, the FDTD implementation is given. The staircase approximation is used to model non-flat longitudinal terrain profile in both simulators. It is achieved by forcing all tangential electric field components (E_x at vertical edges, E_z at horizontal edges for boundary cells and both for the inner cells) to be zero during the TDWP simulation. In TLM-WP, it is performed by using the short-circuit (SC) scattering matrix in the irregular terrain regions [20].

Although PEC boundary approximation is mostly adequate at VHF and above (i.e., frequencies higher than 100 - 150 MHz), accurate impedance boundary simulations are essential at HF frequencies and below. A propagator should be capable of handling imperfect surface effects including multi-mixed path propagation (e.g., land - sea transition). This is achieved by surface impedance modeling as explained in [42]. The first magnetic field component above the impedance boundary (i.e., the magnetic field component at first 2D-FDTD cell over the surface) is calculated from its value at previous time-step plus two neighboring vertical and one horizontal electric fields by using surface impedance relation in terms of the resistance R_s and reactance L_s components [42].

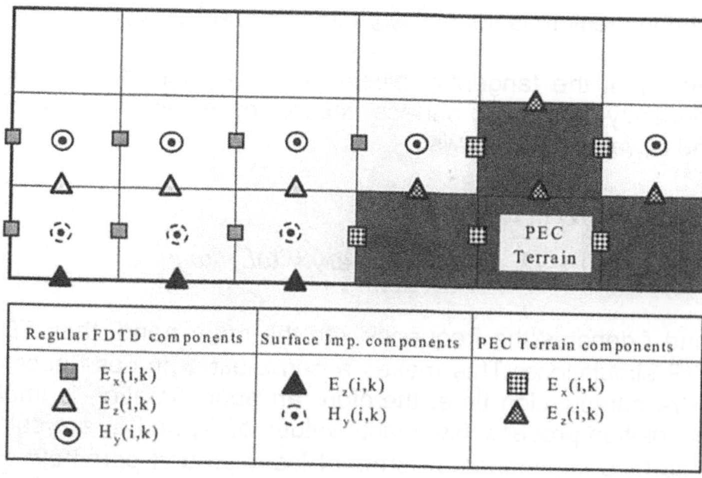

Figure 1.15: *Non-flat terrain and surface impedance modeling*

In TDWP implementations, the first magnetic field component above the impedance boundary is calculated from

$$H_y^{n+1/2}(i,k) = \frac{1 - R_s\Delta t / 2\mu_0\Delta x - L_s / \mu_0\Delta x}{1 + R_s\Delta t / 2\mu_0\Delta x - L_s / \mu_0\Delta x} H_y^{n-1/2}(i,k)$$
$$- \frac{\Delta t}{\mu_0\Delta z(1 + R_s\Delta t / 2\mu_0\Delta x - L_s / \mu_0\Delta x)}\left[E_x^n(i,k+1) - E_x^n(i,k)\right] \quad (1.85)$$
$$+ \frac{\Delta t}{\mu_0\Delta x(1 + R_s\Delta t / 2\mu_0\Delta x - L_s / \mu_0\Delta x)}\left[E_z^n(i+1,k)\right]$$

instead of (1.82), while E_z on the surface is set to zero. For a highly conducting medium (i.e., $\sigma >> \omega\varepsilon$) surface parameters may be approximated as

$$R_s = \sqrt{\frac{\omega\mu}{2\sigma}}, \qquad L_s = \sqrt{\frac{\mu}{2\omega\sigma}}, \qquad (1.86)$$

but

$$R_s = \frac{\sqrt{\mu}}{\sqrt[4]{\varepsilon^2 + \frac{\sigma^2}{\omega^2}}} cos\left[\frac{1}{2} arctan(\frac{\sigma}{\omega\varepsilon})\right] \tag{1.87a}$$

$$L_s = \frac{1}{\omega}\frac{\sqrt{\mu}}{\sqrt[4]{\varepsilon^2 + \frac{\sigma^2}{\omega^2}}} sin\left[\frac{1}{2} arctan(\frac{\sigma}{\omega\varepsilon})\right] \tag{1.87b}$$

should be used if the medium is not highly conducting and $\sigma \approx \omega\varepsilon$.

It should be noted that the tangential electric and magnetic field components are related via a frequency-dependent surface impedance given in terms of frequency-dependent R_s and L_s values, as follows:

$$E_z(\omega) = Z_s(\omega) H_y(\omega) \tag{1.88}$$

$$Z_s(\omega) = R_s(\omega) + jX_s(\omega), \quad X_s(\omega) = \omega L_s(\omega) \tag{1.89}$$

where $\omega = 2\pi f$ and f denotes the frequency. On the other hand, their TD values are required in TDWP simulations. This makes a convolution process necessary in TD, which means extra computation time; therefore an approximation is implemented to eliminate the convolution process. Numerical values of R_s and L_s are calculated from the definition of surface impedance in terms of lossy ground parameters ε and σ, at the center frequency of the band of the source pulse and are assumed to be constant [42]. As mentioned before, the same surface impedance routine is also used in TLM-WP by means of using a connection (transition) cell [24] between TLM computation space and surface boundary (It should be noted that non-flat PEC terrain and surface impedance effects are modeled separately in this study. Implementation of a non-flat imperfect terrain is quite straightforward, yet not implemented.)

1.4 APPLICATIONS WITH CANONICAL TEST CASES

All the methods presented here are tested with characteristic scenarios and comparisons are done among them within range of their validity. Ducting, antiducting, elevated ducting, etc., atmospheric conditions are used together with different non-flat terrain profiles.

In the simulators TDWP and SSPE, both earth's curvature and standard atmosphere condition can be included by using $n(x) = n_0 + x/a_e$, where $a_e = 4a/3 = 8504$ km is the effective earth's radius. Refractive index is either defined as $n = n_0(x)+x/a$ (where $n_0(x)$ is the refractive index height profile for flat earth), in order to take into account the effects of earth's curvature for all variations of atmospheric refractivity index, or an equivalent fictitious medium is introduced, where N as defined in (1.1) is replaced by the modified refractivity M as given in (1.3), where height is given in km. For the standard atmosphere (i.e., for a vertical linearly decreasing refractive index), N decreases by about 40 N unit/km while M increases by about 117 N unit/km. Sub-refraction (super-refraction) occurs when the absolute value of the rate of change is less (more) than 40 N unit/km (79 N unit/km) for the refractivity, N. When the N lapse rate is less than -157 N unit/km ducting occurs. Sub, super, and standard atmosphere, and ducting conditions are pictured in Fig. 1.16.

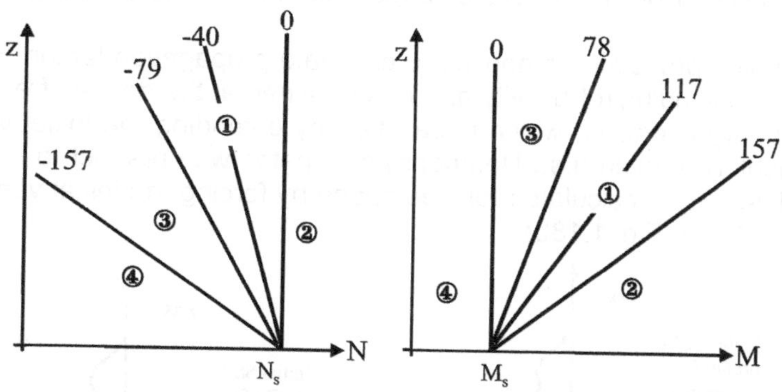

Figure 1.16: *Vertical refractivity variations and sub, super, standard atmosphere and ducting condition definitions*

All the comparisons given in this section are presented either propagation factor or path loss vs. height or vs. range. Propagation factor is defined as the ratio of electric fields computed under realistic surface and atmospheric conditions and fields computed in free space (E/E_0). This is straightforward in analytical computations. In TDWP, propagation factor is simulated as pictured in Fig. 1.17. First, the simulation is run for propagation over ground including refractivity conditions, and $E(t)$ vs. range (at constant height) or height (at a given range) is stored For this run, top, left and right of the computation space are terminated with PML blocks (see the top figure).

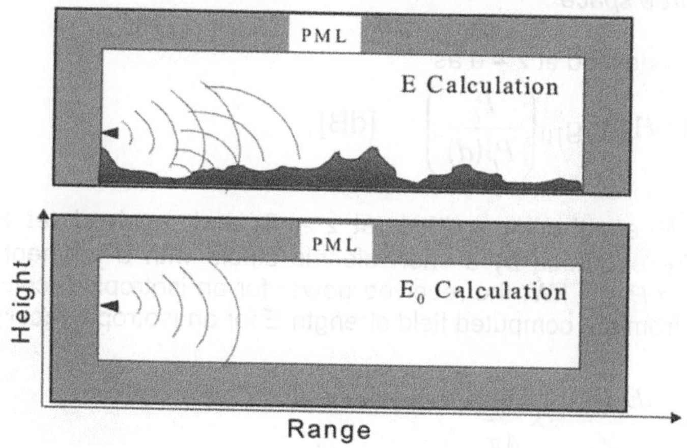

Figure 1.17: *Scenario for numerical propagation factor simulations in TDWP*

Then, the bottom boundary is removed and simulation is repeated assuming free-space propagation ((see the bottom figure).). In this case, $E_0(t)$ vs. range (at constant height) or height (at a given range) is stored. In the second run, the whole computation space is terminated by PML blocks. Moreover, source and observer locations are shifted up until mid heights, so that unwanted reflections from the bottom PML block does not affect the results. Finally, they are Fourier transformed and frequency variations are obtained. This procedure is tested to remove a good amount of noise components, since errors introduced by imperfect PML blocks cancel each other.

In SSPE, a similar procedure is applied to calculate propagation factor. First, SSPE is run over ground including refractivity and E is obtained at the desired frequency for the required boundary condition, which is satisfied by extending the initial vertical profile odd (i.e., Dirichlet) or even (i.e., Neumann) symmetric with respect to $x = 0$ axis (see Fig. 1.18a). Then, E_0 is calculated for free space by forcing the initial vertical profile to be zero for $x < 0$ (see Fig. 1.18b).

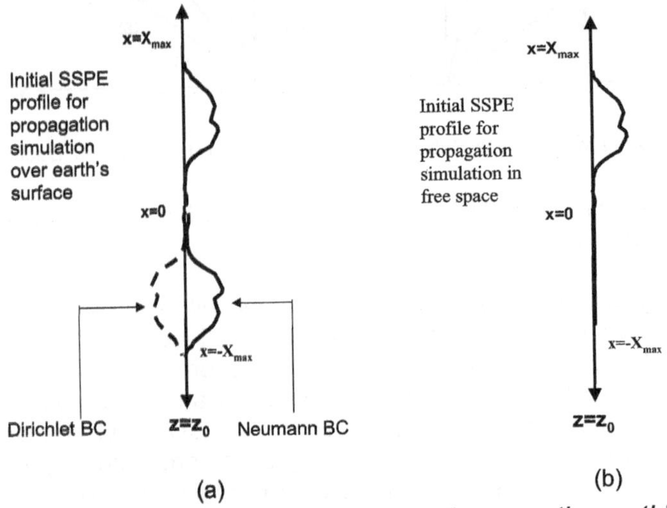

Figure 1.18: *Initial vertical profiles for (a) propagation over the earth's surface and (b) propagation in free space*

Path loss $L_p(z)$ is defined at z = d as

$$L_p(d) = 10\log_{10}\left(\frac{P_t}{P_r(d)}\right) \quad \text{[dB]}, \tag{1.90}$$

where P_t and P_r are the transmitted (at z = 0) and received (at z = d) powers, respectively. For excitation by a short electric dipole with a moment of $M = 5\lambda/2\pi$, corresponding to P_t = 1 kW, the received power for an isotropic receiver at z = d can be determined from the computed field strength E for an isotropic receiver via

$$P_r(d) = \frac{E_r(d)^2}{Z_0} \times \frac{\lambda^2}{4\pi} \tag{1.91}$$

This leads to the expression

$$L_p(d) = 142.0 + 20\log_{10}(f_{MHz}) - 20\log_{10}(E_{\mu V/m}) \quad \text{[dB]} \tag{1.92}$$

with the frequency and field strength measured in MHz and dBμV/m, respectively. Frequency-domain characteristics can be obtained from the TD simulations. The algorithms store necessary information (such as range, terrain profile, transverse and/or longitudinal propagation characteristics at chosen observation points) while propagating from left to right. For example, a vertical field distribution (or propagation factor) at a given range can be obtained as follows:

- During the TDWP and/or TLM-WP simulation, time domain propagation data such as $E_x(x_i,t_j)$ at discrete x_i (i = 1,2, ...) observation heights and discrete t_j (j = 1,2, ...) simulation instants is accumulated at a desired range.
- The accumulation continues until all transients fade away.
- After TDWP simulation, height field distribution $E_x(x_i,f_A)$ at a given frequency f_A is directly obtained by off-line DFT analysis.

1.4.1 Propagation Under Ducting Conditions

The TD propagators are first tested over smooth, spherical earth. The propagation region is characterized by longitudinally homogeneous, bi-linear vertical refractive index, n, with the gradient of $|dn/dx| = 10^{-3}$ [n unit/m] (i.e., decreasing between ground and 25 m and increasing over 25 m). The vertical distribution of the source is Gaussian with spatial extent ≈15 m and maximum at 26 m height (i.e., the vertical profile is non-zero between 18.5 m and 33.5 m). The temporal distribution is a once-differentiated Gaussian pulse with 200-MHz bandwidth centered at 200 MHz. A 1000 × 500 FDTD/TLM computation (corresponding to 100 m × 50 m physical) space is used. A 1000 × 250 virtual window circulates 20 times as if the longitudinal number of cells in FDTD/TLM computation space is 5000. Instant snapshots are taken at different simulation times and are plotted as field profiles at different ranges in Fig. 1.19 (each plot is normalized to its maximum value).

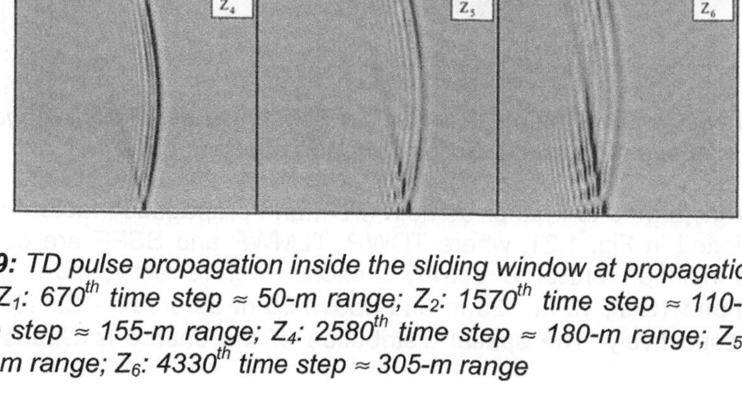

Figure 1.19: *TD pulse propagation inside the sliding window at propagation ranges Z_1, Z_2, ..., Z_6. Z_1: 670th time step ≈ 50-m range; Z_2: 1570th time step ≈ 110-m range; Z_3: 2250th time step ≈ 155-m range; Z_4: 2580th time step ≈ 180-m range; Z_5: 3070th time step ≈ 215-m range; Z_6: 4330th time step ≈ 305-m range*

The relevant wave constituents are the direct, ground-reflected and surface waves whose wavefronts and their interference (wave maxima and minima) appear as 2D images. As shown in the figure, the wave propagates to the right and is above the ground (at $z = Z_1$). While it is propagating it spatially extends because of cylindrical spreading and reaches the bottom surface (at $z = Z_2$ and $z = Z_3$). Then it reflects from the surface, giving rise to interference between the leading direct and trailing ground-reflected waves (as observed at $z = Z_4$ through $z = Z_6$).

Time histories of the pulse during propagation are accumulated at different ranges and along two constant heights; they are plotted in Fig. 1.20, where the results of both TDWP and TLM-WP are almost indistinguishable. In Fig.s 1.20a and 1.20b, the receiver heights are chosen to be 13 m and 35 m above the ground, respectively. Three plots correspond to the 94 ns time histories (i.e., signal vs. time) at three different ranges. Since the scale in each plot is normalized to its maximum value in order to reveal the detailed pulse shapes along the entire trajectory, relative field strengths with respect to the first window are also included as dB values in the plots. At z = 23.5 m, only the initial pulses appear inside the 94 ns propagation windows because the delay of the signal caused by the path difference (distance between direct and ground reflected pulses) exceeds the window length. As the distance increases, the path difference decreases and the ground-reflected pulse also appears inside the 94 ns propagation windows, as shown in the second windows. Inside the last windows the direct and ground-reflected pulses are almost indistinguishable.

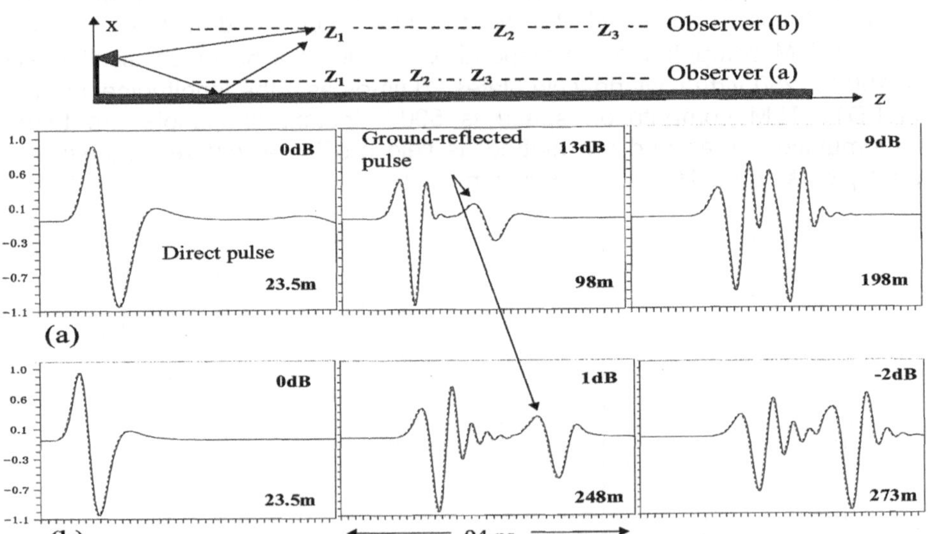

Figure 1.20: *Time histories of the pulse in Fig. 1.19 observed at different ranges along (a) 13 m, (b) 35 m above the surface. Solid: TLM-WP; dashed: TDWP*

TD results are transformed to Fourier domain (FD), and propagation factor vs. height variations are plotted in Fig. 1.21, where TDWP, TLMWP and SSPE are compared. Here, a fictitious tri-linear vertical refractivity is chosen with refractive index gradients between ground and 15 m, 15 m - 25 m and above 25 m as dn/dx=-1.2×10^{-5}, 5×10^{-6} and -1.2×10^{-5}, respectively. The spatial distribution of the source is Gaussian at a height of 15 m, with spatial extent of ≈12 m. Temporal distribution is a once-differentiated Gaussian pulse in TLM-WP and TDWP. This short pulse has 200-MHz bandwidth centered at 200 MHz. The SSPE propagator, with the same spatial source distribution, is run separately for each frequency. With these parameters, energy is mostly trapped along the surface and the pulse reaches to PML blocks at larger distances compared to standard atmosphere. Therefore, numerical dispersion effect is dominant in this example.

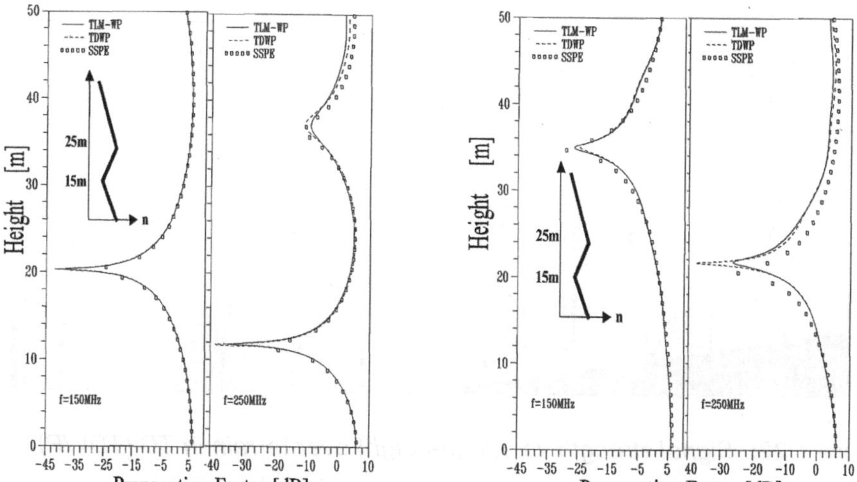

Figure 1.21: *$|E/E_0|$ vs. height after 25 and 45 sliding windows*

The first two plots belong to the results at a range of 625 m (311λ at 150 MHz and 518λ at 250 MHz). At these ranges the propagating pulse is still confined in the sliding window and the agreement is very good. On the other hand, the last two figures belong to a range of 1121.5 m reached at 45th window (561λ at 150 MHz and 935λ at 250 MHz), where the sliding window is hardly capable of holding the propagating pulse. Although the sliding window hardly bounds all the forward-propagated energy, very good agreement may still be obtained in frequency domain as long as most of the dominant contribution is traced.

1.4.2 Propagation Over Non-Flat Terrain

The TDWP is then compared against SSPE in propagation scenario that includes non-flat terrain. A scenario with strong forward scatter (and negligible backward propagation) is chosen in order to make comparisons with SSPE possible. A downward sloping terrain (200 m high at the source point and 0m at 5.5 km) plus smooth surface over which exists a vertical tri-linear, longitudinally homogeneous refractivity profile is taken into account.

A 1500 × 1000 TDWP computation space corresponding to 2250 m height and 1500 m range (with 1.5 m spatial discretization) is chosen. The initial field distribution is a once-differentiated Gaussian function with 15 MHz bandwidth at 10 MHz center frequency. A tri-linear vertical refractive index with n_0=1.0003 at the surface, dM/dx = 120 M/km up to 250 m, dM/dx = -50 M/km between 250 m and 400 m and dM/dx =

118 M/km above 400 m (with respect to the bottom of FDTD computation space) is chosen. Here, *dM/dx* is the gradient of atmospheric refractivity modulus. The spatial distribution of the source is Gaussian at an altitude of 300 m over the terrain. A 1500 × 500 virtual window circulates 13 times and extends the longitudinal number of cells in FDTD computation space to 6500. In Fig. 1.22, instant snapshots of field profiles taken at different simulation times are plotted at ranges Z_1 to Z_6, respectively. Pulse propagation (Z_1), surface reflection (Z_2), energy split (Z_3 and Z_4), and surface and elevated duct formation (Z_5 and Z_6) are clearly observed in Fig. 1.22.

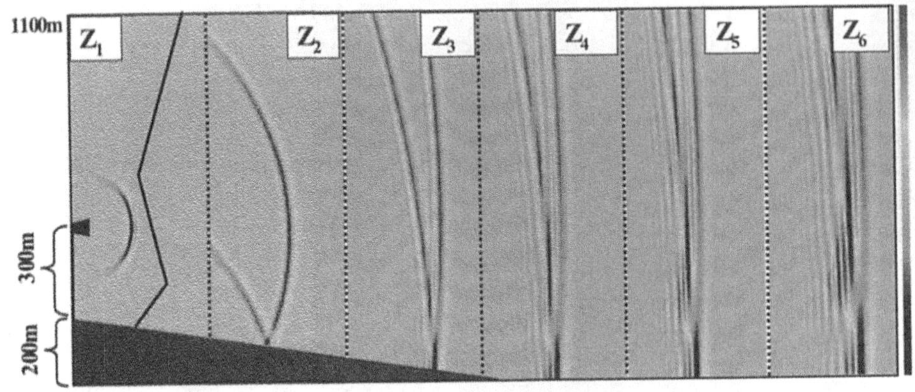

Figure 1.22:. *Signal strength vs. range-height over terrain in TD (TDWP)*

Same source pattern is fed into the SSPE algorithm and is propagated through the same medium. In Fig. 1.23, the range-altitude field distributions at 10 MHz are plotted.

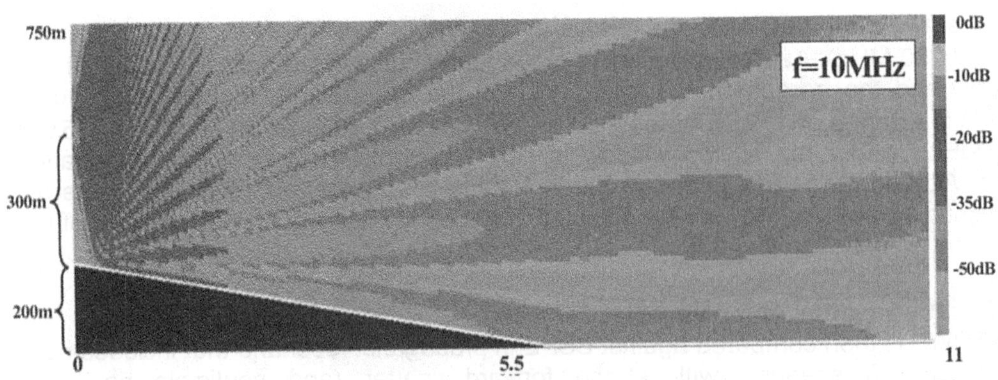

Figure 1.23:. *Signal strength vs. range-height over terrain in TD (SSPE)*

Propagation factor vs. height for the same environment as in Figs. 1.22 and 1.23 are plotted in Fig. 1.24. Neumann boundary condition is applied for SSPE to model VED excitation as in TDWP. A good agreement between TDWP and SSPE is clearly observed. Propagation factor vs. height over a PEC level terrain with the same tri-linear refractivity profile, computed via SSPE, is also given in the figure (dashed curves) to emphasize the effects of the downward sloping terrain portion.

In Fig. 1.25 propagation factor vs. range computed with these three simulators are plotted in Fig. 1.25 for a more realistic scenario. Here, a 1.5-km-long non-flat terrain, with maximum height of 40 m is taken into account (see Fig. 1.25a). The source and observation points are 5 m above the ground and the mesh sizes are taken as 10 cm.

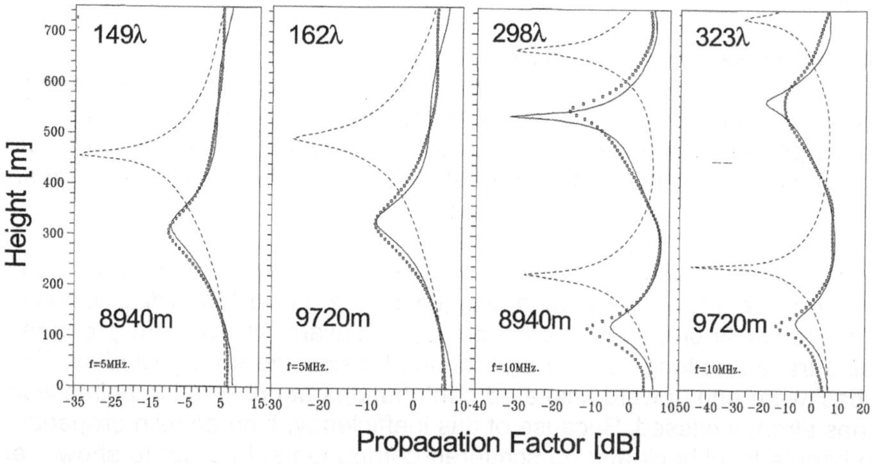

Figure 1.24:. *|E/E$_0$| vs. height at 10 MHz and 15 MHz and at two observation ranges (solid: SSPE; dots: TDWP, dashes: SSPE over level surface).*

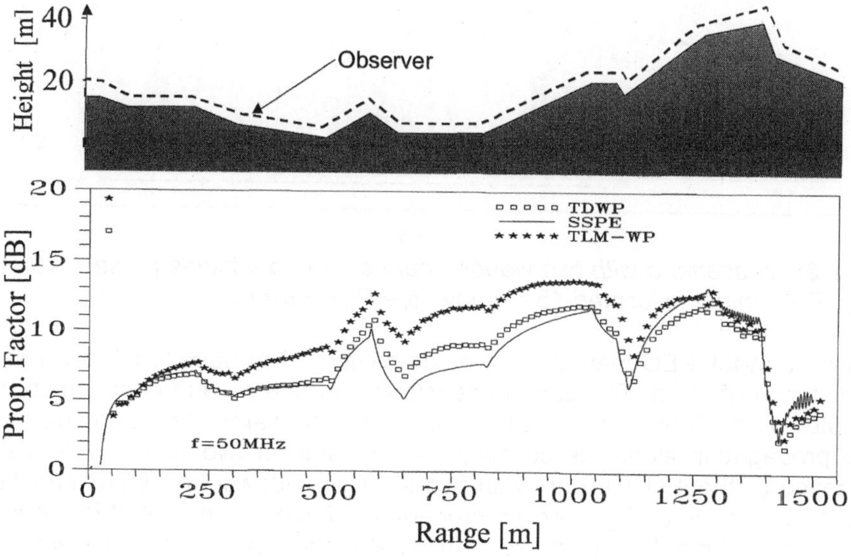

Figure 1.25: *|E/E$_0$| vs. range for the typical non-flat terrain*

Although SSPE does not include the back-scatter effects, TDWP and TLM-WP results contain local back-scatter contribution as long as it is within the sliding window. This explains the discrepancy between SSPE and TD propagators. As seen in Fig. 1.25, there is also a discrepancy between the magnitudes of TDWP and TLM-WP, which is not observed for propagation over smooth ground. This can be because of numerical dispersion or differences in abilities of the FDTD and TLM implementations themselves, resulting in different responses to the scattering mechanisms, such as reflection, edge/tip diffraction, multi-path interference, and surface waves, which is under investigation.

1.4.3 Urban Wireless Communication

TD propagators are also applicable in wireless communications when the depths of the buildings are assumed large enough to reduce the problem to a 2D case. The power of the TD propagators arises from their applicability in wireless communication in urban regions, such as along a narrow street with tall buildings in different sizes. In this section this is presented on various scenarios.

In most of the practical problems, maximum ranges of interest in urban areas cannot be covered by one or a few sliding windows. Therefore, while the sliding window is tracing the longitudinally propagating component, backward and upward propagated waves are also traced as long as they are confined inside the sliding window. There may still be backward and upward scattered contributions at observation points where the window has already passed. Because of this inefficiency, time domain propagators are called to handle local back and up scattered components. In order to show these effects, a scenario in Fig. 1.26 is taken into account.

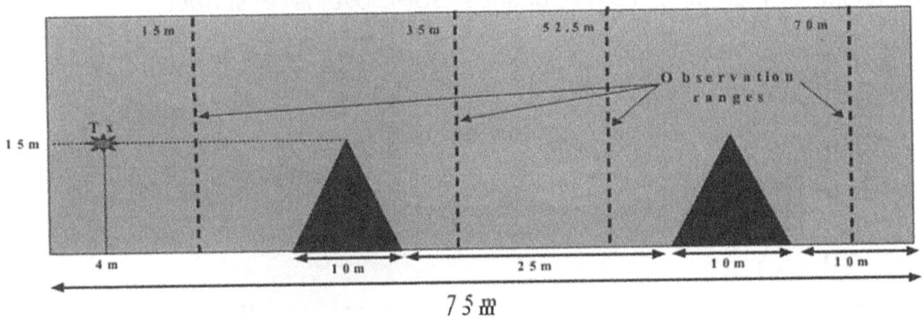

Figure 1.26: *A scenario with two wedge-shaped PEC obstacles (base = 10 m, height = 15 m). Transmitter is located 15 m above the PEC ground*

Here, two 15-m-tall PEC triangular buildings having 10 m base lengths are separated with a distance of 25 m. The other dimensions and source and observation locations are denoted in the figure. The mesh size is 1 cm. The same Gaussian source is used and TD propagation along the buildings is simulated in two ways: (i) with a unique FDTD space (without sliding), (ii) with two sliding windows via TDWP and TLM-WP. The results are given in Fig. 1.27, as propagation factor vs. height at four observation ranges (it should be noted that these results belong to a single time-domain simulation plus multiple DFT application). In the figure, observation ranges from left to right are numbered from 1 (at 15 m) to 4 (at 70 m). Almost indistinguishable solid and dashed lines correspond to TDWP and TLM-WP computations, respectively, while the dashed line represents single window FDTD computations. It is clearly observed in the figure that, although TDWP and TLM-WP results have indistinguishable agreement, they do not agree very well with the FDTD, which may be assumed as a reference solution.

Moreover, unacceptable discrepancies appear at some heights and ranges. Here, FDTD results represent reference solutions (since the numerical dispersion ratio is 60 at 50 MHz), and the reason of the discrepancies may easily be understood from the scenario given in Fig. 1.26. The strongest back- and up-scattered components are expected to be at the first observation range, at the bottom and at heights above the source (because of the inclined wall of the wedge type building). Not so strong as the dominant scatter, back- and up-scattered components are also expected in between the buildings. On the other hand, at the last observation range the forward propagating wave is dominant, therefore, as expected, the best agreement is obtained at this range.

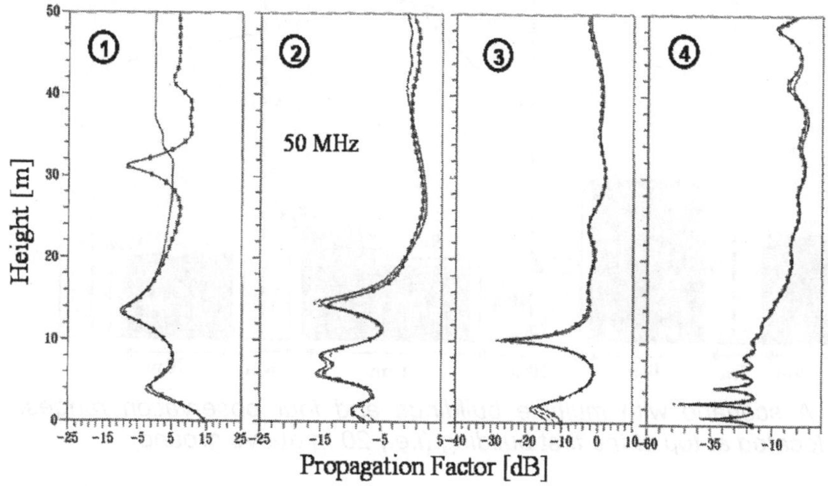

***Figure 1.27:** |F| vs. height at 50 MHz at four observation ranges mentioned in Fig. 1.26. Solid, TDWP; Dashed, TLM-WP; Solid with dots, FDTD*

Longitudinal variations of field strength are also calculated via TDWP and TLM-WP and are presented in Fig. 1.28 at 15-m height and for 50 MHz.

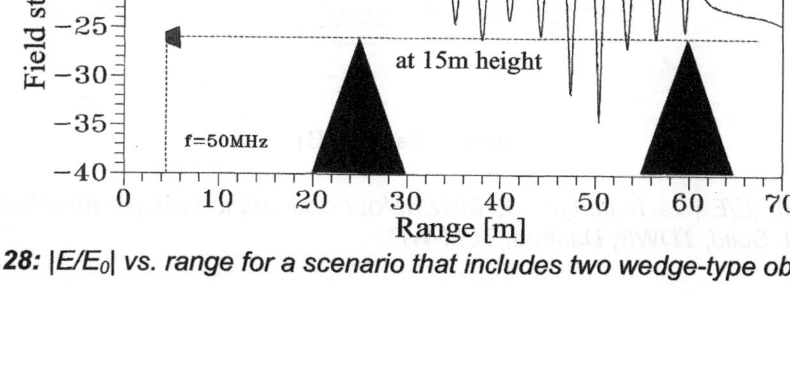

***Figure 1.28:** $|E/E_0|$ vs. range for a scenario that includes two wedge-type obstacles*

As seen, although range variations are highly oscillatory almost perfect agreement is obtained between TDWP and TLM-WP results. It is also interesting to observe tip-diffracted components (two peaks at ranges 25 m and 60 m, which correspond to the tips of the buildings) in the 50 MHz figure. Finally, a complex scenario, where four different-shaped buildings with different sizes are used, is taken into account as given in Fig. 1.29.

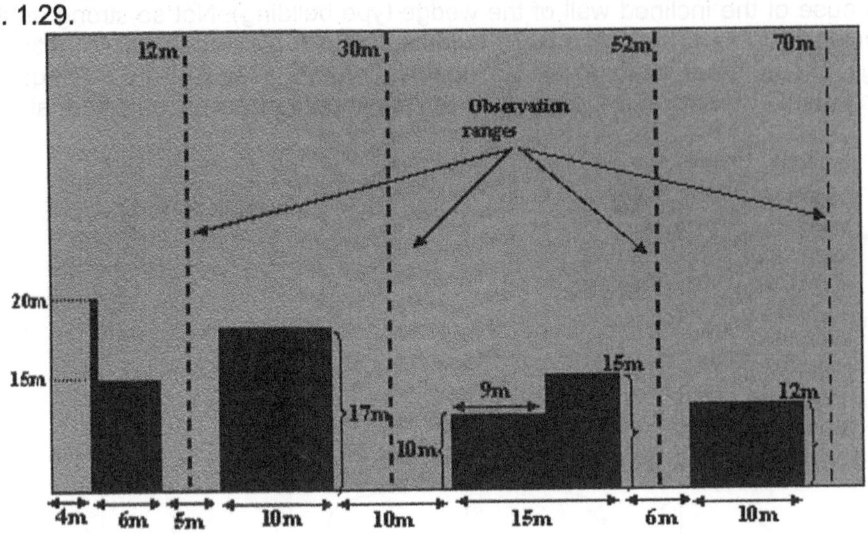

Figure 1.29: *A scenario with multiple buildings and four observation ranges. The transmitter is located at top of the first building (i.e., 20 m above ground)*

All dimensions and source/observation locations are mentioned in the figure. The buildings are assumed to be PEC. The results are presented in Fig. 1.30 as propagation factor vs. height at given four observation ranges. Both TDWP and TLM-WP computations are carried out within a unique FDTD/TLM computation space (to account for all scattered contributions).

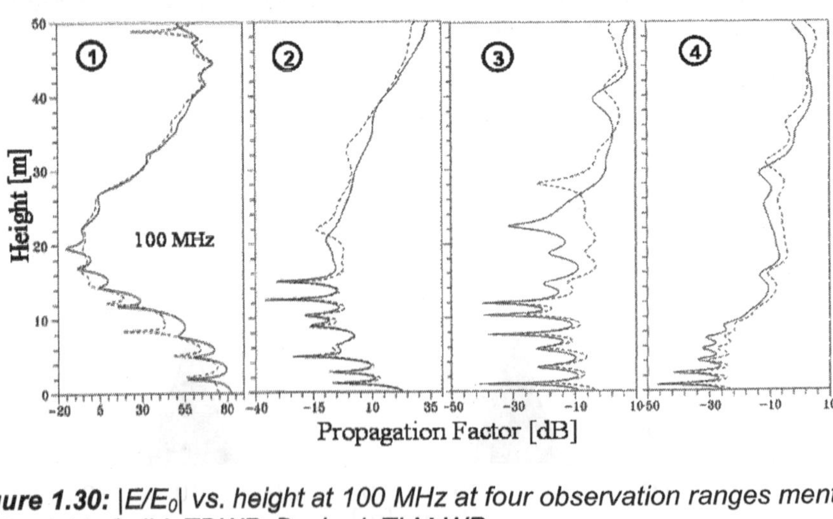

Figure 1.30: *|E/E$_0$| vs. height at 100 MHz at four observation ranges mentioned in Fig. 1.29. Solid, TDWP; Dashed, TLM-WP*

The results do not agree very well as observed in the previous examples. There is no discretization error in this example, since all the buildings are chosen to be rectangular. Possible sources of these discrepancies are (i) numerical dispersion errors (it may require numerical dispersion ratios higher than 30 at 100 MHz), (ii) PML termination (since the scattering mechanisms, such as reflection, edge/tip diffraction, multi-path interference, surface waves, etc., become complex in this example), or (iii) differences in abilities of the FDTD and TLM implementations themselves.

The screen captures of TD pulse scattering at different time instants for the same scenario is given in Fig. 1.31 as signal strength vs. range/height. Since very similar results are obtained, only TDWP results are pictured in this figure. The multi-reflections, edge and tip diffraction and ringing in between buildings are clearly observed in the figure.

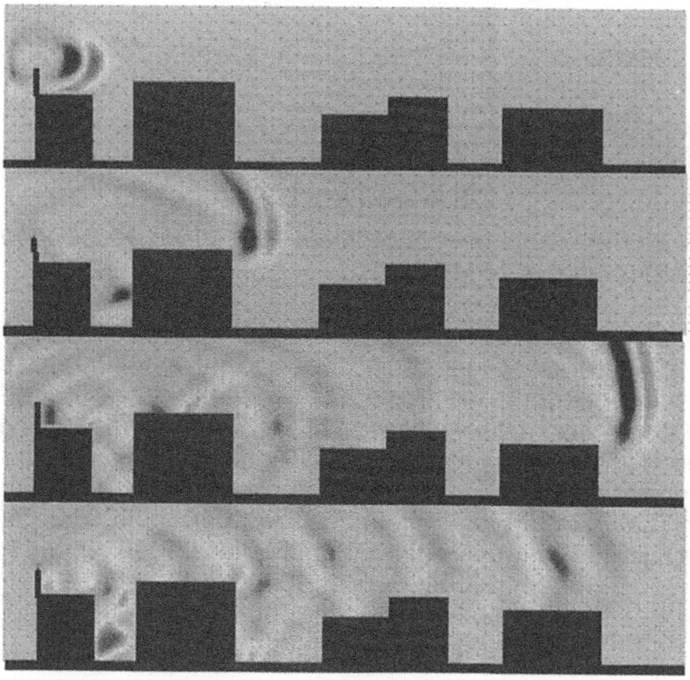

Figure 1.31: *A 3D picture of pulse scattering along the street*

1.4.4 Surface Waves and Mixed-Path Propagation

Although a perfectly reflecting boundary assumption provides in general sufficient approximation at VHF and above (i.e., frequencies higher than 100 - 200 MHz), the use of impedance boundary condition becomes essential at HF frequencies and below. This is especially required for the simulation of long-range marine communication and/or ocean surveillance systems using HF frequencies. A challenging problem is to predict surface wave path loss variations over mixed paths, such as sea - land or sea - land - sea (island) transitions. A sharp decrease occurs in signal strength along sea - land transition and the signal recovers itself after land - sea transition (beyond the island), known as the Millington (recovery) effect [5]. A propagator should be capable of handling these effects in mixed-path simulations.

Data generated via approximate analytical (ray-mode) solutions is used for the validation of the TDWP augmented with the implementation of the impedance boundary condition on the ground surface. Two typical scenarios and the results obtained with TDWP simulations are given in Figs. 1.32 and 1.33. In Fig. 1.32, a sea - land transition problem for a standard atmosphere including earth's curvature (i.e., dM/dx = 118 M/km) is taken into account and range versus surface wave path loss (when both transmitter and receiver are at the surface) are plotted at 15 MHz.

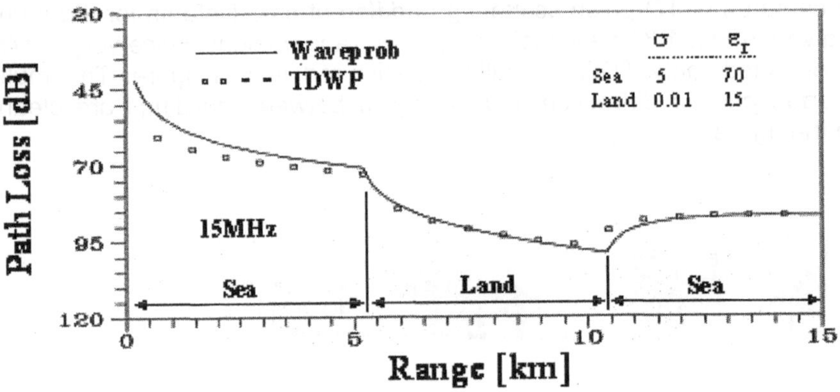

Figure 1.32: *SW path loss vs. range over mixed-path (sea - land transition, the first 8.2 km is sea and the rest is land) at 15 MHz*

The first 8.2 km is assumed to be sea and the rest is land (the parameters of the land and sea are taken as σ_g = 0.01 [S/m], ε_g = 15 and σ_g = 5 [S/m], ε_g = 70, respectively). Ray - mode results are also shown in this figure and serve as reference solutions. A vertical short electrical dipole (with a dipole moment of $5\lambda/2\pi$ Am, corresponding to a 1 kW transmitter) is used in analytical solution and TDWP results are normalized to this value. A very good agreement between the results is clearly seen in the figures. It should be noted that TDWP results at both frequencies belong to a single pulse propagation simulation followed by two off-line DFT applications.

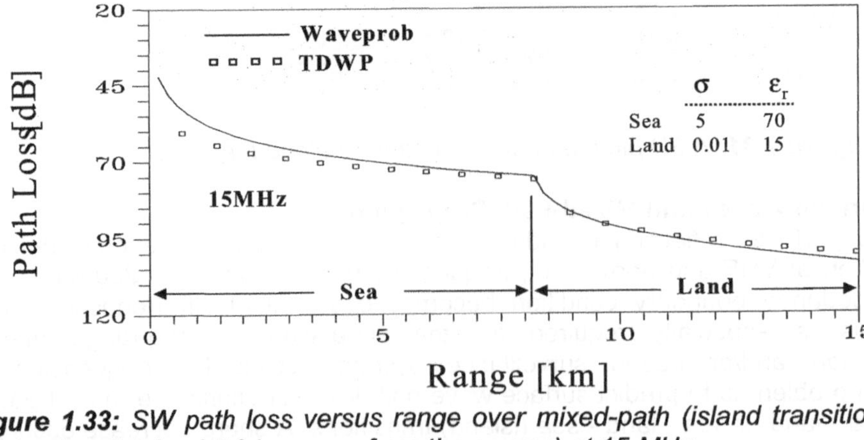

Figure 1.33: *SW path loss versus range over mixed-path (island transition, a 5.2 km length island 5.2 km away from the source) at 15 MHz*

In Fig. 1.33, the results of a scenario corresponding to ground wave propagation over a path including an island transition under the assumption of standard atmospheric conditions are plotted. A 5.2-km length island is assumed to be 5.2 km away from the transmitter. Again, range versus surface wave path loss are plotted at 15 MHz with the same land and sea parameters. Also in this case the results agree very well.

1.5 NUMERICAL SIMULATORS

Five simulator packages are prepared for the reader who wants to exercise different ground wave propagation scenarios. Three of them (GRWAVE, HFMIX, and KNIFE) are based on ray-mode approach and the other two are SSPE in FD and TDWP in TD.

Ray-mode formulations in (1.44) - (1.46) and (1.58) can be combined in an intelligent way to extend ranges of their individual validity and uniformizes the solution in interference, intermediate and diffraction regions [WAVEPROB]. Here, simple sample simulators are presented which can be used in these regions, separately. GRWAVE is valid over lossy spherical earth's surface, only in interference region, where transmitter and receiver is within LOS. HFMIX is prepared for surface-wave path loss calculations which can handle multimixed-paths beyond the diffraction region. Transmitter and receiver height effects are augmented into HFMIX as height gain functions (as given in (1.58)). KNIFE uses four rays to account for single knife-edge diffraction effects as described in [43]. Surface impedance and roughness are also taken into account in KNIFE.

1.5.1 GRWAVE: LOS Ray Code

The flow chart of GRWAVE is given in Fig. 1.34. The simulator reads user-supplied input parameters from the input file GRW.INP, runs and outputs the results in two different output files: GR-R.DAT as range vs. propagation factor [dB] and GR-H.DAT as height vs. propagation factor [dB].

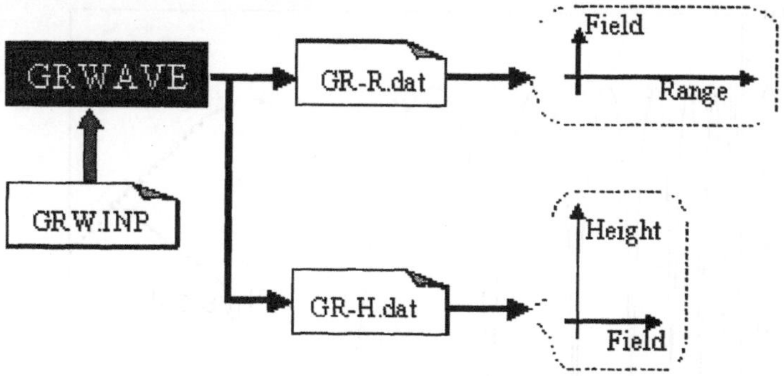

Figure 1.34: *The structure of GRWAVE*

The GRW.INP input file contains five rows of parameters, which are listed in Table 1.2. The fifth line has four real numbers A, B, C, and D, which correspond to range [km], first height [m], last height [m] and height increment [m] when scenario type is 1, and to height [m], first range [km], last range [km] and range increment [km] when scenario type is 2, respectively.

Two typical examples obtained via GRWAVE are pictured in Figs. 1.35 and 1.36. In Fig. 1.35, propagation factor vs. range is plotted and the input parameters are as follows:

GRW.INP (Fig. 1.35)	LINE 1:	500
	LINE 2:	15, .01
	LINE 3:	500
	LINE 4:	2
	LINE 5:	100, 20.0, 120.0, 0.5

Table 1.2: The GRW.INP file and required parameters

Row No.	Parameter
1	Frequency [MHz]
2	Relative permittivity, conductivity
3	Transmitter height [m]
4	Scenario type (1, Field vs. height: 2, field vs. range)
5	A, B, C, D

Similarly, in Fig. 1.36, propagation factor vs. height is plotted. Here, GRWAVE is run three times at three different ranges. Input parameters are as follows:

GRW.INP (Fig. 1.36)	LINE 1:	500
	LINE 2:	15, .01
	LINE 3:	500
	LINE 4:	2
	LINE 5:	A, 0.0, 400, 0.5 (A=40, 80 and 120)

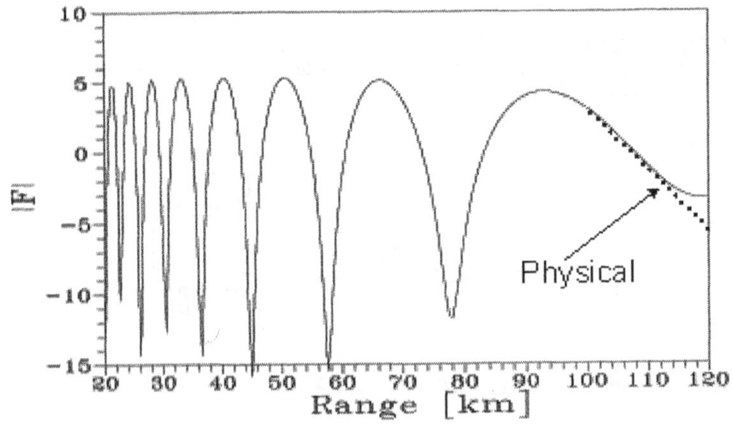

Figure 1.35: *$|E/E_0|$ vs. range at 100 m obtained with GRWAVE*

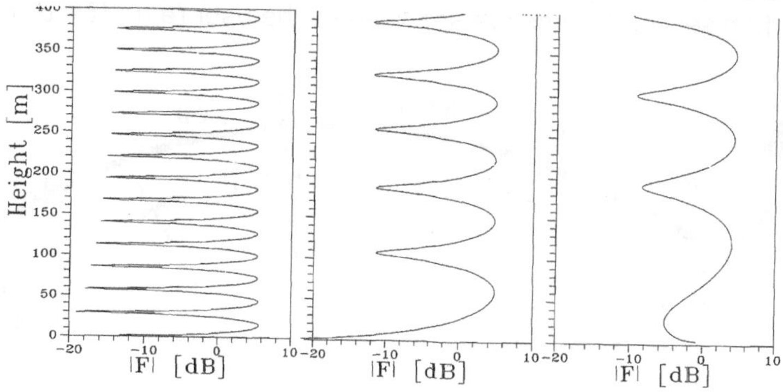

***Figure 1.36 :** |E/E$_0$| vs. height at 40, 80 and 120 km, respectively*

1.5.2 HFMIX: Surface-Wave Path Loss Predictor

The flow-chart of HFMIX is given in Fig. 1.37. The simulator reads user-supplied input parameters from the input file HFMIX.INP, runs and outputs the results in two different output files; LMIX.dat as range vs. path loss [dB] and EMIX.DAT as height vs. field values [dBμV/m].

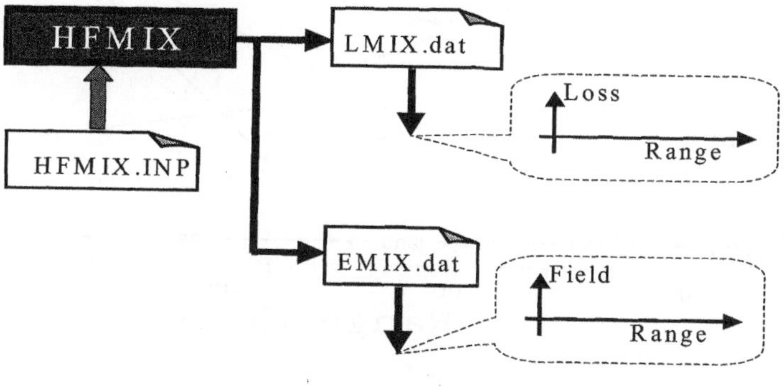

***Figure 1.37:** The structure of HFMIX*

The HFMIX.INP input file contains a minimum of five rows of parameters, which are listed in Table 1.3. The user supplies a number of propagation sections in the fourth line in HFMIX.INP. Depending on this number path length, conductivity, and relative permittivity of each section is given in the fifth, sixth, seventh, etc., lines.

Table 1.3: The HFMIX.INP file and required parameters

Row No:	Parameter
1	Frequency [MHz]
2	Range increment [km]
3	Transmitter and receiver heights [m]
4	Number of propagation paths [Integer]
5	Path length, conductivity, relative permittivity

The scenario of HFMIX is pictured in Fig. 1.38. Total propagation path is automatically calculated when section lengths are specified by the user (e.g., $R = d_1 + d_2 + d_3 + d_4$)

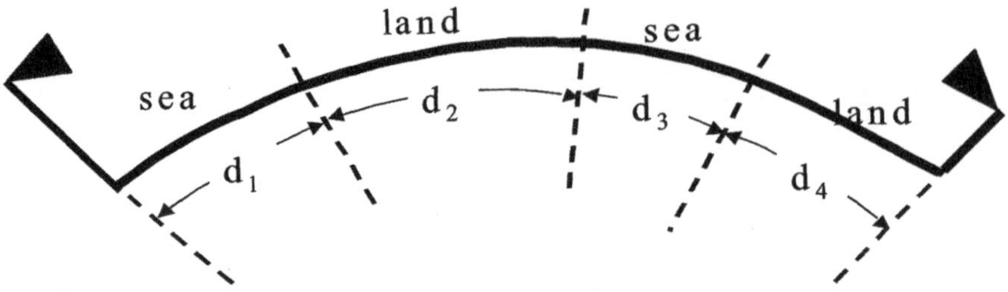

Figure 1.38: *HFMIX scenario definitions*

A typical HFMIX result is presented in Fig. 1.39, as path loss vs. range for a three-section propagation path (first 100 km sea, then an island of 50 km, and another 150 km of sea) at three different frequencies are plotted.

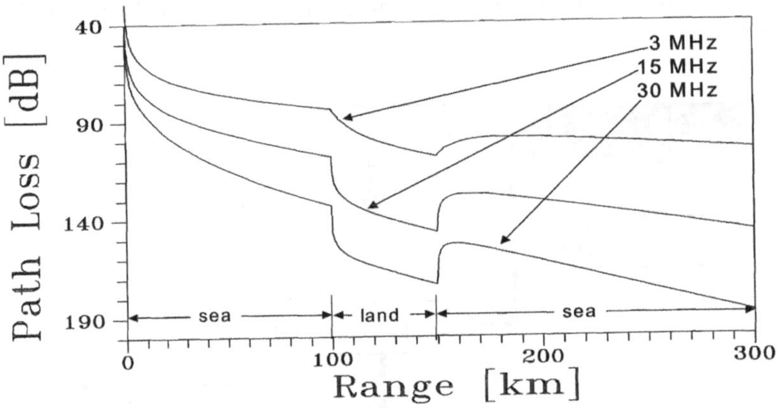

Figure 1.39: *Path loss vs. range over a three-section path*

Some remarks on HFMIX are listed as follows:

- Standard atmosphere over smooth spherical earth is assumed, therefore no refractivity variations are allowed.
- Only surface wave contribution is included. Transmitter and receiver heights are included as height gain functions.
- HFMIX can be best used from a few hundred kHz up to 40-50 MHz. In this frequency region, transmitter and receiver heights better be less than a hundred meters.
- Ground parameters are taken into account, but only two types of grounds are allowed. Mixed-path effects (Millington recovery) can be calculated for up to 10 different propagation sections.
- Homogeneous (single-section) ground, such as land or sea, can be given by giving "1" in the line 4 and total propagation length in the 5th line.

- Codes are calibrated against SSPE and CCIR [44] recommendations. There are no "shortcuts" while coding.
- Since this is an analytical-based code, no improvements are expected parallel to increase in computers' capabilities. Current PC's are fast enough to do computations.

1.5.3 KNIFE: Four-Ray Knife-Edge Code

Knife is a ray method prepared for a single PEC obstacle with zero thickness between a transmitter and a receiver. It uses four rays: the direct ray between the transmitter and receiver, the tip-diffracted ray from the top of the obstacle, te ground-reflected rays, and, finally, the ground-reflected ray behind the obstacle that reaches there via tip diffraction. KNIFE accounts for polarization and surface impedance, including a surface roughness factor. The details of this program can be found in [43].

The structure of this code is illustrated in Fig. 1.40. As usual, the package accepts user defined input parameters from the input file KNIFE.INP. The KNIFE.INP input file contains seven rows of parameters, which are listed in Table 1.4. A typical example is plotted in Fig. 1.41, for PEC surface at three different frequencies.

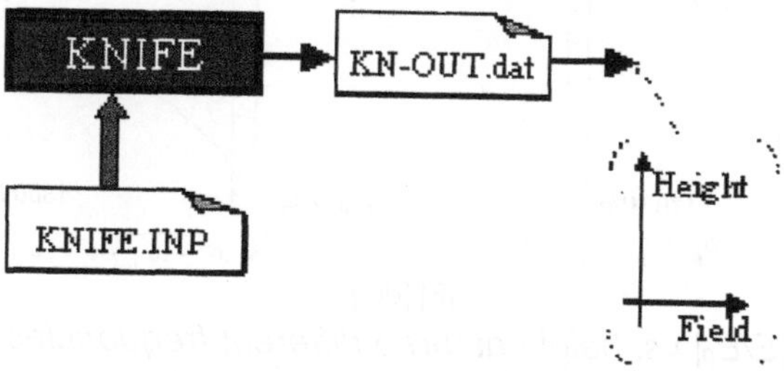

Figure 1.40: *The structure of KNIFE*

The KNIFE.INP file for this example is as follows:

KNIFE.INP (Fig. *1.41*)	LINE 1:	170, 600 and 1300
	LINE 2:	100
	LINE 3:	100, 10000, 5000
	LINE 4:	1, 250, 1
	LINE 5:	4.0, .0001, 4.0, .0001
	LINE 6:	0.0, 0.0
	LINE 7	1

Although relative permittivity and conductivity values are supplied in LINE 5, surface roughness for sections both on the left and right of the obstacle are set to zero in LINE 6. The polarization is horizontal for all these three plots.

Table 1.4: The KNIFE.INP file and required parameters

Row No:	Parameter
1	Frequency [MHz]
2	Transmitter height [m]
3	Obstacle height [m], TR to Obstacle distance [m], Obstacle to Rx distance [m]
4	First height [m], last height [m], height increment [m]
5	Relative permittivity, conductivity [Left part] Relative permittivity, conductivity [Right part]
6	Roughess factor (real number between 0-1)
7	1 [H polarization], 2 [V polarization]

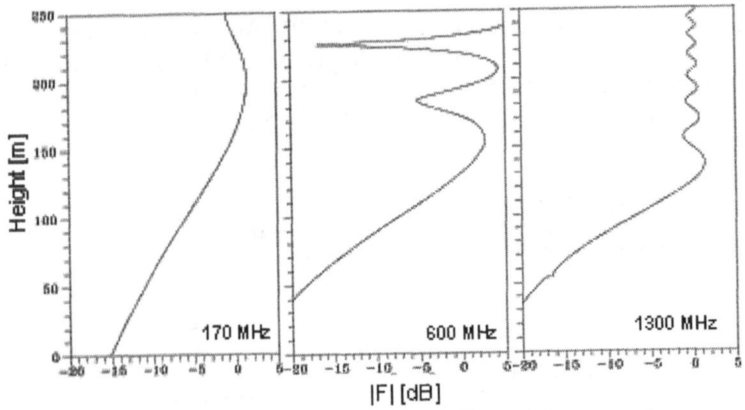

***Figure 1.41:** $|E/E_0|$ vs. height at three different frequencies*

1.5.4 SSPE: Frequency- Domain Wave Propagator

SSPE package is capable of calculating propagation factor and path loss as a range and/or height profile. The structure is given in Fig. 1.42. Some remarks on the package are as follows:

- It uses Dirichlet-type boundary conditions at the surface, therefore source should be minimum 10 m above the surface.
- Dirichlet-type boundary condition for the scalar wave function correspond to TE_z type EM wave problem, therefore only horizontal polarization is possibly chosen.
- The package is better used for frequencies from 200 MHz up to 1 GHz. Maximum height is limited as 500 m, which is divided into 256 height steps.
- Because of the Nyquist criteria, the maximum frequency that can be used with this height step size is limited to approximately 1 GHz for accuracy up to propagation angle of θ = 4.5° with respect to horizontal axis z. For higher propagation angles, the operating frequency should be lower.

Again, the program requires that the user first supply the input file SSPE.INP. It runs and yields two output files RANGE.DAT (first column range [km], second column propagation factor [dB]), and HEIGHT.DAT (first column height [m], second column propagation factor [dB]). Optionally, CONTOUR.DAT may also be obtained, where 3D field data are used in a MATLAB file (PLOT3D.M) for plots as propagation factor vs. range/height is available. Necessary input parameters are listed in Table 1.5.

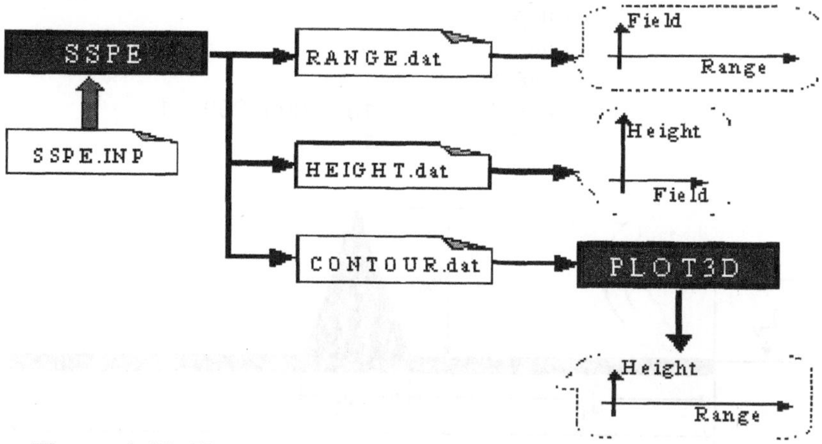

Figure 1.42: *The structure of the SSPE sample package*

Table 1.5: The SSPE.INP file and required parameters

Row No:	Parameter
1	Frequency [MHz]
2	Transmitter height [m]
3	Maximum range of interest [m]
4	Observation height [m]
5	Contour map (1: Yes, 0: No)
6	Scenario Type (1, single obstacle; 2, double obstacle)
7	**If Scenario Type is 1 (Single Obstacle)** Obstacle height [m], transmitter-to- obstacle distance [m], obstacle base length [m], obstacle type (1:wedge, 2: block)
7 – 8	**If Scenario Type is 2 (Double Obstacle)** Obstacle height [m], transmitter-to-obstacle distance [m], obstacle base length [m], obstacle type (1,wedge; 2, block) Obstacle height [m], distance between first and second obstacles [m], obstacle base length [m], obstacle type (1, wedge; 2, block)
8 (or 9)	Refractivity profile (1,2 or 3) 1: Standard atmosphere 2: Sub-refraction 3: Super-refraction

For example the SSPE.INP file given below belongs to Fig. 1.43. The parameters read are 400 MHz, source height h_v = 50 m, single obstacle, obstacle height h_{ter} = 100 m, transmitter-to-obstacle distance d_1 = 5 km, obstacle base w_1 = 2000 m, maximum range of interest R_{max} = 10 km, observation height h_{obs} = 50 m, obstacle: wedge, vertical refractivity profile: standard atmosphere and no contour map data is required.

SSPE.INP (Fig. *1.43*)	LINE 1:	400
	LINE 2:	50
	LINE 3:	10000.
	LINE 4:	50
	LINE 5:	0
	LINE 3:	1
	LINE 4:	100,5000, 2000, 1
	LINE 5	1

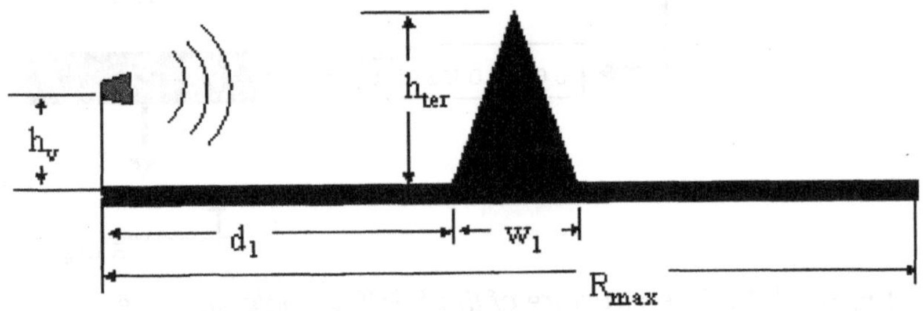

Figure 1.43: *The scenario of the SSPE.INP file in SSPE.INP given above*

Similarly, the scenario in Fig. 1.44 can be obtained with an input file given below (250 MHz, source height h_v = 10 m, double obstacle, wedge and block, wedge height h_{ter1} = 5 m, source-to-wedge distance d_1 = 500 m, wedge base w_1 = 5 m, block height h_{ter2} = 10 m, wedge-to-block distance = 50 m, block base w_2 = 5 m, maximum range of interest R_{max} = 1 km, observation height h_{obs} = 10 m vertical refractivity profile: standard atmosphere and contour map data are required).

SSPE.INP (Fig. *1.44*)	LINE 1:	250
	LINE 2:	10
	LINE 3:	1000
	LINE 4	10
	LINE 5:	1
	LINE 6:	2
	LINE 7:	5, 500, 5, 1
	LINE 8	10,50, 5, 2
	LINE 9	1

SSPE is applied to two typical propagation scenarios, where a wedge shaped PEC obstacle is located in between the transmitter and receiver. The scenario is pictured in Fig. 1.45. Here, the obstacle has a 50-m base length and is 250 m away from a 150-MHz source, which is 25 m above the ground. The height of the obstacle is a parameter.

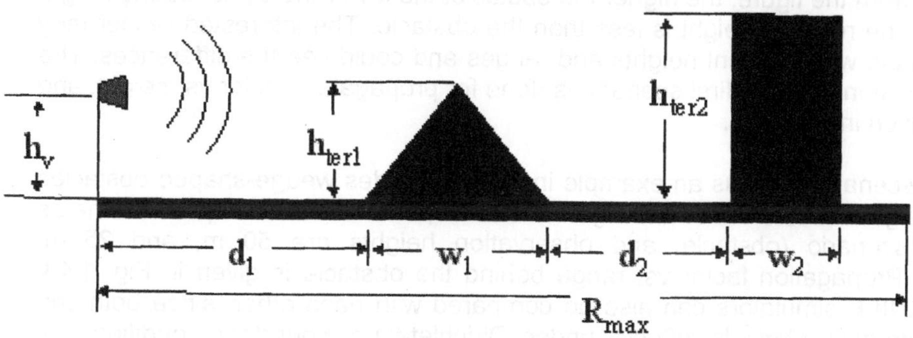

Figure 1.44: *The scenario of the SSPE.INP file in SSPE.INP given above*

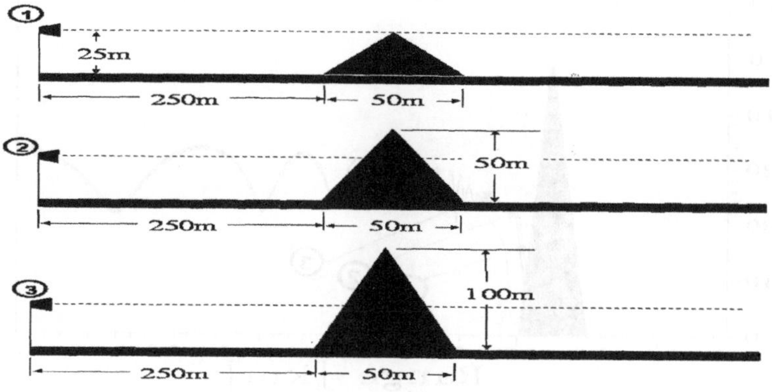

Figure 1.45: *A (first) scenario with single obstacle*

SSPE is run with this scenario and propagation factor vs. both range and height behind the obstacle are calculated. In Fig. 1.46, propagation factor vs. range for the three cases mentioned in Fig. 1.45, are plotted.

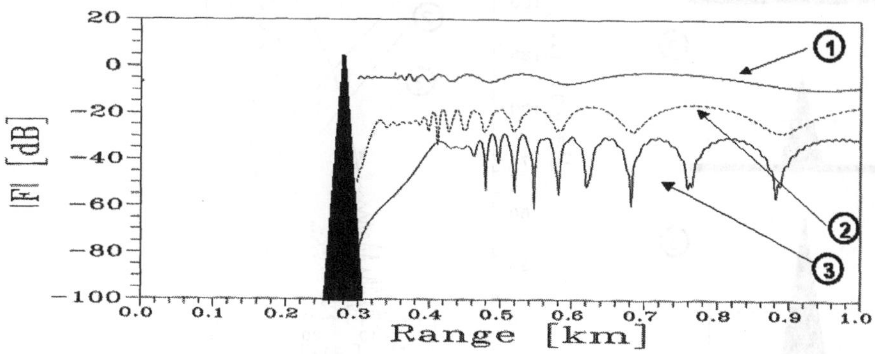

Figure 1.46: *|E/E$_0$| vs. range for the first scenario (observation height = 25 m)*

As observed from the figure, the higher the obstacle, the lower the signal strength right behind when the receiver height is less than the obstacle. The interested reader may run this scenario with different heights and ranges and could see the differences. The second calculation with this first scenario is done for propagation factor vs. height, and results are given in Fig. 1.47.

The second scenario used as an example in SSPE includes wedge-shaped obstacles with same height but different base lengths. All other parameters are kept the same as in the first scenario (obstacle, and observation heights are 50 m, and 25 m, respectively. Propagation factor vs. range behind the obstacle is given in Fig. 1.48. SSPE and KNIFE simulators can also be compared with each other, since both can handle wedge-type obstacle effects under Dirichlet-type boundary conditions. A comparison at 500 MHz for a 100-m height sharp wedge-type PEC obstacle is given in Fig. 1.49. The scenarios are also pictured in the figure.

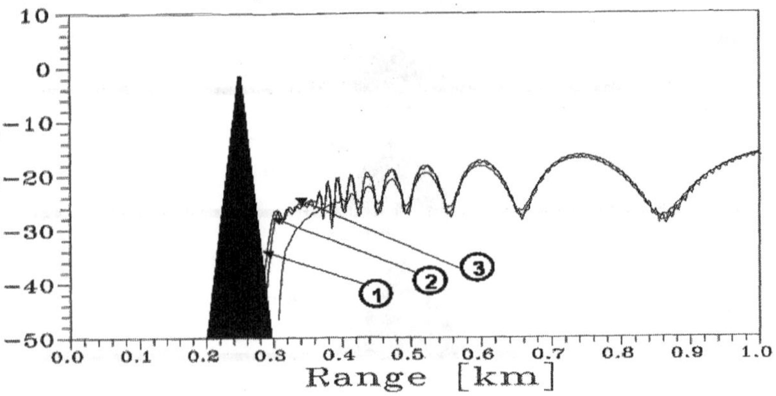

Figure 1.47: *$|E/E_0|$ vs. height for the first scenario (observation range = 25 m)*

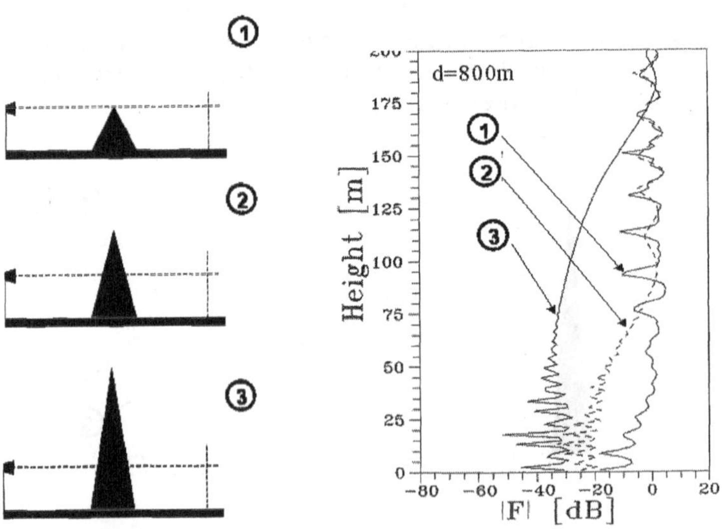

Figure 1.48: *$|E/E_0|$ vs. range for the first scenario (observation height = 25 m) with different base length (1, 50 m; 2, 100 m; 3, 200 m)*

Although the examples starting from Fig. 1.43 to Fig. 1.48 are generated via the SSPE package, some modifications in default parameters (e.g., maximum height) are done to satisfy the accuracy requirements. The contour plots (i.e., propagation factor vs. range/height) are shown in Figs. 1.50 and 1.51 in order to emphasize the importance of parameter selection. In both figures, operating frequency, transmitter-to-wedge distance, and wedge and transmitter heights are selected as 300 MHz, 1000 m, 100 m and 50 m, respectively. In order to show the effect of terrain slope, the wedge base lengths are chosen differently, 2000 m and 100 m for the scenarios shown in Figs. 1.50 and 1.51, respectively. For both scenarios, maximum height is kept as given inside the SSPE package. The color bar at the right side of the figures scales the amplitude of propagation factor. The effect of the obstacle is clearly observed in Fig. 1.50, the propagating wave is scattered from the left edge, and there exists a shadow region behind the wedge. However, in Fig. 1.51, it seems as if the upper part of the wedge permits the wave to penetrate, which is not physically true. The wedge affects as if it is lower. The reason is the slope angle, which is very high compared to the one in Fig. 1.51. Therefore, smaller height step sizes have to be used to satisfy the Nyquist criteria. Besides, staircase terrain approximation in SSPE has difficulties in modeling sharp slopes and needs more diligent discretization in height.

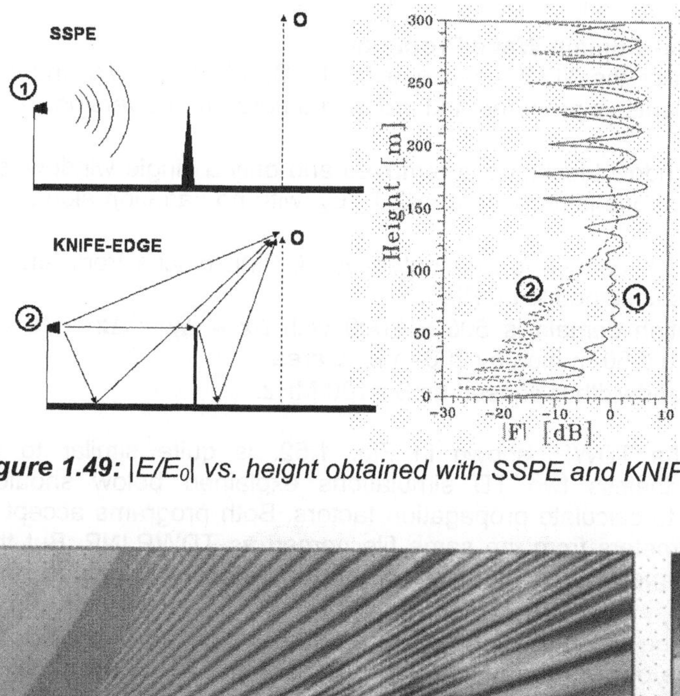

Figure 1.49: *$|E/E_0|$ vs. height obtained with SSPE and KNIFE*

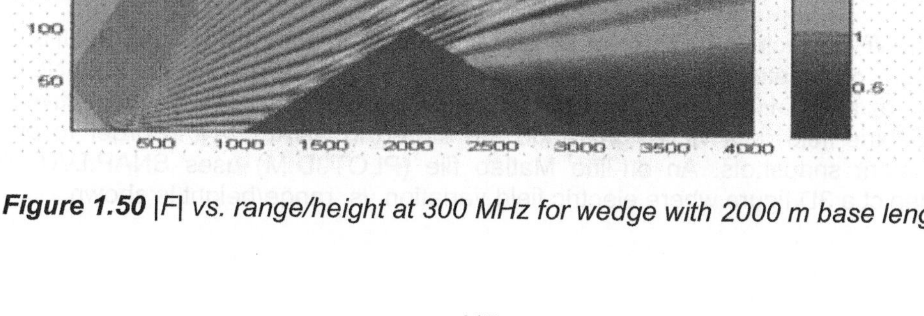

Figure 1.50 *$|F|$ vs. range/height at 300 MHz for wedge with 2000 m base length*

At the upper left of the Figs. 1.50 and 1.51, the amplitude of propagation factor is higher than it should be, because the propagating wave cannot reach that height at that range and numerical error is slightly smaller for free space, causing the denominator in the formula of the propagation factor to be small.

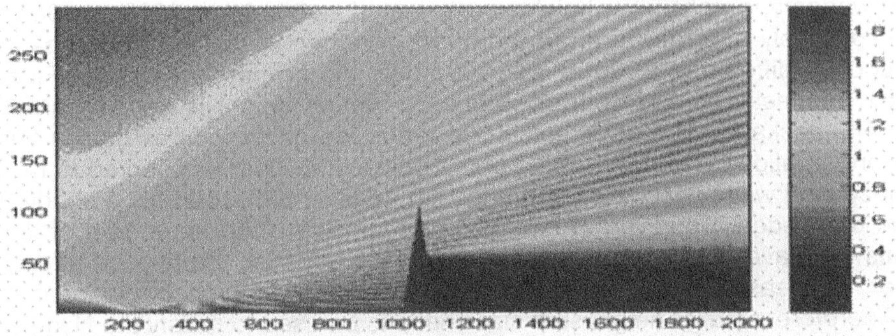

***Figure 1.51**: |F| vs. range/height at 300 MHz for wedge with 100 m base length*

1.5.5 TDWP: Time- Domain Wave Propagator

A simplified version of the TDWP package for HED, where H_x, H_z and E_y are the related field components, is prepared and presented here. In this version:

- The sliding window technique is removed and only a single window is allowed. Therefore, the code is reduced to a 2D TE_z (with no variation along *y*-direction) FDTD code.
- The propagation domain is terminated by 16-PML blocks from left, right, and top. Ground is assumed to be PEC.
- The computation domain is 500 $\times$ 2000 with $\Delta x = \Delta y = \Delta z = 10$ cm, which corresponds to a 50-m-high by 200-m-long area.
- The source bandwidth is chosen as B =100 MHz.

The structure of the TDWP, shown in Fig. 1.52, is quite similar to the other propagators. Nevertheless two TD simulations explained below should be run separately in order to calculate propagation factors. Both programs accept the user-supplied input parameters from the same file named as TDWP.INP. But the for the rest of the input parameters, each program needs a different additional parameter file.

- **TDWP-F**: Free space simulation. The additional input parameter file is TDWP-F. PAR. The outputs are given in T-RANGE-F.dat and T-HEIGHT-F.dat files for range and height profiles, respectively.

- **TDWP-S**: Includes obstacles, PEC ground, and refractivity. The additional input parameter file is TDWP-S. PAR. The outputs are given in T-RANGE-S.dat and T-HEIGHT-S.dat files for range and height profiles, respectively. Electric field data vs. height/range are stored in SNAP.DAT in order to plot instant snapshots. An off-line Matlab file (PLOT3D.M) uses SNAP.DAT to depict a 3D figure where electric field variation vs. range/height is shown.

Off-line DFT is applied to the data in these output files, and propagation factor vs. range or height are calculated for a given frequency. The FD data are given in F-RANGE.dat and F-HEIGHT.dat files. Since short pulse propagation is simulated in TDWP, off-line DFT may be applied as many frequencies as the user desires (within bandwidth of the short pulse injected). Necessary input parameters are listed in Table 1.6.

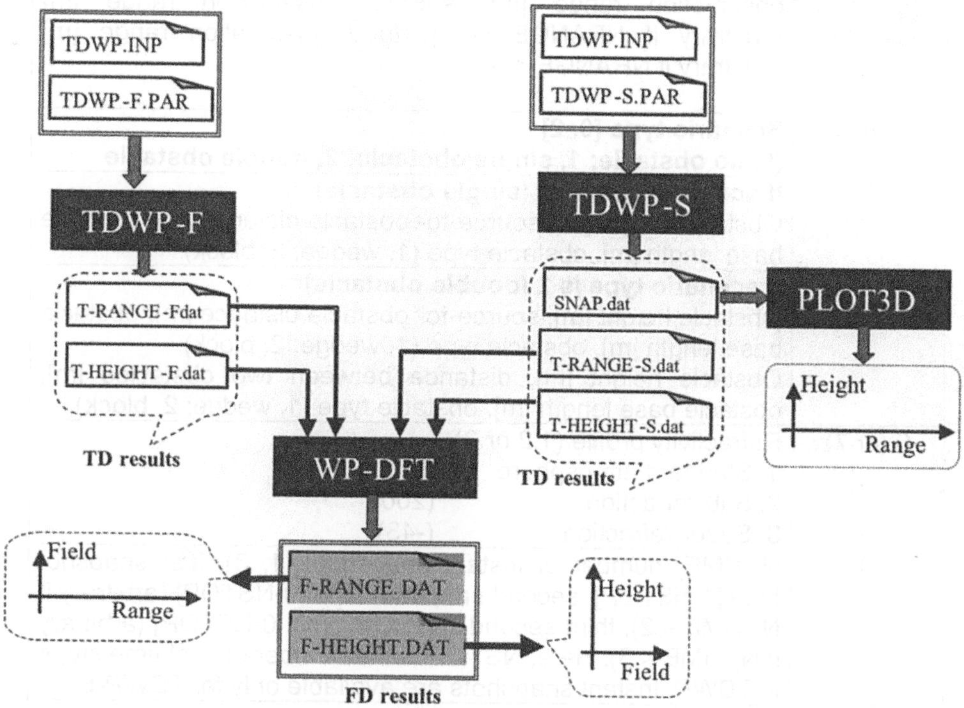

Figure 1.52: *The structure of TDWP package*

For example, the TDWP.INP file given below belongs to Fig. 1.48. The parameters read are transmitter height h_t = 20 m; receiver height h_r = 20 m; single obstacle; observation ranges: 50 m, 100 m, 150 m; obstacle height $h_{ter.}$ = 20 m, source-to-obstacle distance d_1 = 50 m; obstacle base; w_1 = 25 m, obstacle: wedge; vertical refractivity profile: standard atmosphere; time steps for instant snapshots: 500 and 1000.

TDWP.INP (Fig. 1.48)	LINE 1:	10
	LINE 2:	10
	LINE 3:	3, 50, 100, 150
	LINE 4:	1
	LINE 5:	20, 50, 25 ,1
	LINE 5	1
	LINE 6	1
	LINE 7	2, 500, 1000

Table 1.6: The TDWP.INP file and required parameters

Row No:	**Parameter**
1	Transmitter height [m]
2	Receiver (observer) height [m]
3	NRANGE (number of observation ranges: 1 - 3), first observation range [m], second observation range [m] (arbitrary if NRANGE < 2), third observation range [m] (arbitrary if NRANGE < 3)
4	**Scenario type [0, 2]** **(0, no obstacle; 1, single obstacle; 2, double obstacle**
5	**If scenario type is 1 (single obstacle)** Obstacle height [m], source-to- obstacle distance [m], obstacle base length [m], obstacle type (1, wedge; 2, block)
6	**If scenario type is 2 (double obstacle)** Obstacle height [m], source-to- obstacle distance [m], obstacle base length [m], obstacle type (1, wedge; 2, block) Obstacle height [m], distance between two obstacles [m], obstacle base length [m], obstacle type (1, wedge; 2, block)
6 (or 7)	Refractivity profile (1,2 or 3) 1: Standard atmosphere (117) 2: Sub-refraction (200) 3: Super-refraction (-43)
8	N_TIME (number of instant snapshots: 1, 3), first snapshot time [1-NSTOP], second snapshot time [0-NSTOP] (arbitrary if N_TIME < 2), third second snapshot time [0-NSTOP] (arbitrary if N_TIME < 3). Here, NSTOP is the number of total time steps in TDWP. Instant snapshots are available only for TDWP-S.

Like SSPE in FD, TDWP may also be used to obtain 3D plots of propagation directly in TD. An example of this is given in Fig. 1.53, where four different snapshots are plotted.

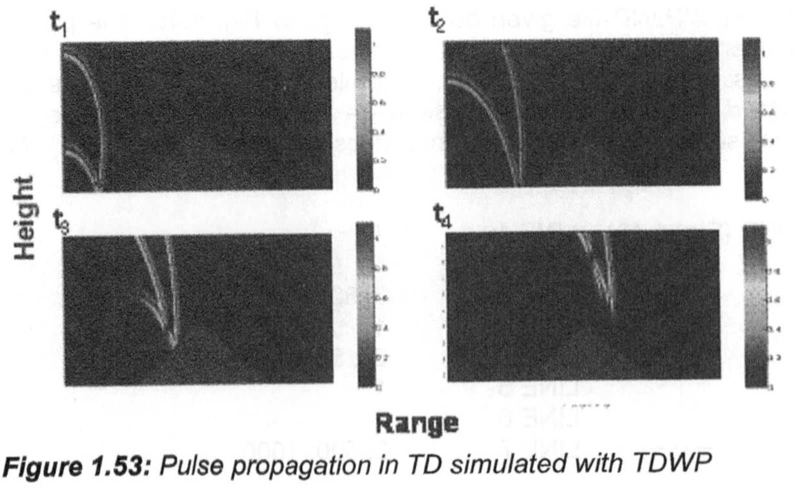

Figure 1.53: *Pulse propagation in TD simulated with TDWP*

1.6 CONCLUSIONS AND DISCUSSION

The problem of ground-wave propagation over a non-flat, imperfect earth's surface through a non-homogeneous atmosphere is discussed in this chapter. Three methods are presented; analytical approximate (ray-mode) solutions, the SSPE in FD and the TDWP in TD.

The main conclusions of this chapter are the following:

- Analytical ray-mode techniques (also their hybridized form WAVEPROB) yield sufficiently accurate results only when propagation over a spherical smooth earth (including lossy ground) is considered.
- SSPE is efficient and reliable for modeling narrow-band wave propagation (especially for frequencies VHF and above) over weakly back-scattering rough surfaces, above which exists a vertically and/or horizontally varying refractivity profile.
- The TDWP provides an efficient and reliable method for ground-wave propagation problems involving broadband pulse signals. Although computationally expensive and not yet optimized, this technique is quite flexible and can be used in a variety of applications. By adding decision-makers and parallel processing, the TDWP may be capable of simulating not only 2D but also 3D propagation of pulsed signals over terrain, which may include surface impedance variations.
- Major factors that at present restrict the applicability and/or validity of TDWP are
  (i) computer memory and speed requirements,
  (ii) unwanted reflections from the top PML boundary,
  (iii) numerical dispersion effects.

The restrictions of all these methods are listed in Table 1.7. Also, the comparisons in terms of ground-wave propagation requirements in order to clarify the utilities are listed in Tables 1.8.

Table 1.7: Comparison of the algorithms in terms of operating frequency

Operating Frequency	Ray-Mode (WAVEPROB)	**SSPE**	**TDWP**
MF	Easy	Difficult	Easy
HF	Easy	Difficult	Easy
VHF	Easy	Easy	Difficult
MW	Easy	Easy	Difficult

Most of the numerical results given in this chapter were performed via a personal computer (PCs, Pentium II with 128 MB RAM). The restrictions mentioned in (i) and (iii) above can partly be removed when more capable computers (with larger memory and higher speed) are used. This will allow one to utilize a larger computation space (as well as a larger sliding window) than that we have used in this study and hence extend the simulation range. At this stage, TD simulators are limited to ranges up to few thousands of wavelengths. The effects of unwanted reflections from the top PML boundary dominate the numerical dispersion effects in most of the propagation scenarios considered in this study. Also, PML works better for propagation over irregular terrain, because of the wide spread scattering of the waves. However, in this case, due to direct and ground scattered pulse interference together with numerical dispersion accuracy deteriorates at longer ranges. On the other hand, some signal processing techniques (such as windowing, signal estimation, etc.) can be incorporated to the TDWP to achieve a (slight) improvement in performance.

Table 1.8: Comparison of the algorithms in terms of ground-wave propagation requirements

Propagation Scenario	Ray-Mode (WAVEPROB)	**SSPE**	**TDWP**
Atmosphere	Standard atmosphere with and without earth's curvature	All kinds of vertical and horizontal atmospheric changes including ducts, but for relatively small atmospheric gradient.	No restriction
Non-flat terrain effects	Only rough surface	All terrain types can be included, but for piecewise linear approximation there exists a limitation for slope angles, and for staircase approximation it is difficult to apply Neumann type surface boundary condition.	No restriction
Surface impedance oundary	Easy	Easy, but there are problems for the source located at a highly conducting surface, such as sea.	Easy, but needs extra calculations if recursive convolution algorithm is applied.
Back and up scattering	Not applicable	Not applicable (possible with finite difference PE or vector PE).	Local back and up scatter effects are included.
Long-range propagation	Easy	Easy	Difficult and needs extra precautions.

References

[1] J. R. Wait, "The Ancient and Modern History of EM Ground-Wave Propagation", *IEEE Antennas and Propagation Magazine*, Vol.40, pp. 7 - 24, 1998

[2] N. DeMinco, "Propagation Prediction Techniques and Antenna Modeling (150 to 1705khz) for Intelligent Transportation Systems (ITS) Broadcast Applications, *IEEE Antennas and Propagat. Magazine*, Vol. 42, pp. 9 - 33, 2000

[3] J. R. Wait, *Electromagnetic Waves in Stratified Media*, Pergamon, Oxford, 1962

[4] V. A. Fock, *Electromagnetic Diffraction and Propagation Problems*, Pergamon, Oxford, pp.191 - 235, 1965

[5] G. Millington, "Ground Wave Propagation over an inhomogeneous smooth earth", *Proc. IRE*, Vol. 96 No. 39, pp. 53 - 64, 1949

[6] R. H. Ott, L. A. Berry, "An alternative integral equation for propagation over irregular terrain", *Radio Science*, Vol. 5, pp. 767 - 771, 1970

[7] R. H. Ott, L. A. Berry, "An alternative integral equation for propagation over irregular terrain", 2, *Radio Science*, Vol. 6, pp. 429 - 435, 1971

[8] K. A. Norton, "The Propagation of Radio Waves over the Surface of Earth and in the Upper Atmosphere", *Proc. IRE*, Vol.24, pp.1367 - 1387, 1936

[9] T. Ishihara, L. B. Felsen, "Hybrid Ray-mode Parameterization of High Frequency Propagation in an Open Waveguide with Inhomogeneous Transverse Refractive Index: Formulation and Application to a bilinear Surface Duct", *IEEE Transactions on Antennas and Propagat.* Vol. 39 No. 6, pp. 780 - 788, 1991

[10] T. Ishihara, L. B. Felsen, "Hybrid Ray-mode Parameterization of High Frequency Propagation in an Open Waveguide with Inhomogeneous Transverse Refractive Index: Numerical results and Quality Assessment", *IEEE Transactions on Antennas and Propagat.* Vol. 39 No. 6, pp. 789 - 797, 1991

[11] L. Sevgi, L. B. Felsen, "A new Algorithm for Ground Wave Propagation Based on a Hybrid Ray-Mode Approach", *Int. J. of Numerical Modeling*, Vol.11 No 2, pp. 87 - 103, March 1998

[12] L. B. Felsen, L. Sevgi, "Adiabatic and Intrinsic Modes for Wave Propagation in Guiding Environments with Longitudinal and Transverse Variations: Formulation and Canonical Test", *IEEE Transactions on Antennas and Propagat.* Vol. 39 No. 8, pp. 1130 - 1136, Aug. 1991

[13] L. B. Felsen, L. Sevgi, "Adiabatic and Intrinsic Modes for Wave Propagation in Guiding Environments with Longitudinal and Transverse Variations: Continuously Refracting Media", *IEEE Transactions on Antennas and Propagat.* Vol. 39 No. 8, pp. 1137 - 1143, Aug. 1991

[14] M. Levy, *Parabolic Equation Methods for Electromagnetic Wave Propagation*, IEE, Institution of Electrical Engineers, 2000

[15] G. D. Dockery, J. R. Kuttler, "An Improved Impedance Boundary Algorithm for Fourier Split-step Solutions of the Parabolic Wave Equation", *IEEE Transactions on Antennas and Propagat.* Vol. 44 No. 12, pp. 1592 - 1599, 1996

[16] T. Ishihara, L. B. Felsen, "Hybrid Ray-Parabolic Equation analysis of Propagation in Ocean Acoustic Guiding Environments", *J. Acoust. Soc. Am.*, Vol. 83, pp. 950 - 960, 1988

[17] W. Marcus, "A Hybrid (Finite Difference-Surface Green's function) method for computing Transmission Losses in an Inhomogeneous Atmosphere Over Irregular Terrain", *IEEE Transactions on Antennas and Propagat.* Vol. 40 No. 12, pp. 1451 - 1458, Dec. 1992

[18] F. Akleman, L. Sevgi, "A Novel Time-domain Wave Propagator", *IEEE Transactions on Antennas and Propagat.* Vol. 48 No. 5, pp. 839 - 841, 2000
[19] F. Akleman, L. Sevgi, "Realistic Surface Modeling in a Time-domain Wave Propagator ", IEEE Transactions on Antennas and Propagation, (to appear) July 2003
[20] M.O. Özyalçın, F. Akleman, L. Sevgi, "A TLM Based Novel Time Domain Ground Wave Propagator", *IEEE Transactions on Antennas and Propagat.* (to appear) June 2003
[21] J. Deygout, "Correction Factor for Multiple Knife-edge Diffraction", *IEEE Trans. Antennas and Propagat. Vol.* 39, pp. 1256 - 1258, Aug. 1991
[22] R. J. Luebbers, "Propagation Prediction for Hilly Terrain Using GTD Wedge Diffraction", *IEEE Transactions on Antennas and Propagat.* Vol. 32, pp. 951 - 955, 1984
[23] J. H. Whitteker, "Fresnel-kirchhoff Theory Applied to Terrain Diffraction Problems", *Radio Science*, Vol. 25, pp. 837 - 851, 1990
[24] J. T. Hviid, J.B. Andersen, J. Toftgard, J. Bojer, "Terrain-based Propagation Model for Rural Area – An Integral Equation Approach", *IEEE Transactions on Antennas and Propagat.*, Vol. 43, pp. 41 - 46, Jan. 1995
[25] L. Sevgi, F. Akleman, L. B. Felsen, "Ground Wave Propagation Modeling: Problem-Matched Analytical Formulations And Direct Numerical Techniques", *IEEE Antennas and Propagation Magazine*, Vol. 44 No.1, pp. 55 - 75, 2002
[26] M. P. Shatz, G. H. Polychronopoulos, "An Algorithm for the Evaluation of Radar Propagation in the Spherical Earth Diffraction Region", *IEEE Transactions on Antennas and Propagat.*, Vol.38 No. 8, pp.1249 - 1252, 1990
[27] D. E. Kerr (ed.), *The Propagation of Short Radio Waves*, Radiation Lab. Series, Vol. 13, McGraw-Hill, Ne York, 1951
[28] M. P. M. Hall, L. W. Barclay, M. T. Hewitt (ed.), *Propagation of Radiowaves*, IEE Press, London, 1996
[29] E.C. Jordan, K. G. Balmain, *Electromagnetic Waves and Radiating Systems*, Prentice-Hall, Englewood Cliffs, NJ, 1968
[30] M. Abramowitz, I.A. Stegun, *Handbook of Mathematical Functions*, Dover, 1969
[31] F.D. Tappert, The Parabolic Approximation Method in *Wave Propagation and Underwater Acoustics*, pp. 224-287 (Eds.) J. B. Keller and J. S. Papadakis Springer-Verlag, New York, 1977.
[32] F.R Di Napoli and R.L. Daevenport, Numerical Methods of Underwater Acoustic Propagation in *Numerical methods of Underwater Acoustic propagation*, J. A DeSanto (Ed.), Springer-Verlag, New York, 1977.
[33] J. R. Kutler and G. D Dockery,. "Theoretical Description of the Parabolic Approximation / Fourier Split Step Method of Representing Electromagnetic Propagation in the Troposphere", *Radio Science*, Vol. 26, pp. 381 - 393. 1991
[34] L. Sevgi, E. Topuz, "Split step Parabolic Equation analysis of Coupled Dielectric Waveguides", ELEKTRIK, Turkish J. of Electronics and Communication, Vol. 3 No 2-3, pp. 85 - 92, 1995
[35] L. Sevgi, "Split Step Parabolic Equation Solutions in Surface Duct-to-Elevated Duct Transition", Turkish J. of Physics, Vol. 19 No. 3, pp. 541 - 551, 1995
[36] D. J. Thomson and N. R Chapman "A Wide-Angle Split-Step Algorithm for the Parabolic Equation", *J. Acoust. Soc. Am.*, Vol. 74, pp. 1848 - 1854, 1993
[37] M. D Collins, "A Split Step Pade Solution For The Parabolic Equation Method", *J. Acoustic. Soc. Am.*, Vol. 94, pp.1736-1742, 1993

[38] F. Akleman, *Time and Frequency Domain Numerical Modeling of Ground Wave Propagation*, Ph.D Thesis, Istanbul Technical University, July 2002, Istanbul, Turkey

[39] K. S. Yee, "Numerical Solution of Initial Boundary Value Problems Involving Maxwell's equations," *IEEE Transactions on Antennas and Propagat*, Vol. 14 No. 3, pp. 302 - 307, May 1966

[40] C. Christopoulos, *The transmission-line modeling method*, Oxford University Press, Oxford, 1995

[41] C. Eswarappa, W. J. R. Hoefer, "Implementation of Berenger Absorbing Boundary Conditions in TLM by Interfacing FDTD Perfectly Matched Layers", *Electronic Letters*, Vol. 31 No. 15, pp. 1264 - 1266, 1995

[42] J. H. Beggs, R. J. Luebbers, K. S. Yee, K. S. Kunz, "Finite-Difference Time-Domain Implementation of Surface Impedance Boundary Conditions", *IEEE Transactions on Antennas and Propagat*, Vol. 40 No. 1, pp. 49 - 56, 1992

[43] M. L. Meeks, *Radar Propagation at low Altitudes*, Artech House, Boston, MA, 1982

[44] CCIR, "Ground Wave Propagation Curves for Frequencies Between 10 kHz and 30 MHz", *CCIR Rec*. 368-6, 1990

2

Antenna Design and Array Processing

2.1 INTRODUCTION

This chapter covers antenna fundamentals, design principles of complex antenna systems, electronic beam forming and steering, numerical simulation, and finally design examples. The primary purpose is to emphasize understanding of principles and development techniques as well as to give examples on how numerical simulation approaches are used in designing complex arrays, which are of interest in many practical applications. The reader is assumed to be familiar with antenna engineering, at least with concepts and mathematical background. EM terms such as Poynting vector, plane wave, wave impedance, phase and group velocity, permittivity, conductivity, permeability, reciprocity, duality, free space, lossy and lossless media, etc., are assumed known (see Appendix A for a brief summary). Some fundamental books on antennas are listed at the end of the chapter [1-11].

2.2 ANTENNA FUNDAMENTALS

Since EM waves have been employed for communication, through the use of radio, TV, satellite, radar, etc., radiowave transmit and receive devices that are called antennas have been in use. An antenna is a device that provides a transition from a *guided wave* on a transmission line to a *free-space wave* [1] (or vice versa). The IEEE definition of an antenna is given as "*that part of a transmitting or receiving system that is designed to radiate or receive EM waves.*" Most antennas are reciprocal devices and behave the same on transmit and on receive. Simple radiators and some antenna concepts are pictured in Fig. 2.1.

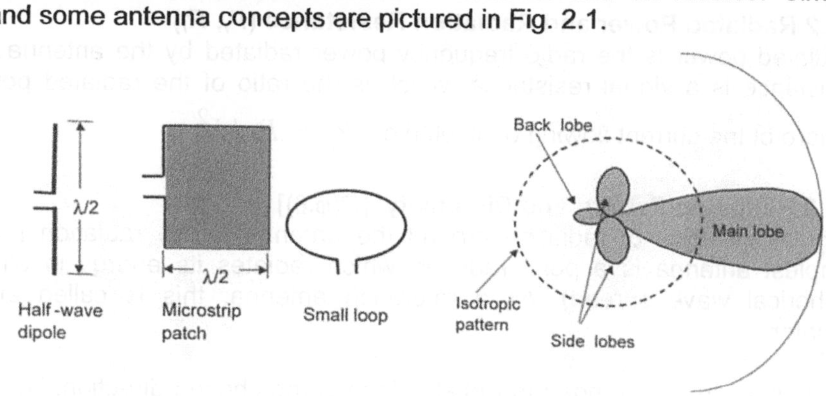

Figure 2.1: *Basic radiators and a radiation pattern*

A basic antenna may be assumed as a one-port device as illustrated in Fig. 2.2. When transmitted, high-frequency energy from the source with voltage V_s and impedance Z_s is transferred to the complex load $Z_{in} = (R_r+R_l)+jX_A$ via a transmission line with characteristic impedance Z_0. Radiation characteristics and antenna losses are represented in terms of radiation resistance R_r and R_l, respectively. Reactive power storage is represented by X_A. When receive, the flow is in opposite direction; the captured EM energy is transferred to the input port of the receiver-processor via the same transmission line.

Basic parameters are then given as follows:

2.2.1 Antenna input power (P_{in}) and input impedance ($Z_{in} = R_l + R_r + jX_A$)

It is the power that is fed to the antenna from the power source and equal to the antenna losses, plus radiated power.

Antenna input impedance is the (complex) impedance of an antenna at the input port at the radiating frequency. It is determined by input voltage divided by the input current.

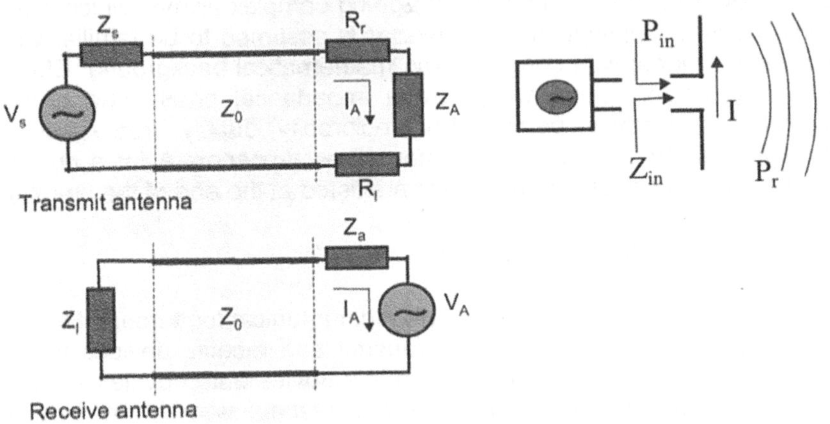

Figure 2.2: *Network equivalent of an antenna*

2.2.2 Radiated Power and Radiation Resistance (P_r, R_r)

Radiated power is the radio frequency power radiated by the antenna. Radiation resistance is a virtual resistance, which is the ratio of the radiated power to the square of the current flowing on itself (i.e., $R_r = P_r / I_A^2$).

2.2.3 Radiation Pattern and Directivity [$D(\varphi,\theta)$]

Angular variation of radiation around the antenna is the radiation pattern. The simplest antenna is a point radiator, which radiates its energy in all directions (spherical wave spread). As a reference antenna, this is called an *isotropic* radiator.

Antennas can be arranged to radiate along some chosen direction; i.e., the power of the antenna can be directed. Directivity is the ratio of power density in the direction of the pattern maximum to the average power density at the same distance from the antenna. Power density that an isotropic source radiates with the same input power is called average power density.

2.2.4 Antenna Efficiency (e) and Gain [$G(\varphi,\theta)$]

Antenna efficiency is a parameter between zero and one, which is defined as the power radiated by the antenna divided by the power delivered to the antenna. Efficiency is 1 (100 %) for ideal (loss-free) antennas.

Directivity reduced by the losses on the antenna (it should be noted that directivity and gain often misused in antenna theory. Directivity can be calculated and/or measured but, gain of the antenna can only be measured, since it is related to cable material, connector losses, etc.). The relation of directivity and gain is given as $G(\varphi,\theta) = e \times D(\varphi,\theta)$.

2.2.5 Antenna Lobes and Beamwidth (HPBW and FNBW)

Directive antennas direct majority of its power to a desired direction. This is called main lobe (main beam). Unwanted lobes at the sides and back are called side lobes and back lobe, respectively. The ratio of the power in the main lobe to the back lobe is the front-to-back ratio (FBR) of the antenna.

Half-power beamwidth (HPBW) is the angular distance between half power (3 dB) on each side of beam maximum of the main beam. First null beamwidth (FNBW) is the angle between first null points of the main lobe.

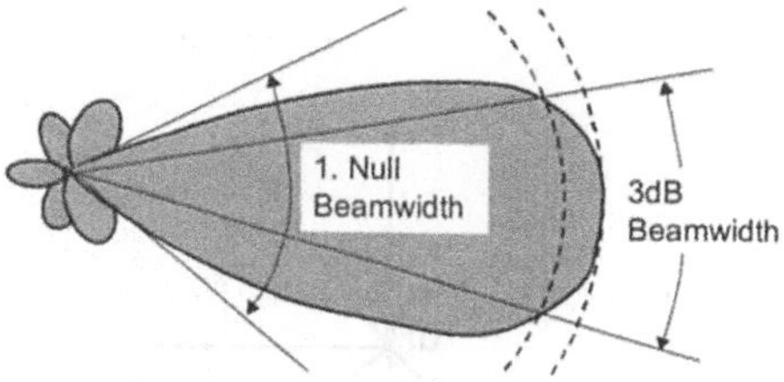

2.2.6 Effective Aperture (A_e)

When receive, total power incident on the receive antenna is the sum up of power density over the "area" of the antenna, which is called effective aperture. It is the ratio of the available power at the terminals of the antenna to the power flux density of a plane wave incident upon the antenna, which is polarization matched to the antenna. Maximum gain and effective aperture are related via $G = (4\pi A_e/\lambda^2)$.

2.2.7 Polarization

The figure traced out with time by the instantaneous electric field vector associated with the radiation from an antenna when transmitting (e.g., linear, circular, or elliptical polarization).

2.2.8 Standing Wave Ratio (SWR)

The ratio of maximum voltage to the minimum voltage along the feeding transmission line that connects power source to the antenna input terminals. It is a parameter that specifies reflected power at the antenna input. SWR = 1 means the impedance of the line is matched to the antenna input impedance Z_{in}, therefore no reflection occurs at the antenna terminals.

2.2.9 Bandwidth (B)

The range of frequencies over which performance of the antenna is assumed almost constant.

2.2.10 Antenna Factor (AF)

Antenna is a transducer; it converts voltage to the electric field (or vice versa). A receiver antenna factor is the ratio of electric field to the voltage read at the receiver screen, i.e., AF = E/V [1/m].

2.2.11 Near-Field/Far-Field

An infinitesimal current element is called an ideal radiator (or Hertzian dipole, doublet, electric dipole) and its electric and magnetic fields at a radial distance r can be obtained analytically [1] as shown in Fig. 2.3, where I, dl, k_0 and Z_0 are the current amplitude, length, free-space wavenumber, and free-space impedance (120π = 377 Ω), respectively. Characteristics of this element are important, since the field radiated by any complex antenna can be represented as a superposition of the fields due to the infinitesimal current elements on the antenna surface.

Field of an infinitesimal current element

$$\mathbf{E} = \frac{jZ_0\mathbf{I}dl}{2\pi k_0}\cos\theta\left(\frac{jk_0}{r^2}+\frac{1}{r^3}\right)e^{-jk_0 r}\mathbf{e}_r - \frac{jZ_0\mathbf{I}dl}{4\pi k_0}\sin\theta\left(-\frac{k_0^2}{r}+\frac{jk_0}{r^2}+\frac{1}{r^3}\right)e^{-jk_0 r}\mathbf{e}_\theta$$

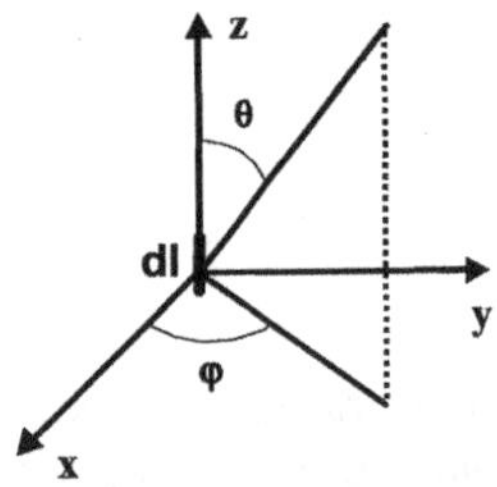

In far field

$$\mathbf{E} = jZ_0\mathbf{I}dlk_0\sin\theta\,\frac{e^{-jk_0 r}}{4\pi r}\mathbf{e}_\theta$$

$$\mathbf{H} = j\mathbf{I}dlk_0\sin\theta\,\frac{e^{-jk_0 r}}{4\pi r}\mathbf{e}_\varphi$$

Figure 2.3: *Ideal current element and electric field at a distance r*

Because of azimuthal (φ) symmetry (see Fig. 2.3), electric field has only radial and vertical (r,θ) components in the spherical coordinate system. Similarly, magnetic field has radial and horizontal (r, φ) components. Both electric and magnetic fields have terms with $1/r$, $1/r^2$ and $1/r^3$. When the distance r is much larger than the wavelength (i.e., $r >> \lambda$) all terms except $1/r$ can be ignored. In this case only E_θ and H_φ components with $1/r$ terms are present and they propagate radially outwards. The distance at which fields assumed to have only $1/r$ terms is called *far field*. Very close to antenna none of the above terms can be ignored and field components are very complex. This region is called *near field*. Distances much larger than the wavelength ($r >> \lambda$) usually satisfy this condition.

Near-field/far-field terms are also used according to mutual interaction distance in an antenna system. This is explained in Fig. 2.4. Here, distance variation of wave impedances of electric and magnetic dipoles are plotted.

As shown in the figure, wave impedance of an electric (magnetic) dipole is high (low) in the close proximity of the source and decreases (increases) with distance. At a wavelength or more, it asymptotically reaches the value of free-space wave impedance of 377 Ω. In multi-element complex systems, mutual interaction also becomes negligible at these distances. Therefore, the second definition of the far field is that of a distance at which wave impedance reduces to the free-space value and mutual interaction is negligible.

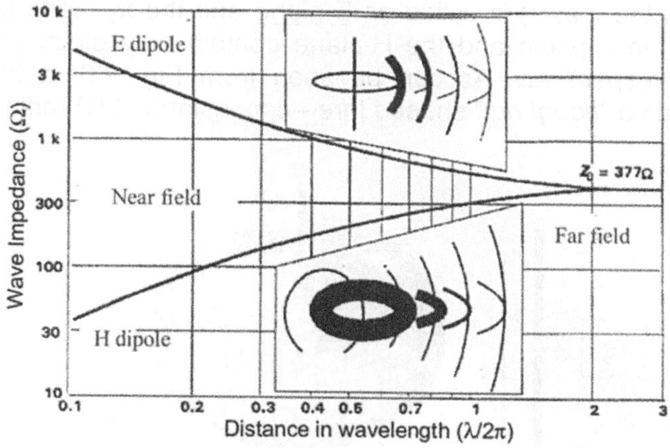

Figure 2.4: *Wave impedance vs. distance of ideal dipoles*

The third definition of the far field is related to equi-phase surfaces. An isotropic radiator has a spherical wavefront (see Fig. 2.5). As the distance *r* increases the wavefront locally along a given direction may locally be assumed plane. If the longest dimension of an antenna is *D* and wavelength is λ, then far field is defined as

$$r > \frac{2D^2}{\lambda} \text{ [m]} \tag{2.1}$$

a distance where phase error of the plane wave assumption is less than 22.5° (i.e., λ/16) [1]. In practice, whichever the greatest (eq. (2.1) or $r >> D$ or $r >> \lambda$) should be taken as the far-field distance.

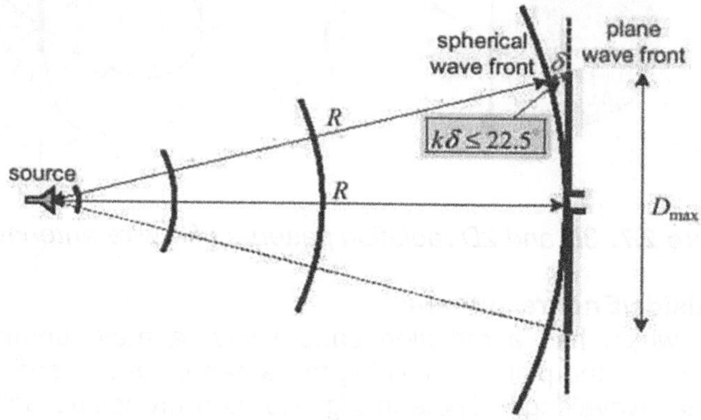

Figure 2.5: *Plane wave assumption distance*

Spherical coordinates are mostly used in antenna engineering as shown in Fig. 2.6. Here, a half-wavelength dipole antenna and its radiation pattern are plotted. As shown in the figure, electric field vector is parallel to the axis of the wire, which is longitudinally located along the z axis. In antenna theory, any vertical plane (i.e., plane parallel to the z axis) is called an E-plane and the xy-plane is called an H-plane. That is, the E-plane and the H-plane contain only electric field and only magnetic field, respectively. As can be seen from Fig. 2.6, a half-wavelength vertical dipole has a "*doughnut*" shaped three-dimensional (3D) radiation pattern.

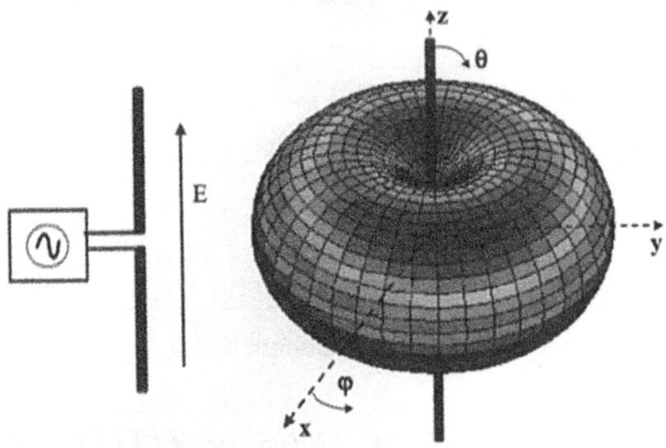

Figure 2.6: *A half-wavelength dipole and its radiation pattern*

Wire antennas are also called resonant antennas, since they resonate and give maximum radiated power at certain discrete frequencies, where the length of the antenna is multiple of the wavelength. Variety of different radiation patterns can be obtained by changing the frequency. A typical example is given in Fig. 2.7.

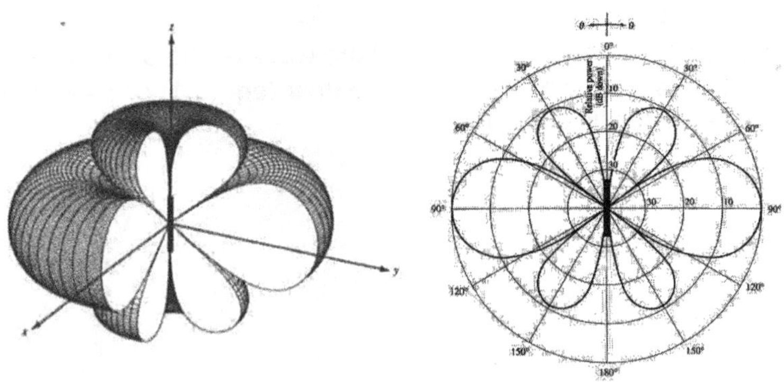

Figure 2.7: *3D and 2D radiation patterns of a wire antenna (l = 1.25 λ)*

2.2.12 Broadside/Endfire antenna

An antenna, which has a radiation pattern with a main beam maximum in a direction normal to the plane containing the antenna, is a broadside antenna. For example, a half-wavelength dipole in Fig. 2.6, is a broadside antenna. An endfire antenna is the one for which the main beam is in the plane containing the antenna. Typical patterns of broadside and endfire antennas are pictured in Fig. 2.8.

There are three antennas that are commonly used as a reference; isotropic point radiator, ideal (Hertzian) dipole and a half-wavelength dipole. Table 2.1 lists directivity gains of these reference antennas.

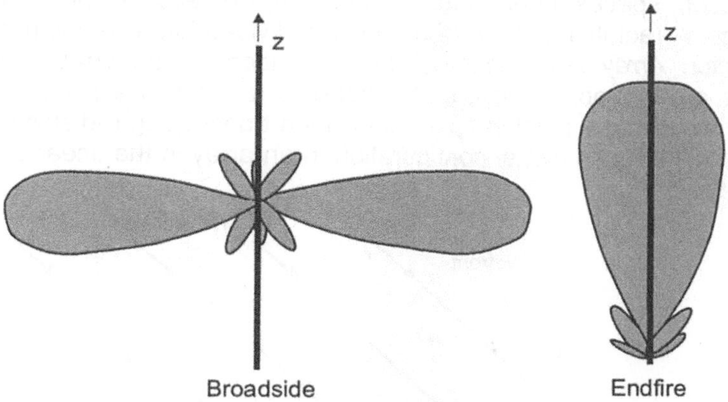

Figure 2.8: *Broadside and endfire antenna patterns*

In practice, dBi, dBs, and dBd are used to refer isotropic, ideal (short) and half-wavelength radiators, respectively.

Table 2.1: Directivity gains of reference antennas

Antenna	Gain	Gain dB]
Isotropic radiator	1.0	0.0
Hertz (short) dipole	1.5	1.75
Half wavelength dipole	1.64	2.15

2.3 ELECTRONIC BEAM STEERING AND DIRECTION FINDING (DF)

Several antennas can be arranged in free space and interconnected to produce a directional radiation pattern. Such a configuration or an arrangement of multiple radiators is referred to as an *antenna array*. The direction of the directional radiation pattern can also be changed, which is called beam forming or *beam steering*. Beam steering can be achieved by either mechanical or electronic scanning. The antenna is turned in mechanical scanning, but phases (and sometimes amplitudes) of excitation currents of array elements are changed in electronic scanning. The latter is then called *phased arrays*. Phased arrays have many applications, especially in HF, VHF, and UHF communication, radars and direction finders.

Arrays have many geometrical configurations such as linear arrays, planar arrays, and conformal arrays, where antenna element centers (individual radiators) lie along a straight line located over a plane and element locations conform a non-planar surface, respectively.

The radiation pattern of an array is determined by the type of individual elements, their location, spaces among them, orientation, excitation amplitudes and phases. In practice, A radiation pattern of an array is divided into two: element pattern and array factor. *Array factor* is the radiation pattern of an array, where all array elements are isotropic radiators. An isotropic radiator is a hypothetical, lossless antenna occupying a point in space and when transmitting radiates uniformly in all directions. The fundamental configuration in an array is the linear array as shown in Fig. 2.9.

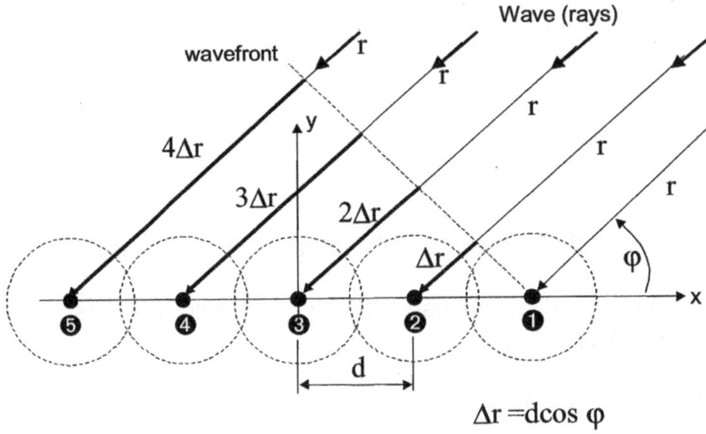

Figure 2.9: *A linear array with isotropic radiator elements*

Here, a 5-element linear array located along the *x*-axis on the *xy*-plane, with isotropic radiators separated by a distance *d* is pictured. The dashed circles are the element radiation patterns. The angle φ is that of an incoming plane wave relative to the axis of the receiving array. When φ = 90° (i.e., when the angle of incidence is from +*y*-axis) all array elements receive the signal with the same phase. When φ is different than 90°, the second array element has a path difference of $d \times \cos(\varphi)$ with respect to element one. Similarly, a third element has a path difference of $d \times \cos(\varphi)$ with respect to the second and $2 \times d \times \cos(\varphi)$ with respect to the first elements. The path difference of $d \times \cos(\varphi)$ correspond to a phase difference of ($\beta = k_0$)

$$Phase\ \ difference = \beta \times d \times \cos(\varphi) = \frac{2\pi}{\lambda} \times d \times \cos(\varphi) \tag{2.2}$$

at the given frequency. Therefore, one can electronically steer the beam when transmitted, by adjusting elemental phases according to this rule. The beamwidth is inversely proportional with the array length (i.e., with the distance between the first and last array elements located along the *x*-axis).

An example is given in Fig. 2.10. Here, broadside and endfire radiation characteristics obtained with the same 5-element array is shown [1]. The radiators are 1 m high and vertically located in free space. The elements are separated 0.45 m apart. At 300 MHz the array corresponds to $\lambda/2$ dipoles, 0.45λ apart from each other. When they are equally fed the boresight pattern is obtained with a gain of 9.8 dB and beamwidth of 26°. The endfire characteristic is obtained with the same array when they are fed with a phase difference of 0.9 π (corresponding to $\varphi = 0°$), with a gain of 9.2 dB and beamwidth of 74°. While FBR is zero when boresight, it is 4.2 dB when endfire.

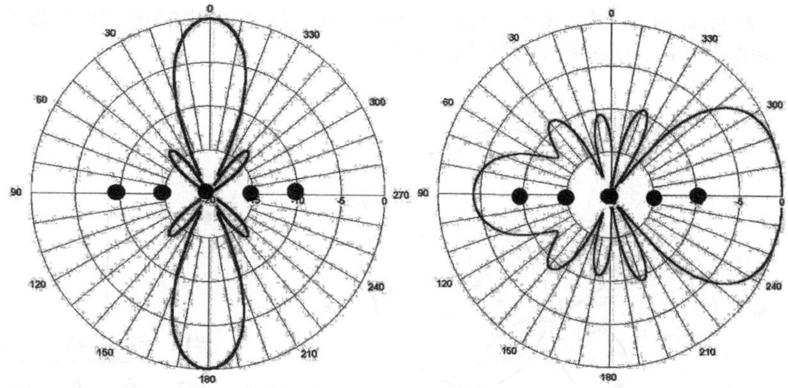

Figure 2.10: *Five-element broadside and endfire antenna array*

A one-dimensional (linear) array is symmetrical with respect to array axis (*x*-axis in Fig. 2.9). This means, main beam points both positive and negative *y*-directions and the array has zero FBR. However, in most of the practical problems the array is designed to cover only one direction (e.g., positive *y*-direction in Fig. 2.9) and only one-way beam steering is required. This necessitates a two-dimensional (planar) array. Examples of a linear and a 2D planar arrays are given in Fig. 2.11.

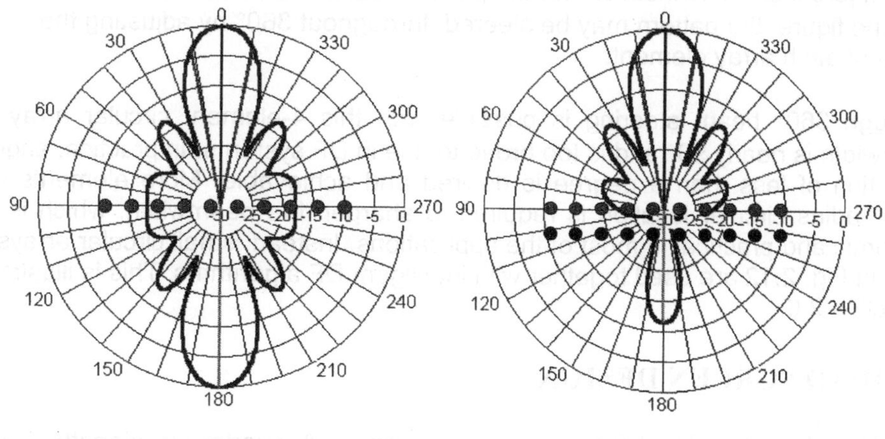

Figure 2.11: *Coverage and beam forming with 1 × 10 and 2 × 10 arrays*

Here, 1 × 10 and 2 × 10 arrays and their horizontal radiation patterns are shown. On the left, all center fed, a 10-m-long linear array with 5-m inter-element distance is given. Its beamwidth, gain, and front-to-back ratio are 22°, 11 dBi, and 0 dB, respectively, at 15 MHz. By locating parasitic elements quarter wavelength back of the first row, changes these values to 22°, 14 dBi, and 11 dB, respectively, at the same frequency. If the 2 × 10 planar array (on the right) is assumed a 10-channel array then it has a broad-side array, but endfire channel characteristics.

Electronic beam steering is generally used in direction finding (DF). A DF antenna may be a planar array if a sector is to be covered, or circular if 360° of coverage is of interest. A typical 4-element circular array, its feeder phases, and its horizontal radiation patterns are presented in Fig. 2.12.

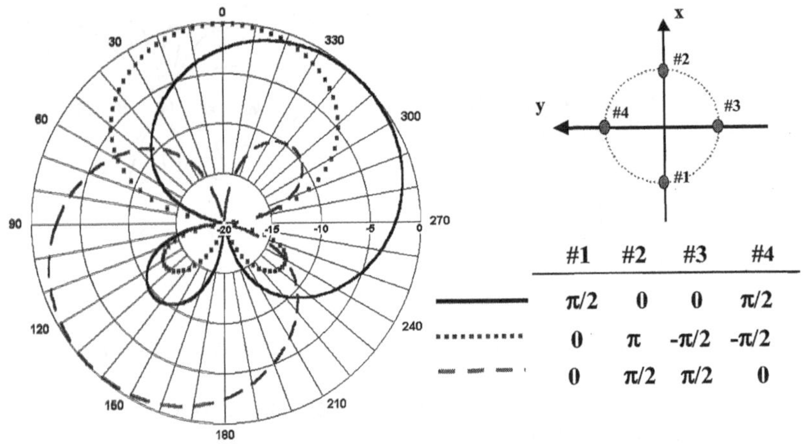

***Figure 2.12:** 4-element circular DF antenna and electronic beam forming*

Here, four quarter-wavelength-long (center-fed) vertical wire antennas are located on a circle with a radius slightly less than quarter wavelength and are fed as given in the figure inset. The inset shows the phase differences of each element. As seen from the figure, the pattern may be steered throughout 360° by adjusting the phase of each array element.

Although 360° beam steering is possible with this 4-element circular array, its beamwidth is nearly 90° and is too broad to use in DF systems. In practice, angular resolution of less than a degree is desired and achievable. More elements with larger radius circular location is required to sharpen the beamwidth, which is not economic and practical in most of the applications. Instead, small circular arrays as given in Fig. 2.12 are used together with intelligent DF algorithms. This is illustrated in Section 2.6.

2.4 GROUND SCREEN DESIGN

A PEC surface has ideal 0-Ω surface impedance. A quarter-wavelength vertical radiator over PEC surface acts as a half-wavelength element in free space.

Therefore, its vertical radiation pattern has a maximum towards horizontal. This is the design goal for most of vertically polarized ground (surface)-wave propagation antenna elements. HFSWR antennas are typical examples, where good ground screen design is almost a must.

A lossy ground with ground parameters conductivity σ_g [S/m] and relative permittivity ε_g has surface impedance, which can be calculated via

$$Z_s = Z_0 \left[\frac{j\omega\varepsilon_0}{\sigma_g + j\omega\varepsilon_g} \right]^{1/2} \left[1 + \frac{j\omega\varepsilon_0}{\sigma_g + j\omega\varepsilon_g} \right]^{1/2} \tag{2.3}$$

where $\omega = 2\pi f$ is the radian frequency and Z_0 is the free space impedance (377 Ω). The surface impedance of the POOR ground (i.e., the ground with σ_g = 0.003 S/m and ε_g = 4.0) is around 150 Ω at 2 MHz, increases exponentially with the frequency and reaches to 300 Ω at 10 MHz. But, the surface impedance of the ocean (i.e., when σ_g = 5.0 S/m and ε_g = 80.0) is in between 5 Ω and 10 Ω in the same frequency region. Therefore, ocean surface may easily be assumed as a PEC surface at, for example, operating frequencies HF and above.

The difference between erecting the antennas over POOR ground or PEC surface may result in a reduction of gain by up to 15 dB. Besides this loss, there is also an extra near-field propagation path loss because of the POOR ground. Table 2.2 lists vertical electrical field strength and path loss at 1 km away from a 1 kW vertical radiator. It is clear from Table 2.2 that, at 1 km distance; propagation loss over POOR ground may be 10 - 15 dB higher than the propagation loss over ocean surface.

Table 2.2: Field strength and path loss values of a vertical radiator with 1kW transmitter power at 1-km distance (d = 1 km, ε_g = 15.0, f = 3 MHz)

Conductivity [S/m]	Field Strength [dBμV/m]	Path Loss [dB]
0.0001	95.2	56.3
0.001	96.6	55.0
0.01	106.4	45.1
0.1	109.3	42.2
1.0	109.5	42.0
5.0	109.5	42.0

Therefore, together with the reduction in antenna gain, there may be a total of 30 dB extra loss just because the antenna elements are erected over POOR ground.

In order to overcome this problem, surface impedance of antenna elements must be reduced to acceptable levels. This is accomplished by one of the following ways:

- Erect antenna elements on top of a PEC surface, which typically extends a quarter-wavelength radially. This is applicable for UHF and above, when sizes are relatively much smaller.
- Select antenna site at the edge of the shore so that ocean water flows under the antenna elements (e.g., for large radar and/or communication arrays in HF and VHF).
- Use *ground screen*.

If the first two choices are not available, then, ground screen shall be used to reduce surface impedance of the antenna elements. Using ground screen

- will stabilize radiator's input impedance, hence increase the antenna gain
- will lower the antenna take of angle (TOA, defined as the vertical angle between beam direction and the ground), hence more energy shall be coupled to ocean surface [5,15].

In practice, three different types of ground screens are available. They are radials, horizontal wires, and rectangular meshes, or combinations thereof. It is simple to evaluate the effects of ground screen via approximate analytical approaches. The impedance of a horizontal wire screen can be given as [5,11]

$$Z_{scr} \approx \frac{j\mu\omega d}{2\pi} \ln\left[\frac{d}{2\pi a}\right] \tag{2.4}$$

where *a* is the wire length and *d* is the wire separation. The same equation may also be used for a mesh-type screen if the mesh sizes are equal (i.e., square mesh) and equal to *d*.

Using the ground screen with impedance Z_{scr} over lossy ground with surface impedance Z_s will cause an equivalent surface impedance, which can be given as

$$Z_{eq} = \frac{Z_{scr} \times Z_s}{Z_{scr} + Z_s}, \tag{2.5}$$

the parallel equivalent of the two impedances.

2.5 ANTENNA SIMULATION

Antenna simulation is one of the earliest applications in numerical EM. The well-known numerical EM (NEC) code has been in use for more than four decades. More recently (i.e., after 1980s) TD techniques, such as FDTD and TLM, have also begun to be used in antenna simulations. These three simulators are discussed in this section.

2.5.1 MoM-Based NEC2 Code

The Numerical Electromagnetics Code (NEC) is an outgrowth of a program developed in the 1970s, called the Antenna Modeling Program (AMP). The NEC has been developed at the Lawrence Livermore Laboratory, Livermore, California, under the sponsorship of the Naval Ocean Systems Center and the Air Force Weapons Laboratory. There are at least four versions of NEC, with NEC2 emerging in 1981 and NEC4 appearing in 1992. NEC2 is the highest version of the code under public domain. NEC4 remains proprietary with the Lawrence Livermore National Laboratory and the University of California. It requires a separate license for use.

NEC in all its forms is a computer code for the analysis of the EM response of antennas and other metal structures that uses MoM techniques. It is a numerical solution to integral equations for the currents induced on a metallic structure by sources or incident fields. The approach has no theoretical limit and may be used for very large arrays or for the very fine subdivision of smaller arrays.

NEC2 [12] is a public domain code and may be found in many Internet sites [13]. The reader may reach hundreds of related sites including source codes written in Fortran or C languages by just doing a quick search in any Internet Search Engine with keywords "NEC," numerical electromagnetic code," "antenna simulator," etc. NEC2 with excellent, user-friendly GUI may also be purchased from different companies [14].

The most important task in using NEC2 package is the preparation of the input file *.NEC, where the geometry and control commands are used to discretize the antenna under investigation in terms of segments. Some of the basic NEC2 commands are listed in Table 2.3

Table 2.3: Basic NEC2 geometry commands

Geometry Commands			
Comment End	CE	Generate Helix	GH
Comment	CM	Generate Cylinder	GR
Generate Wire	GW	Coordinate transfrom	GM
Generate Arc	GA	Scale Dimensions	GS

The NEC input file *.NEC can be prepared just as writing a computer code. Each line is a unique command and starts with a two-letter NEC2 command. The commands CM and CE are for the user and they are not taken into account by the NEC2 compiler. Any wire, arc, helix and/or cylinder can be generated by GW, GA, GH, and GR commands, respectively. The GW command generates a wire between the points (x1,y1,z1) and (x2,y2,z2). For example, to locate a 1 m vertical wire along the *z*-axis with 5 mm radius from the origin, a command line

GW wire #, # of segments, x1, y1, z1, x2, y2, z2, radius

GW 1, 21, 0., 0., 0., 0., 0., 1., 0.005

is given. Here the numbers from left to right are wire label (1), number of segments (21), initial *x,y,z* coordinates (0,0,0), final *x,y,z* coordinates (0,0,1), and the wire radius (0.005), respectively. The wire may be duplicated and/or coordinate translated/rotated by the command GM and the wire and/or the whole geometry may be scaled by the command GS. These three commands are usually sufficient to build as a wire mesh equivalent of the geometry. The geometry definition in NEC input file is terminated with GE command. The following lines are filled with the control commands, such as listed in Table 2.4. For example frequency is given with FR, excitation is supplied by EX, and ground parameters (conductivity and permittivity) are introduced with GD. The requested output values, such as near electric field, near magnetic field and vertical and/or horizontal radiation patterns are mentioned by NE, NH and RP commands, respectively. The last line of the NEC input file must be EN command.

Table 2.4: Basic NEC2 control commands

Control Commands			
Frequency	FR	End of run	EN
Execute	XQ	Ground parameters	GD
Excitation	EX	Near *E* Fields	NE
Radiation pattern	RP	Near *H* fields	NH

A typical NEC input file is given in Table 2.5. Here, three 1-m vertical wires (with 0.00 5 radius) are located along the *y*-axis, separated by a distance of 0.5 m. Each wire is represented by 11 segments. They are all equally bottom-fed and the frequency is 300 MHz. The spatial sampling (segment length - wavelength) is $\lambda/11$ at 300 MHz. The antenna system is located above POOR ground. The program computes horizontal radiation pattern with a degree resolution (i.e., $\theta = 90°$, $0° \leq \varphi \leq 360°$) for sky wave component.

Table 2.5: Typical NEC input file for a 3-element antenna array

```
CM
CM A sample NEC input file for a 3-element array
CM
CE
GW 1  11  0.0 -0.5 -0.5 0.0 -0.5 0.5 0.005
GW 1  11  0.0  0.0 -0.5 0.0  0.0 0.5 0.005
GW 1  11  0.0  0.5 -0.5 0.0  0.5 0.5 0.005
GE 1
GN 2 0 0 0 4.0 .003
FR 0 1 0 0 300 5
EX 0 1 1 00 1.0 0.0
EX 0 1 1 00 1.0 0.0
EX 0 1 1 00 1.0 0.0
RP 0 1 361 1000 90 0 1 1
EN
```

Any geometry can be modeled in NEC input file such as a loop antenna, helical antenna, a monopole over a rectangular PEC box, an array over the ground with complex ground screen layout, a radio antenna on a car, a radar antenna beneath an aircraft, etc., as long as their mesh representation are built with the GW, GM, GS, etc., commands. This is the most difficult part in NEC simulations. Two typical example geometries are pictured in Figs. 2.13 and 2.14. In Fig. 2.13, a vertical wire antenna above the ground with underlying ground screen radials is shown. Antennas over ground option is directly mentioned by the GD command. On the other hand, the user should be careful when generating a mesh for a PEC surface and a ground screen underneath the antenna.

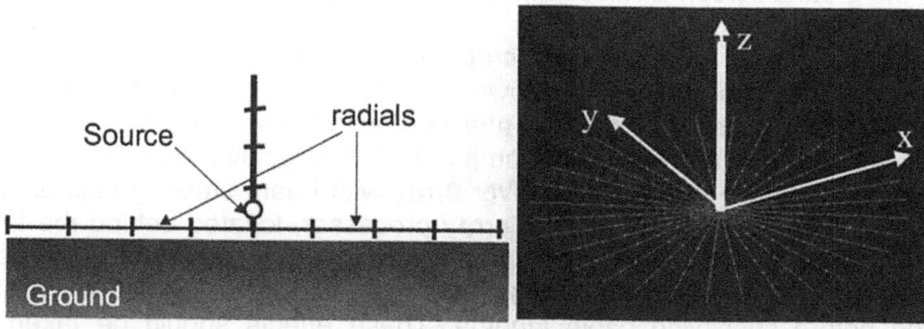

Figure 2.13: *A vertical wire antenna above the 30-radial ground screen*

Table 2.6: A Fortran code to generate ground screen radials

```
C  PROGRAM      : MN-SCR.FOR
C  DATE         : OCTOBER 1998
C  AUTHOR       : L. SEVGI
C  AIM          : TO LOCATE GROUND SCREEN UNDER A MONOPOLE
   PARAMETER (PI = 3.1415927)
   OPEN (UNIT = 22, FILE = 'MN-SCR.NEC')
   READ(*,*)  'ANTENNA COORDINATES (X,Y) = ?', X0,Y0
   READ(*,*)  'NUMBER OF RADIALS = ?', NRAD
   READ(*,*)  'RADIAL LENGTH [m] = ?', RL
   ACI =  2.*PI / NRAD
   READ(*,*) 'RADIAL RADIUS [m] = ?', RAD
   READ(*,*) 'RADIAL SEGMENTS = ?', NSEG

    II = 100                              ! WIRE NUMERATOR
   Z1 = 0.005                             ! 5 mm ABOVE THE GROUND
   Z2 = Z1
   DO KR = 1, NRAD
       PHI = (KR-1)*ACI
       DX = RL*COS(PHI) ;   DY = RL*SIN(PHI)
       X2 = X0 + DX ; Y2 = Y0 + DY
        WRITE(22,55) 'GW',II,NSEG,X1,Y1,Z1,X2,Y2,Z2,RAD
        I=II+1
   END DO
55 FORMAT (A2,I4,',',I2,',',2(F9.4,',',F9.4,',',F9.6,','),F7.6)
   STOP ; END
```

A wire with any length can be given with one GW command, since it can be divided as many segments as required to satisfy at least 10 to 20 segments per wavelength. Any two wires are connected if one of the ends of a wire has the same (*x*,*y*,*z*) coordinates with one of the ends of the second wire. Therefore, the length of a wire can only be given within the mesh between the crossing points and wires can not cross each other. A simple Fortran code is given in Table 2.6, where any number of radials are automatically generated underneath a vertical antenna located at the given (*x*,*y*) coordinate (although this ground screen can be obtained with GM command easily, it is better to write down a simple code for automatic radial generation). Since NEC2 does not allow wires located parallel on the surface, the ground screen is raised 5mm above the ground as shown in Fig. 2.13.

In Fig. 2.14, top view of a ground screen under a triplet array (three vertical monopoles) above the ground is shown. When the radials of the monopoles intersect with the others they are cut and connected. This structure is a typical channel of an antenna array for, for example, HFSWR receiver antenna system. A number of these channels form a receiver array with beam steering capability. All radiators are connected to the receive array processor, located behind the triplet, via signal cables. Obviously, each signal cable has different length. Moreover, signal velocity inside the cable is less than the velocity of light and must be measured before specifying cable lengths. These effects should be taken into account when digital beam forming is applied.

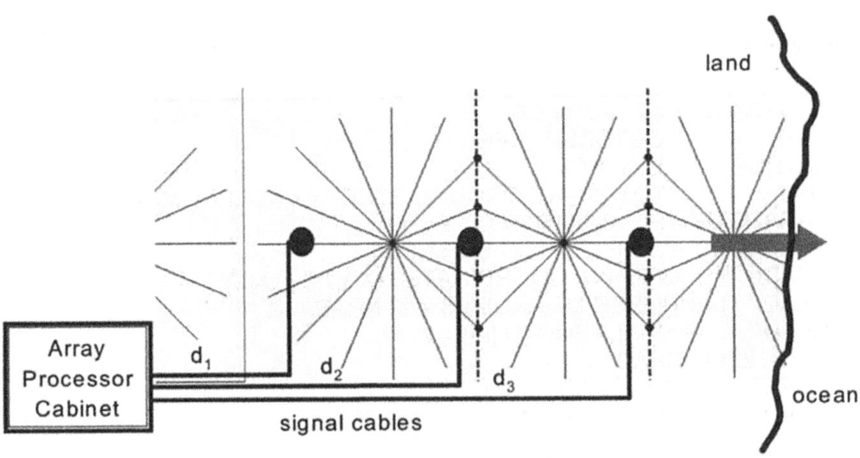

Figure 2.14: *A design of a 16-radial ground screen*

A Fortran code which generates the ground screen layout as shown in Fig. 2.14 under a 2 × *N* planar array is given in Table 2.7. The user supplies the number of array elements *N*, and the program automatically generates the ground screen layout of the 2 × *N* planar array as pictured in Fig. 2.14

Another important geometry in NEC simulations is a PEC box given in Fig. 2.15. Here, a PEC box with dimensions BW, BL, and BH is shown. The box is generated with six meshed surfaces with a mesh size that satisfies the required segment wavelength relation.

Table 2.7: A Fortran code for ground screen generation under a 2×N array

```
C  PROGRAM    : DU-SCR.FOR
C  DATE       : OCTOBER 28, 1998
C  AIM        : TO LOCATE GROUND SCREEN FOR DOUBLET ARRAY
C               ARRAY IS ALONG Y-AXIS, +X AXIS POINTS BORESIGHT

    PARAMETER(PI=3.1415927)
    OPEN(UNIT=22,FILE='DU-SCR.NEC')
    READ(*,*) 'NUMBER OF ANTENNA ELEMENTS = ?', NN
    READ(*,*) 'NUMBER OF RADIALS = ?', NRAD
    READ(*,*) 'RADIAL LENGTH [m] = ?', RL
    ACI=2.*PI/NRAD
    READ(*,*) 'RADIAL RADIUS [m] = ?', RAD
    READ(*,*) 'RADIAL SEGMENTS = ?', NSEG
    READ(*,*) 'ANTENNA SPACING WITHIN DOUBLET [m] = ?', DQ
    READ(*,*) 'SPACING BETWEEN DOUBLETS [m] = ?', DA

    ALENGTH=(NN-1)*DA ; QLENGTH=DQ
     II=100                                     ! WIRE NUMERATOR
     DO L=0,1                                   ! DOUBLET LOOP (X)
      DO K=1,NN                                 ! ARRAY LOOP (Y)
       X1=-QLENGTH/2+L*DQ                       ! X-COORDINATE
       Y1=-ALENGTH/2+(K-1)*DA                   ! y-COORDINATE
       Z1=0.005 ; Z2=Z1                         ! 5 mm ABOVE THE GROUND
       DO KR=1,NRAD
        PHI=(KR-1)*ACI
        DX=RL*COS(PHI) ;  DY=RL*SIN(PHI)
        X2=X1+DX ;  Y2=Y1+DY
        IF(X2.GT.(X1+DQ/2)) THEN
         X2=X1+DQ/2 ; DY=DQ/2.*SIN(PHI) ; Y2=Y1+DY
        END IF
        IF(X2.LT.(X1-DQ/2)) THEN
         X2=X1-DQ/2 ; DY=DQ/2.*SIN(PHI) ;  Y2=Y1+DY
        END IF
        IF(NN.EQ.1) GOTO 77
        IF(Y2.GT.(Y1+DA/2)) THEN
         Y2=Y1+DA/2 ;  DX=DA/2.*COS(PHI) ;  X2=X1+DX
        END IF
        IF(Y2.LT.(Y1-DA/2)) THEN
         Y2=Y1-DA/2 ; DX=DA/2.*COS(PHI) ;    X2=X1+DX
        END IF
77       WRITE(22,55) 'GW',II,NSEG,X1,Y1,Z1,X2,Y2,Z2,RAD
        II=II+1
       END DO
      END DO
    END DO
55  FORMAT(A2,I4,',',I2,',',2(F9.4,',',F9.4,',',F9.6,','),F7.6)
    STOP
    END
```

In NEC, one wire with multi-segmentation can be generated to satisfy segment wavelength relation. Unfortunately, this kind of PEC surface generation in terms of wire meshes must be generated on a one-wire/one-segment basis. Therefore, depending on the mesh size, as many as thousands of wires (segments) may be required to generate a six-faced PEC box, as given in Fig. 2.15. A short Fortran Code is given in Table 2.8 for this purpose.

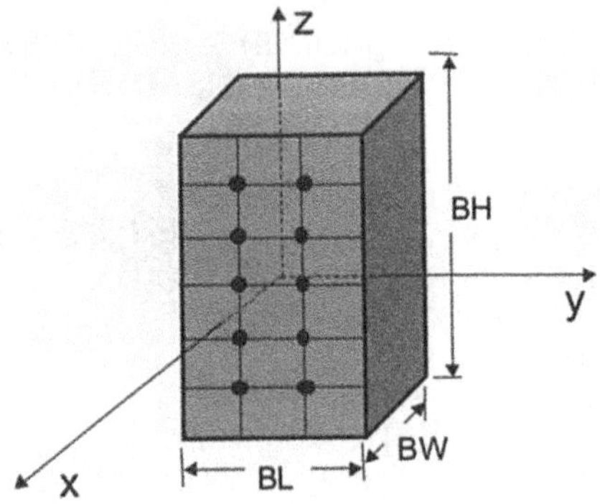

Figure 2.15: *A PEC box and mesh generation*

As important as geometry generation in NEC is the array excitation. Suppose a 24-channel quadlet (a 4×24) array is of interest and electronic beam steering effects is to be simulated (this is a typical receiver array for a HFSWR). Since all bottom-fed monopoles are used, $4 \times 24 = 96$ EX command lines will be used for one beam steering direction. As will be discussed in Section 2.5, beam steering along different direction requires to change these 96 EX lines accordingly. It is again very practical to generate these excitation lines automatically. A simple Fortran code for this purpose is given in Table 2.9. In this routine, supplying the element distances, the frequency and the beam direction is sufficient to produce 96 line of excitation, where connection cable lengths, propagation velocity in this cable, and voltage drop because of the cable losses are all taken into account.

Table 2.8: A Fortran code for PEC box mesh generation

```
C  PROGRAM      : BOX.FOR
C  AUTHOR       : L. SEVGI
C  NOTE         : THIS PART CONSTRUCTS ONE OF SIX FACES OF THE BOX

   OPEN(UNIT=22,FILE='BOX.NEC')
   READ(*,*) 'BOX WIDTH [m] = ?', BW
   READ(*,*) 'BOX LENGTH [m] = ?', BL
   READ(*,*) 'BOX HEIGHT [m] = ?', BH
   READ(*,*) 'BOX GRID [m] = ?',  BG
   READ(*,*) 'NUMBER OF SEGMENTS = ?', NSEG

   RA=.001 ; II=10
C**************************************************************
C     HORIZONTAL WIRES AT XMIN SURFACE                        *
C**************************************************************
   DO ZZ=-BH/2,BH/2,BG
```

```
        X1=-BW/2 ;  X2=X1 ; Z1=ZZ ; Z2=Z1
        DO YY=-BL/2,BL/2-BG,BG
          Y1=YY ;  Y2=YY+BG
          WRITE(22,55) 'GW',II,NSEG,X1,Y1,Z1,X2,Y2,Z2,RA
          II=II+1
        END DO
      END DO
C****************************************************************
C     VERTICAL  WIRES AT XMIN SURFACE                          *
C****************************************************************
      DO YY=-BL/2,BL/2,BG
        X1=-BW/2 ;  X2=X1 ; Y1=YY ;  Y2=Y1
        DO ZZ=-BH/2,BH/2-BG,BG
          Z1=ZZ ;  Z2=Z1+BG
          WRITE(22,55) 'GW',II,NSEG,X1,Y1,Z1,X2,Y2,Z2,RA
          II=II+1
        END DO
      END DO
55    FORMAT(A2,1X,I4,',',I2,',',2(F9.4,',',F9.4,',',F9.6,','),F7.6)

      WRITE(22,*) 'GS 0 0 1'
      WRITE(22,*) 'GE 0'
      WRITE(22,*) 'GN -1'
      WRITE(22,*) 'FR 0 1 0 0 5.0 .25'
      WRITE(22,*) 'EX 1 1 1 0 90 -45 0 1 1'
      WRITE(22,*) 'RP 0 181 1 1500 -90 0 1 1 1E6 0.'
      WRITE(22,*) 'RP 0 1 361 1500 90 0 1 1 1E6 0.'
      WRITE(22,*) 'EN'
      CLOSE(22) ;  STOP ; END
```

Table 2.9: A Fortran code for a 24-channel excitation in NEC

```
C   PROGRAM                : QU-FEED.FOR
C   AUTHOR                 : L. SEVGI
C   AIM                    :TO PREPARE EXCITATION FOR QUADLET ARRAY
C
      PARAMETER(PI=3.1415927,C=2.997925e+08,vp=0.84)
      REAL AL(24), UY(24)

      OPEN(UNIT=22,FILE='qu-feed.nec')
      READ(*,*) 'NUMBER OF ANTENNA ELEMENTS = ?', NN
      READ(*,*) 'ANTENNA SPACING WITHIN QUADLET [M] = ?', DX
      READ(*,*) 'FREQUENCY [MHz] ? = ?', FREQ
      READ(*,*) 'BEAM ANGLE FROM BORESIGHT [Deg] ? = ?', ANG

      ANG=ANG*PI/180. ; FARK=DY*SIN(ABS(ANG))
      DO K=1,24
        AL(K)=0. ; UY(K)=0.
      END DO

C   Cable lengths
      Al(1)=4.0525128e+002;  al(2)=3.7682316e+002;   al(3)=3.5028252e+002;
      Al(4)=3.2320008e+002;  al(5)=2.9453256e+002;   al(6)=2.6277384e+002;
      Al(7)=2.3755116e+002;  al(8)=2.0984376e+002;   al(9)=1.7955588e+002;
      Al(10)=1.5334284e+002; al(11)=1.2473916e+002;  al(12)=9.5967480e+001;
      Al(13)=6.7925760e+001; al(14)=9.6365640e+001;  al(15)=1.2366396e+002;
      Al(16)=1.4880600e+002; al(17)=1.8088644e+002;  al(18)=2.0916588e+002;
      Al(19)=2.3266824e+002; al(20)=2.6014128e+002;  al(21)=2.8866180e+002;
```

```
   Al(22)=3.1590300e+002;  al(23)=3.4311984e+002;        al(24)=3.7521204e+002
C  Times MW LMR-240 Cable Attenuation [dB/100m]
   DD=DX ; WAVE=300*vp/FREQ ; WFREE=300/FREQ ; PHASE=DD*2.*PI/WAVE
   FARK=FARK*2.*PI/WFREE
   ATT=0.79426*SQRT(FREQ)+0.00108*FREQ

C  Feeder voltages
   DO K=1,24
     ALOSS=AL(K)*ATT/100 ; UY(K)=10.**(-ALOSS/20)
   END DO
   MM=1
   BASLA=3*PHASE
   DO J=0,3                              ! Quadlet loop
     AA=BASLA-J*PHASE
     DO K=1,NN                           ! Channel loop
       IF(ANG.GT.0.) THEN
         ACI=AA+(NN-K)*FARK
       ELSE
         ACI=AA+(K-1)*FARK
       END IF
       ACI=MOD(ACI,2*PI) ; UR=ABS(UY(K)*COS(ACI)) ; US=ABS(UY(K)*SIN(ACI))
       IF(ACI.GT.0..AND.ACI.LE.PI/2.) THEN
         UR=UR ; US=US
       ELSEIF(ACI.GT.PI/2.AND.ACI.LE.PI) THEN
         UR=-UR ; US=US
       ELSEIF(ACI.GT.PI.AND.ACI.LE.3*PI/2) THEN
         UR=-UR ; US=-US
       ELSEIF(ACI.GT.3*PI/2.AND.ACI.LE.2*PI) THEN
         UR=UR ; US=-US
       END IF
       WRITE(22,88) 'EX 0',MM,' 1 00 ',UR,US
       MM=MM+1
     END DO
   END DO
88 FORMAT(A4,1X,I3,A6,F9.5,',',F9.5)
   STOP ; END
```

2.5.2 FDANT and TLMANT Codes

FDTD- and TLM-based antenna simulators, FDANT and TLMANT, use classical YEE and SCN-TLM cell structures, respectively, and the total field formulation. Same auxiliary routines are used in both TD simulators, and their input and output parameters are also same. Therefore, explanations given below for FDTD-based antenna simulation are also valid for TLM-based simulations. The main routine of the FDTD based antenna simulation is listed in Table 2.10.

YAKUZ, HESAP, POYNTING, ABSORBED, and VA are the subroutines that are used in antenna simulations besides the regular FDTD subroutines. YAKUZ and HESAP are used in NTFF transformation and are explained in Part I. YAKUZ is called at every time step (inside the time loop), and HESAP is called once when the time simulation is ended. POYNTING is a power computation routine which calculates the radiated power.

Table 2.10: FDTD BASED TD antenna simulator's structure

```
C  PROGRAM        : ANTENNA SIMULATOR
C  METHOD         : FDTD
C  DATE           : FEBRUARY 1995

   INCLUDE 'fdant.par'
   OPEN(UNIT=30,FILE='FD-FAR.DAT')
   OPEN(UNIT=40,FILE='POYNTING.DAT')
   OPEN(UNIT=50,FILE='PABS.DAT')

1  WRITE(*,*) 'ANTENNA PATTERN ON WHICH PLANE= ?'
   WRITE(*,*) 'ENTER (0) FOR HORIZONTAL (H) PLANE (0),
  *        (1) FOR VERTICAL (E) PLANE (1)'
   READ(*,*) KEH
   IF(KEH.EQ.0) THEN
      READ(*,*) 'THETA [Deg] FOR FAR FIELD OBSERVATION = ?', TETA
      TETA = TETA * PI / 180.
      READ(*,*) 'ANGULAR INCREMENTS [Deg] = ?', DELTA
      DELTA = DELTA * PI / 180.
      MAX = 2. * PI / DELTA
   ELSEIF(KEH.EQ.1) THEN
      READ(*,*) 'PHI [Deg] FOR FAR FIELD OBSERVATION = ?', PHI
      PHI = PHI * PI / 180.
      READ(*,*) 'ANGULAR INCREMENTS [Deg] = ?', DELTA
      DELTA = DELTA * PI / 180
      MAX = 2. * PI / DELTA
   ELSE
      GOTO 1
   ENDIF

C ***************** MAIN FDTD LOOP ****************
    T = 0.0
    DO 100 N=1,NSTOP
       UPDATE ELECTRIC FIELDS
       CALL RADIATION BOUNDARY SIMULATORS
       T=T+DT/2.
       UPDATE MAGNETIC FIELDS
        T=T+DT/2.
        CALL YAKUZ
        CALL POYNTING
        CALL ABSORBED
        WRITE(40,*) T*1.E9, PYT
        WRITE(50,*) T*1.E9,PABS*DELX
C ****************** END OF MAIN LOOP ***************
100 CONTINUE
101
C *****************  OFFLINE NTFF TRANSFORMATIONS  ***************
    CALL HESAP
    STOP
    END
```

The subroutine POYNTING uses

$$P_{rad} = \oint\!\!\oint_S E(t)\times H(t)\ ds \qquad (2.6)$$

over a closed virtual surface that surrounds the antenna under investigation. Usually, same virtual surface is used for both NTFF and POYNTING calculations. The virtual closed surface is a cubical surface and tangential field components are required in (2.6) for instant power calculations. For example at x = constant surfaces of this virtual surface E_y, E_z and H_y, H_z are the tangential field components. The radiated power in (2.6) is the instantaneous power and is divided by a period T when the source is sinusoidal and average power is of interest.

The gain of the antenna is calculated from

$$Gain = \frac{4\pi\times E(\omega,\varphi,\theta)^2\ /\ Z_0}{P_{in}} \qquad (2.7)$$

where $E(\omega,\varphi,\theta)$ is the far field along a direction determined by the angles (φ,θ) at an angular frequency ω, and Z_0 is the free-space wave impedance. The input power at ω can be calculated from

$$P_{in} = \mathrm{Re}\left[V_{in}(\omega)\times I^*{}_{in}(\omega)\right] \qquad (2.8)$$

where input voltage and current at ω are directly obtained from the TD simulations of $v_{in}(t)$ and $i_{in}(t)$ by applying off-line DFT. When the simulation is run, it first requires selection of (vertical or horizontal plane) radiation pattern (see Fig. 2.16).

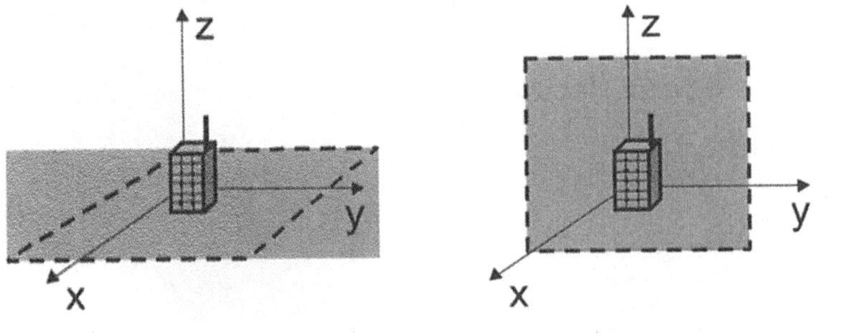

Figure 2.16: *Horizontal and vertical radiation patterns*

Then, angular resolution is supplied. If, for example, horizontal plane radiation pattern is selected with 2° resolution, NTFF transformation is applied at 181 assumed directions (with 2° angular separation) inside the TD loop (although they are same, calculations at 0° and 360° are performed separately, only for graphical illustrations). The results are written to the output file AN-FAR.DAT. There are four columns in this file: These are the angle, time, E_θ and E_φ. If the number of time steps and angle are NSTOP and 181, respectively, then the number of lines will be NSTOP+200 for each angle with a total of 181 × (NSTOP+200). It should be noted that, extrapolated TD far fields are stored in a time array, and dimension of this array must be greater than the number of simulation time steps, NSTOP. This is due to TD far field extrapolation as explained in [19]. Here, 100 time cells are reserved for this transformation in front, and at the back of the array. This is why the total simulation time will be NSTOP+200.

The user may select to use a short pulse or a single frequency sinusoidal excitation in TD simulations. Choosing a short pulse results in obtaining a broadband response with a single TD run. For example, a typical short pulse response of a quarter-wavelength antenna of a mobile phone in *xy*-plane is pictured in Fig. 2.17. In broad band TD simulations simulation period must be long enough to decease all transients as shown in the figure.

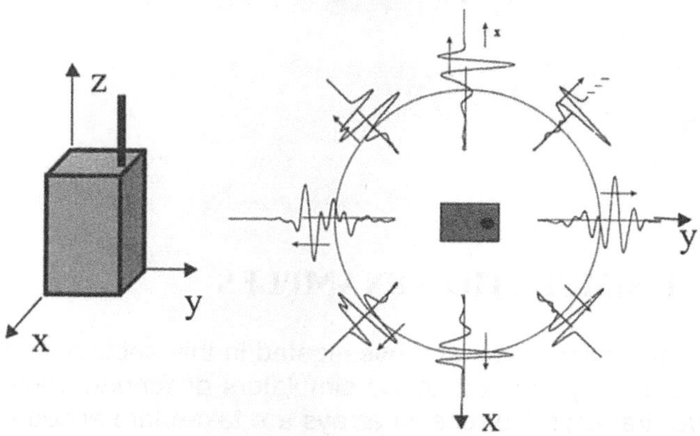

Figure 2.17: *Short pulse far-field responses of an antenna along different directions*

FDTD-based simulation yields broadband response in time domain. Radiation patterns at multiple frequencies are computed by the Fortran program AN_DFT. Table 2.11 lists the code. The program only needs to know the frequency, number of angles and time steps (and whether a Hanning window is applied or not), and the results are given in the output file AN-DFT.DAT; angles are given in the first column, and E_θ and E_φ are given in the second and third columns, respectively.

Table 2.11: An_DFT Fortran code

```
C  PROGRAM   : AN - DFT.FOR
C  AIM       : TO OBTAIN ANTENNA RADIATION PATTERNS FROM
             : AN-FAR.DAT AT A GIVEN FREQUENCY

   REAL FETETA,FEPHI,TIME,ACI,PI,aa(2000),OFF
   COMPLEX CFETETA,CFEPHI,J,cf
   OPEN (UNIT = 22,FILE = 'AN-DFT.DAT')
   OPEN (UNIT = 11,FILE = 'AN-FAR.DAT')
```

```
READ(*,*) 'FREQUENCY [GHz] = ?', FREQ
READ(*,*) 'NUMBER OF ANGLES, TIME STEPS = ?', NACI, NT
READ(*,*) 'APPLY HANNING WINDOW: YES (1), NO (0) = ?', KK
IF(KK.EQ.0) THEN
  OFF = 0.0
ELSE
  OFF = 1.0
ENDIF
PI = 4.0*ATAN(1.0) ; J = CMPLX(0.0,1.0) ; ARG = 2.0*PI
DO I = 1, NACI
   CFETETA = CMPLX(0.0,0.0) ; CFEPHI = CMPLX(0.0,0.0)
   DO K = 1, NT
      READ (11,*) ACI,TIME,FETETA,FEPHI
      CFETETA = CFETETA+FETETA*CEXP(-J*ARG*FREQ*TIME)
*                *OFF*0.5*(1.+COS(PI*FLOAT(K)/NT))
      CFEPHI = CFEPHI+FEPHI*CEXP(-J*ARG*FREQ*TIME)
*               *OFF*0.5*(1.+COS(PI*FLOAT(K)/NT))
   END DO
   b1 = CABS(CFETETA) ; b2 = CABS(CFEPHI)
   WRITE (22,*) ACI,b1,b2
 END DO
 STOP ; END
```

2.6 NUMERICAL SIMULATION EXAMPLES

Typical, but complex structures are investigated in this section, to show the power, efficiency and accuracy of the antenna simulators described above. First, HF and VHF high-power transmit and receive arrays are taken into account. The examples presented here are actual radar antennas, which are designed, built and tested for HFSWR [15 - 17].

HFSWR uses vertically polarized surface waves that follow the curvature of the earth and capable of over the horizon coverage. Long-range surveillance is possible because of the relatively low attenuation propagation over the highly conductive ocean surface. HFSWR can provide continuous, all-weather, 24-hour coverage in regions up to 500 km in range and 120° in azimuth [15,16]. Antenna systems for HFSWR should therefore be designed to satisfy a number of criteria. Also, when predicting the performance of both the transmit and receive arrays the effect of local site topography and ground conditions must be considered.

2.6.1 HFSWR Transmit Array

HFSWR transmit antenna must provide high gain over the designed operational frequency band. Energy must be distributed equally and only over the desired surveillance area. In order to eliminate ionospheric interference, transmit antenna must also provide a deep, broad, null at near vertical incidence (NVI). It is located very close to shore, generally over the ground with poor electrical properties; therefore a good design of a ground screen is essential. A transmit antenna system designed to meet these requirements is given in Fig. 2.18.

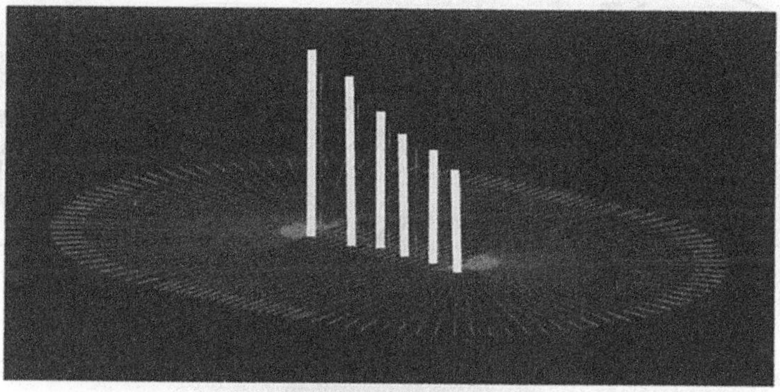

Figure 2.18: *HFSWR transmit array system [16]*

The transmit array designed for the HFSWR is a 6-element log-periodic monopole antenna [17]. Physical dimensions of the antenna sistem is listed in Table 2.12.

Table 2.12: Physical dimensions of the HFSWR transmit antenna

Element No.	Height [m]	Distance [m]
1	24.8	---
2	21.6	6.8
3	18.8	6.0
4	16.3	5.2
5	14.2	4.5
6	12.3	3.9

The array is aligned across the shore (the highest element #1 is at the back and the shortest one #6 is in the front). The array is designed to operate in the frequency band of 3 MHz to 5 MHz, with at least 8 dB gain and 10 dB FBR. The ground screen consists of 144 38.8-m-long (approximately half wavelength at 3.85 MHz) 144 radials located as pictured in Fig. 2.18.

Theoretically, this log-periodic monopole array satisfies gain requirements above a PEC surface (i.e., its gain is between 9.0 and 9.5 dBi in 3.2-5.2 MHz and FBR is higher than 10 dB). The HFSWR transmit array is located on a POOR ground (i.e., $\varepsilon_r = 4.0$, $\sigma = 0.003$ S/m) and this causes a loss of more than 20 - 25 dB in antenna gain. To overcome ground losses, a special ground screen is designed as illustrated in Fig. 2.18.

Horizontal and vertical radiation patterns of this transmit antenna array at 3.5 MHz is given in Fig. 2.19. Solid and dashed lines correspond to with and without ground screen layout. As observed from the figure, ground screen increases the gain at least 10 dB at this frequency, but decreases overhead nulling.

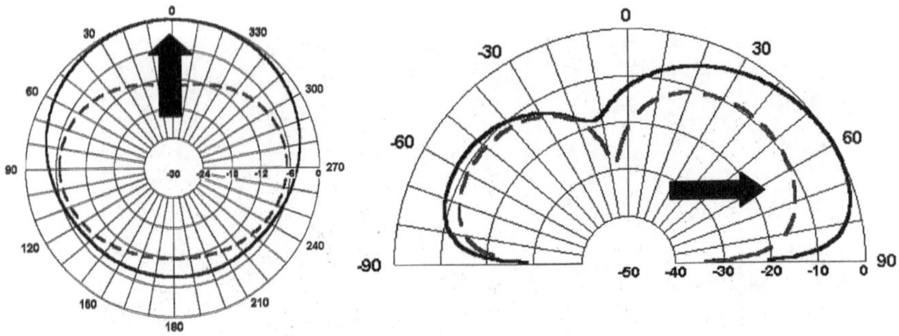

Figure 2.19: *Horizontal and vertical radiation patterns of the HFSWR transmit antenna system*

2.6.2 HFSWR Receive Array

HFSWR receive array is much more complicated than the transmit array. The receive array must be parallel to the shore line and must have high and equal array gain over the entire surveillance area with minimum sensitivity to signals arriving from other directions (especially from the ionosphere). Operating at the lower end of the HF band requires that the receive array occupy a significant shoreline area, with the aperture of the array inversely proportional to frequency. For example, at 3 MHz, a 5° azimuth beamwidth requires an array aperture of approximately 800 m to 1 km. Moreover, the receive array should direct its power towards the ocean surface, with high FBR, so that it neither interfere land based other systems, nor is interfered by them. This is illustrated in Fig. 2.20. To satisfy all these requirements, first array elements (also called as channels) are designed and then array performance is analyzed.

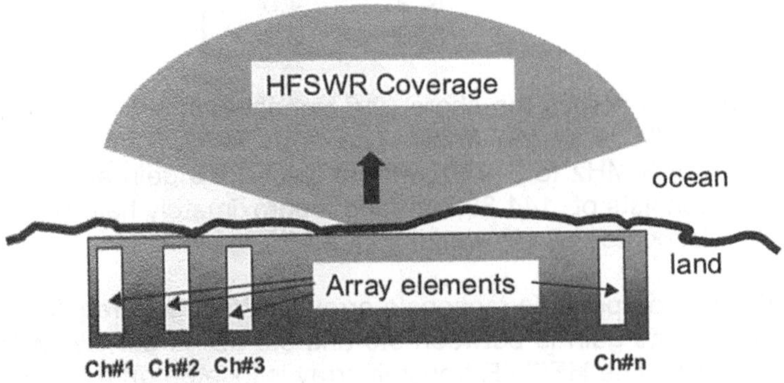

Figure 2.20: *HFSWR receive array and antenna park*

The mission of the HFSWR channel elements can be summarized as to:

- form a maximum radiation along the channel axis with as much azimuth coverage as possible (typically 100° to 120°),
- give as high gain as possible (at least 2 - 3 dB)

- give as much FBR ratio as possible (at least 12 - 15 dB),
- give as deep over-head null as possible (at least 25 - 30 dB)

in the whole operation frequency band. To satisfy these requirements, three different structures are investigated as array channels. These are *quadlet, triplet* and *doublet.* In quadlet, four elements are spaced approximately quarter-wavelength apart and are phased to form an end-fire channel. Similarly, three and two elements are used in triplets and doublets, respectively. A short monopole is used as a channel radiator and a 16-radial ground screen layout is used underneath. Figure 2.21 pictures the top view of the channel design. As shown in the figure, radials of different elements in a channel are connected to each other when intersected. The channels are optimized for the frequency band of 3 - 6 MHz. The parameters of the channels are listed in Table 2.13.

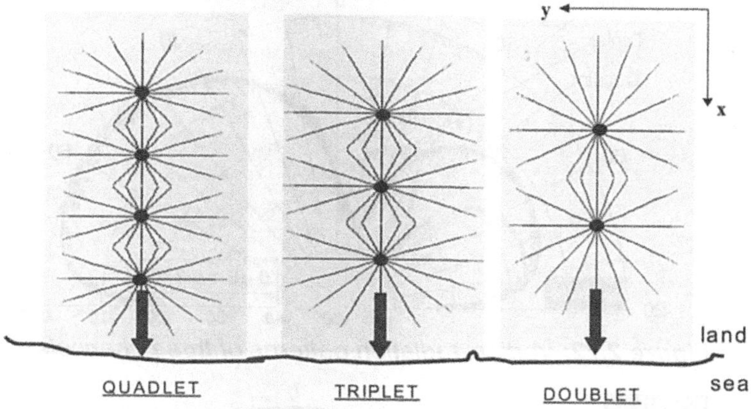

Figure 2.21: *Top view of sample HFSWR receive array channels*

Ground is assumed to be POOR. Parameters for the ocean are assumed to be ε_r = 80.0, σ = 5.0 S/m in NEC calculations, where EM shooting towards ocean surface is simulated.

Table 2.13: HFSWR Receive array channel parameters

Parameter	Value	
Antenna height	9.22	[m]
Element spacing	19.0	[m]
Element wire radius	0.15	[m]
Number of radials	16	
Radial wire lengths	19.0	[m]
Radial wire radius	0.001	[m]
Incremental increase in cable length	19.0	[m]
Phase velocity in electrical cable	0.84 *c*	[m/s]

The channel elements are bottom-fed. The feeder cable losses are represented by voltage drops. The loss of the RG-213u cable is given as 0.55 dB/100 ft at 10

MHz), therefore 0.5 dB/100 ft cable loss is taken into account within the band of 3 - 6 MHz. For a 19.0-m cable length, voltage drop corresponds to 5 V. Therefore, elemental feeding voltages are taken as:

Quadlet:	100 V, 95 V, 90 V, and 85 V
Triplet:	100 V, 95 V, and 90 V
Doublet:	100 V, and 95 V.

Vertical and horizontal radiation patterns are obtained at different operating frequencies and are given in Figs. 2.22 and 2.23. Here, the channels are located along the x-axis. Vertical radiation patterns are obtained at the *xz*-plane ($\varphi = 0°$, $-90° < \theta < 90°$). Horizontal radiation patterns are plotted 5° above the *xy*-plane ($\theta = 85°$, $0° < \varphi < 360°$).

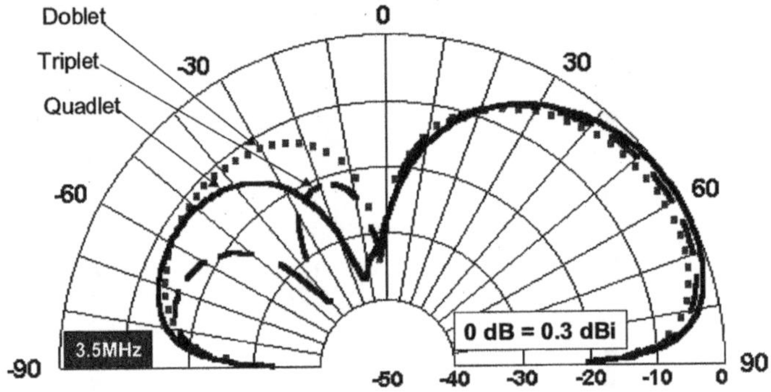

Figure 2.22: *Vertical radiation patterns of three channels*

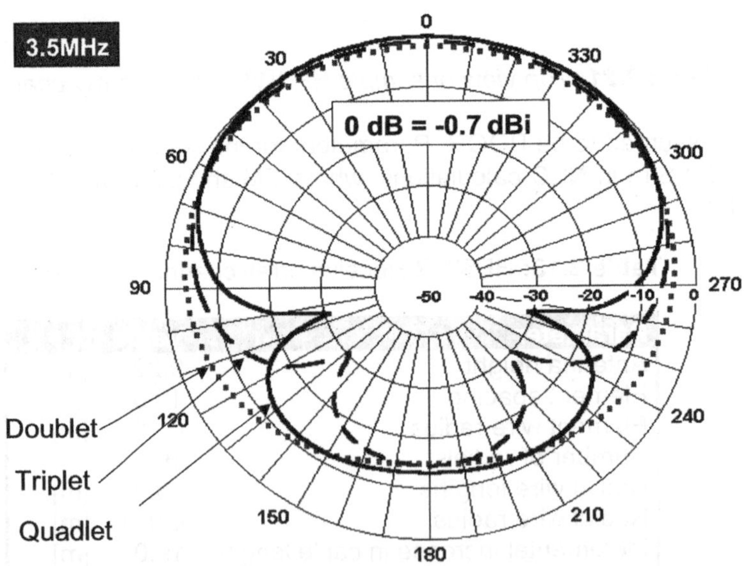

Figure 2.23: *Horizontal radiation patterns of the channels*

The pattern dynamic range is 50 dB and the outer circle corresponds to 0 dB. The actual gain of the outer circle is given in the figure inset as 0.3 dBi. As observed in

Figs. 2.22 and 2.23, quadlet, triplet, and doublet as receive array channel elements have similar channel gain and FBR at this frequency. On the other hand, vertical and horizontal back-lobe shapes and overhead nulling effects are quite different. These tests should be performed throughout the frequency band before deciding the channel element.

The surface over which the array is to be erected is very important in array performance. To show this, the vertical radiation pattern of the quadlet channel over different surface types is given in Fig. 2.24. The quadlet with parameters in Table 2.13 has an 11-dBi channel gain at 4.5 MHz when erected over PEC surface and the beam maximum points the ground. When erected over POOR ground as much as 20-dB gain reduction is observed at this frequency. Approximately half of this loss is compensated when the 16-radial ground screen is used.

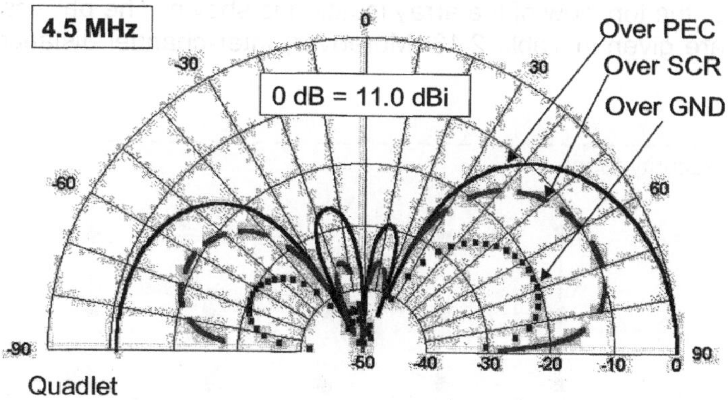

Figure 2.24: *Channel over different type surfaces (SCR: screen, GND: ground)*

It should be noted that only the sky wave effects are shown in these figures, where the vertical beam has a maximum around 30° above the surface. In practice, this vertical maximum is directed towards the ocean surface by using the Brewster angle effect. NEC2 has the capability of taking the land-ocean coupling effect into account in an approximate sense [12]. By using this capability, it can be shown that the vertical beam maximally points the ocean surface. This is illustrated in Fig. 2.25, where ocean parameters are assumed to be the same. In the figure, solid and dashed lines correspond to surface wave coupling and sky wave without surface wave, respectively. A ground screen is used not only to obtain a high antenna gain (by stabilizing the input impedance of the vertical radiators) but also to lower the vertical radiation pattern to cover the sea surface.

To illustrate receive array performance, a 7-channel quadlet array has been modeled (as shown in Fig. 2.26). This provides much of the information related to radiation patterns; it also provides gain and ground screen effects without the excessive computational requirement needed for a larger array.

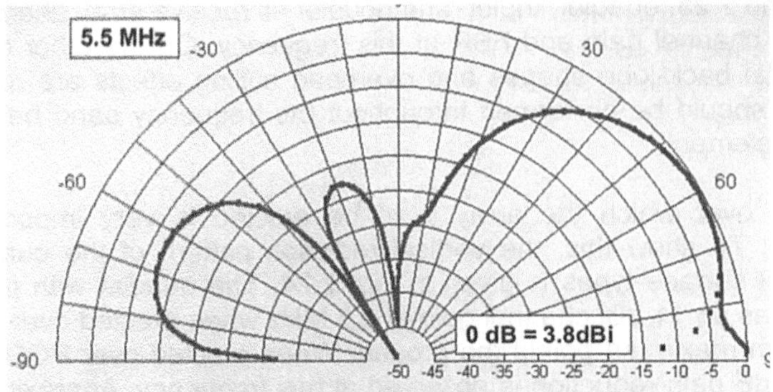

Figure 2.25: *Beam shooting over ocean surface*

In Figure 2.26, the top view of the array location is shown. The physical parameters of the array are given in Table 2.13. Moreover, inter-channel distance is taken as 31 m.

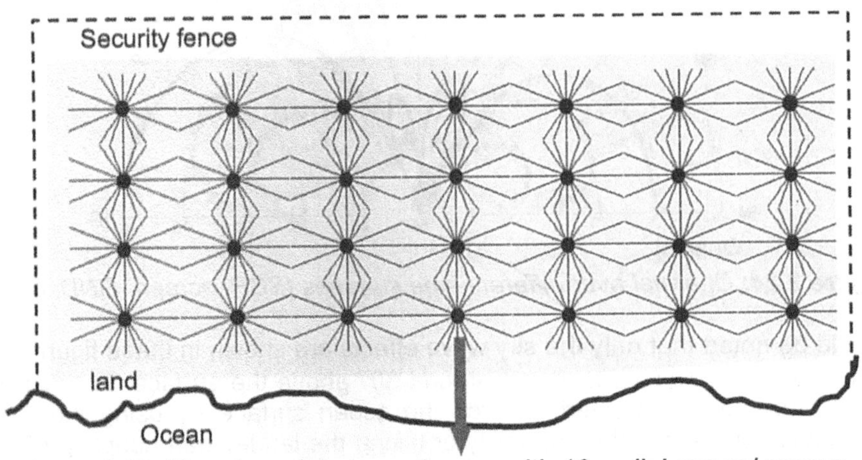

Figure 2.26: *Top view of 7-channel array with 16-radial ground screen*

As pictured in Fig. 2.26, the HFSWR receive array requires a leveled area with hundreds of square meters along the shore line. A security fence surrounds this area. All elements are connected to the receive array processor located behind the array via cables with hundreds of meters of lengths.

Vertical and horizontal radiation patterns of this 7-channel array are given in Fig. 2.27 and Fig. 2.28. The dashed and solid lines in these figures correspond to quarter-wavelength monopoles and fat monopoles, respectively [17]. More than 10 dB loss in over head nulling (with respect to a single quadlet) is observed for this 7-element quadlet array. It may be speculated that the reason for this is the excitation of horizontal field component because of the ground screen layout.

The 4 × 7 array is chosen because of computer requirements. A personal computer with 128-MB RAM memory and 500-MHz speed can handle MoM problems with as many as 2400 - 2500 segments within a reasonable computation time (i.e., less than an hour). This array is sufficient to test operational performances as well as ground screen effects.

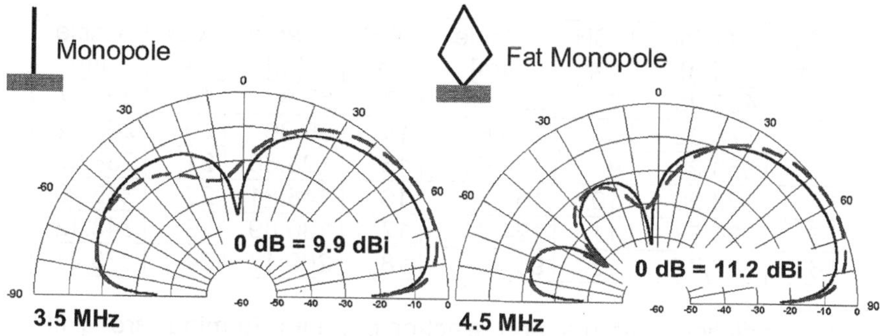

Figure 2.27: *Vertical radiation patterns of a 4 × 7 element array*

Numerical simulations have shown that ground screen directly affects overhead nulling, but has almost negligible effects on horizontal radiation pattern and beam steering. Therefore, 24-channel array of the HFSWR is investigated over POOR ground without ground screen.

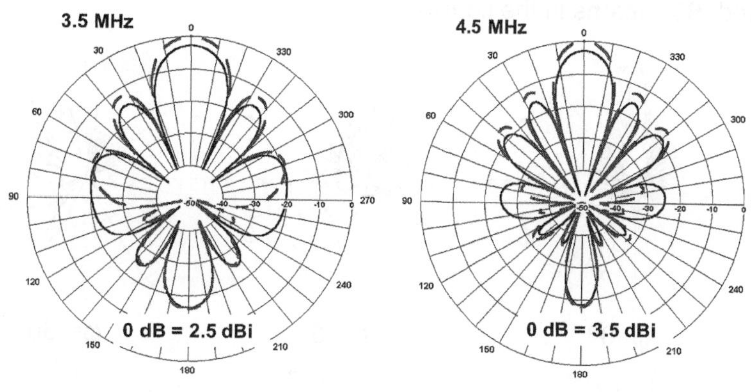

Figure 2.28: *Horizontal radiation patterns of a 4 × 7 element array*

One HFSWR receive array is designed as a 24-channel array by using quadlets [15]. Inter-channel distances are taken as 31 m. With these parameters, a rectangular area of 57 m × 713 m is required only for the 4 × 24 monopoles. With ground screen and a security fence, this area may be as large as 100 m × 1000 m. Behind this area, an operational control center (OCC) is built where receive array channels are connected to receiver processor. The connections between channels and OCC are done via underground cables. The length of connection cables are very important, since they introduce extra signal attenuation as well as phase differences. The location of OCC directly affects the lengths of the connection cables. The best way is to locate the OCC symmetrically so that channels from the left and right have almost equal cable lengths. In Table 2.14, real cable lengths of the designed 24-channels are listed.

Table 2.14: 24-Channel receive array cable lengths

CH #	Cable [m]	CH #	Cable [m]	CH #	Cable [m]	CH #	Cable [m]
1	405.25	7	237.55	13	67.93	19	252.67
2	376.82	8	209.84	14	96.36	20	260.14
3	350.28	9	179.56	15	123.66	21	288.67
4	323.20	10	153.34	16	148.81	22	315.90
5	294.53	11	124.74	17	180.89	23	343.12
6	262.77	12	95.97	18	209.17	24	375.21

Horizontal radiation patterns and electronic beam forming are simulated and illustrated in Fig. 2.29 for different beam pointing angles. Here, patterns with beam angles 0°, 15°, and 55° towards the right and -15°, -30° and -60° towards the left are plotted. As observed in the figure, for the 0° beam the array has nearly 5° beam width, 30-dB FBR, and more than 20 dB side-lobe suppression. These performance parameters are almost satisfied for the beams within ± 30°. Acceptable values are obtained when the beam is within ± 45°. On the other hand, unwanted strong side lobes appear when the beam angle is higher than ± 45° (see the 55° and -60° beams in the figure).

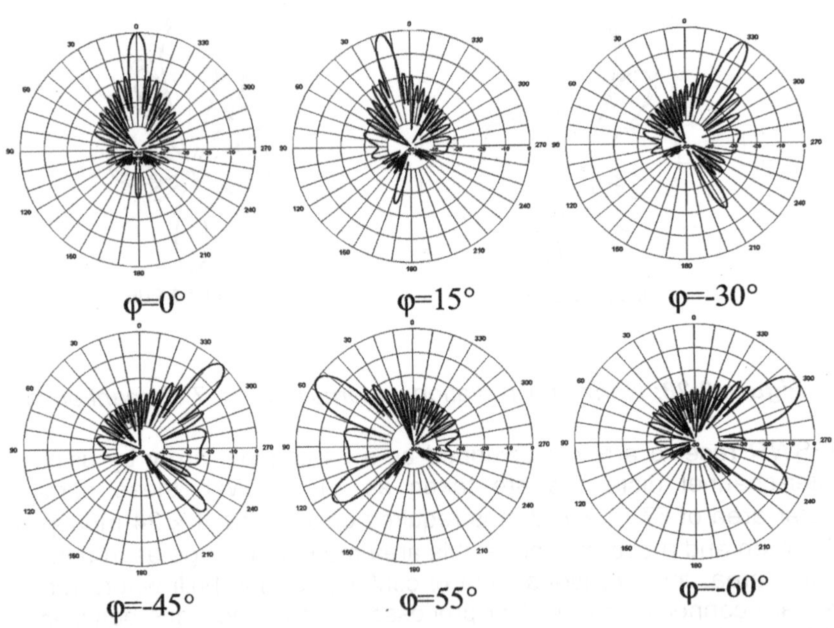

Figure 2.29: *Electronic beam steering in the 4 × 24 array*

It may be concluded that electronic beam steering is possible within ± 45° in this planar array. In practice, a horizontal coverage of 100° to 120° is desired. Simulations in this section show that this may be possible only by using intelligent algorithms, which manages to eliminate undesired side and/or back lobes.

HFSWR is an all-whether, 24-hour continuous system designed to operate a few decades. Very often one or more channels fail to operate mostly because of cable problems. The operator should know what happens when one or more channels are not working properly. This is simulated in NEC and typical results are plotted in Fig. 2.30. In the 24-channel array, the channels are numbered 1 to 24 from left to right. The blanked channels are (left) 1-3-5 and (right) 11-12-13 in the figure. The thin and thick lines correspond to 24-channel beam forming with and without blanked channels, respectively. From an operational point of view, the channel blanking modeled in Fig. 2.30 represents almost the worst case. The probability that three channels fail simultaneously is very low. However, even for these cases, digital beam synthesis is still possible with only minor degradations at the sides of the radar coverage.

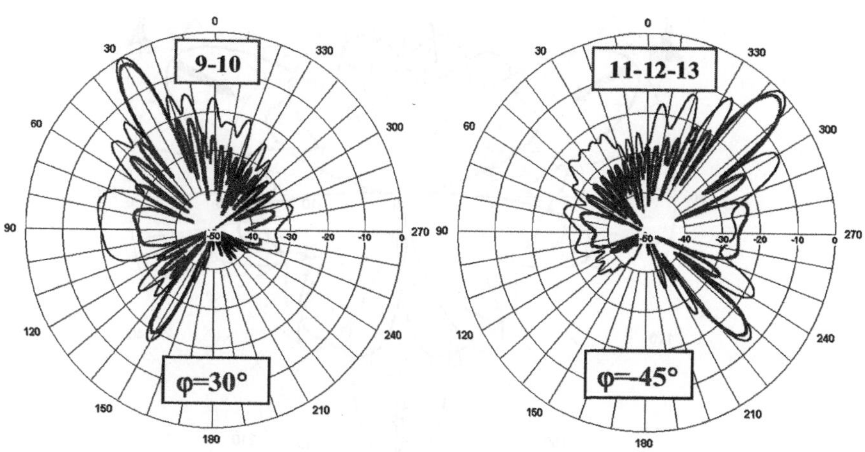

Figure 2.30: *Channel failure effects on digital beam steering*

In order to simulate array performances via NEC package effectively, the number of wires and segments is important. MoM technique is based on numerical solution of a set of equations in a matrix form. The order of the matrix system is proportional to the number of segments. As the number of segments increases linearly, the computation time required for the matrix inversion increases exponentially. With a P-II-based PC having 128-MB RAM memory, segments up to 2400 to 2500 can be handled in a reasonable time. After 2000 segments it takes hours, even days to obtain NEC simulation results. In a multi-channel receive array modeling via NEC, the number of segments mostly depends on the ground screen.

2.6.3 VHF DF Antenna

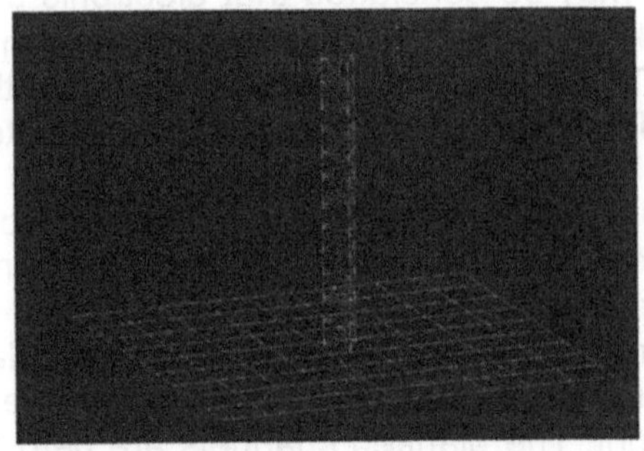

The last MoM simulation example is a direction-finding antenna. It is designed to operate at VHF (118 - 163 MHz), and has 16 half-wavelength dipoles, located circularly with a radius of 2.9 m and mounted on top of a 20-m-high antenna tower [18].

By adjusting elemental excitation (i.e., by changing the feeding phases) of the DF antenna, maximum radiation may horizontally be aligned to a desired direction. This is illustrated in Fig. 2.31. Here, horizontal beam is directed to 60°, 180°, and 300°.

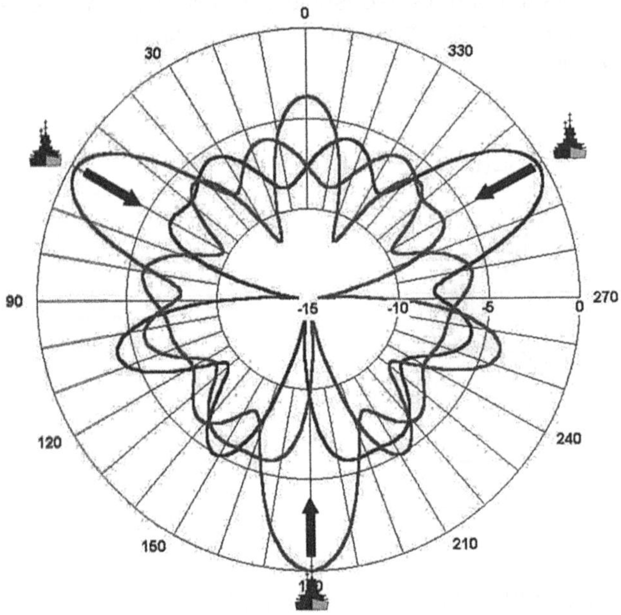

Figure 2.31: *Horizontal pattern of the VHF DF atenna*

As shown in the figure, the beamwidth obtained with this 16-element, 2.9-m-radius DF antenna is more than 20°. Moreover, main beam side lobe difference is less than 5 dB. This requires intelligent algorithms to find out the direction of arrival within 1° accuracy. One technique is to use a correlation process. In practice, the pattern of the receive signal is correlated with previously stored theoretical patterns (with, for example, 1° increments) and the direction is determined from maximum correlation. A typical example is given in Fig. 2.32, where correlation results of four different angles are shown. Maximum correlation occurs at 0°, where the target is present [18].

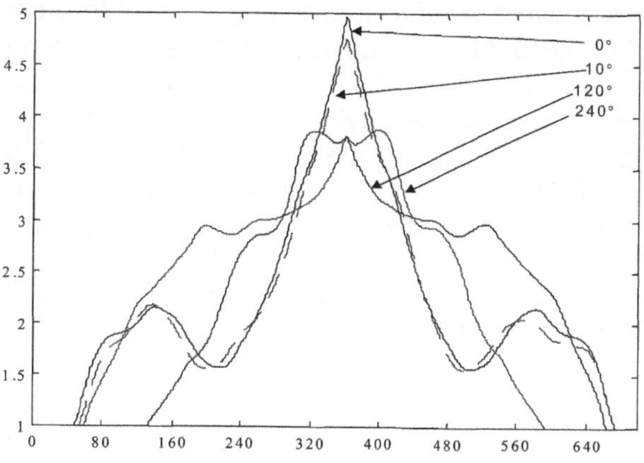

Figure 2.32: *Beam sharpening after correlation process*

2.6.4 Cellular Phone Antenna

Antenna structures that can be represented with meshes of wires are best handled via MoM as long as the frequency and number of segments are within its range of application. On the other hand, structures like cellular phone antennas, where human hand and/or head should be included, are better modeled via TD simulators, such as FDTD and TLM mentioned above. To illustrate this, typical examples are given in this section. First, a simple wire antenna is taken into account. A 30-cm-long vertical wire resonates at 1000 MHz and acts as a half-wavelength dipole. This is a reference antenna structure since its radiation characteristics and gain is known. This wire antenna is discretized in $40 \times 40 \times 70$ FDTD and TLM computation spaces with a cell size of 1 cm. The wire antenna is modeled as $1 \times 1 \times 30$ in *x,y* and *z* coordinates, respectively. The antenna is center-fed with a once-differentiated Gaussian pulse source and far fields are accumulated in time in both horizontal (when $\theta = 90°$) and vertical ($\varphi = 0°$) planes. Off-line DFT is applied to obtain radiation patterns. A typical example is given in Fig. 2.33, where very good agreement is obtained.

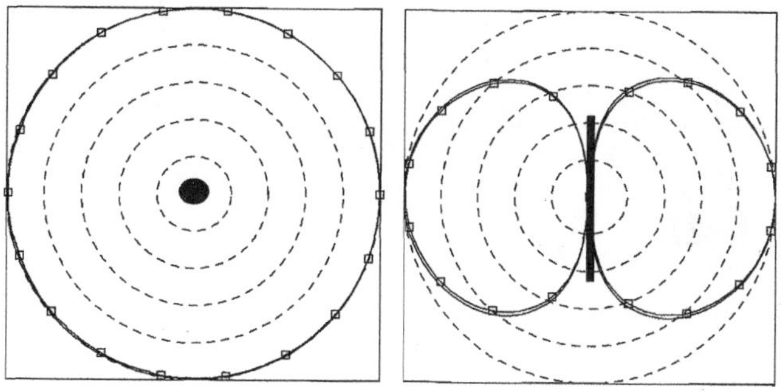

Figure 2.33: *Radiation patterns of a half wavelength dipole (left) horizontal, (right) vertical (E_θ is shown), f = 1000 MHz. Solid, FDTD; dots, TLM*

The second example where FDTD and TLM results are compared is given in Fig. 2.34. Here, a PIFA above a conducting rectangular box (representing a hand held device) is taken into account. The device's dimensions are 5 cm × 8cm × 13cm and is located inside a 50 × 60 × 100 discrete computation space. The cell sizes are $\Delta x = \Delta y = \Delta z = 0.5$ cm and the device occupies a space of 10 × 16 × 26. A 4 × 6 PIFA is excited with a Gaussian pulsed voltage source. Again near fields are simulated inside the FDTD and TLM spaces, and far fields are extrapolated at each time-step. A total of 600 time steps is found to be enough to let all transients be absorbed in the absorbing boundary blocks. In the figure, normalized vertical radiation patterns for both θ and φ components of electric fields, obtained via off-line DFT at 900 MHz, are shown.

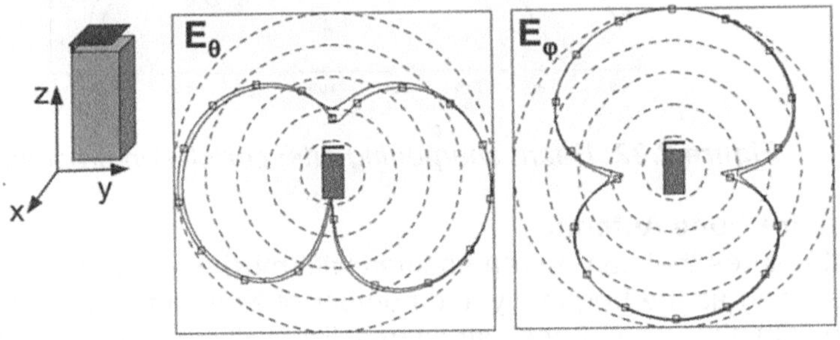

Figure 2.34: *Vertical radiation patterns of a PIFA over PEC box (left) E_θ, (right) E_φ at $\varphi = 0°$, f = 900 MHz. Solid, FDTD; dots, TLM*

The following three figures belong to the FDTD simulations, where mobile phone performance is tested with and without a human head. In Fig. 2.35, a quarter-wavelength monopole is used as the phone antenna. In Fig. 2.36 and Fig. 2.37, two types of PIFA are used [19 - 22].

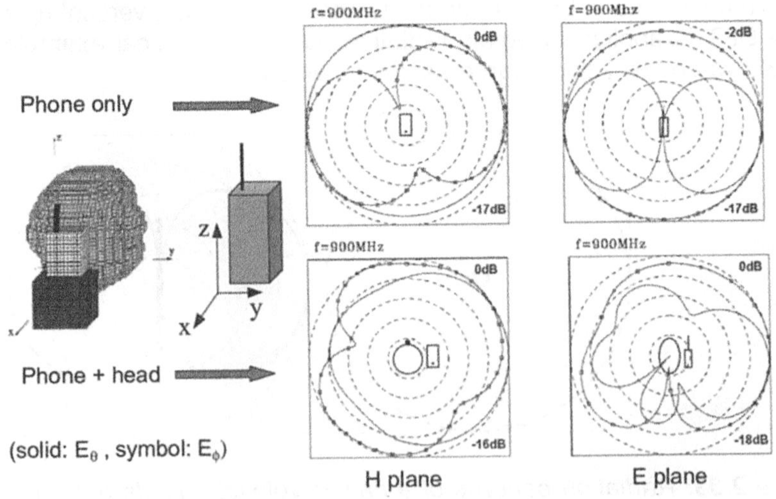

Figure 2.35: *Radiation patterns of a mobile phone (λ/4 monopole)*

In these figures, the patterns are normalized to 0 dB and are drawn with 30 dB dynamic range (i.e., each circle correspond to a 6 dB decrease). Maximum value of each pattern is also given in the figure insets.

Discrete human head model, its tissue types and number of cells used for each tissue are given in Chapter 5 in detail, therefore, neither those details nor computation specifications are included here. Examples are given to demonstrate the capabilities of the TD simulators.

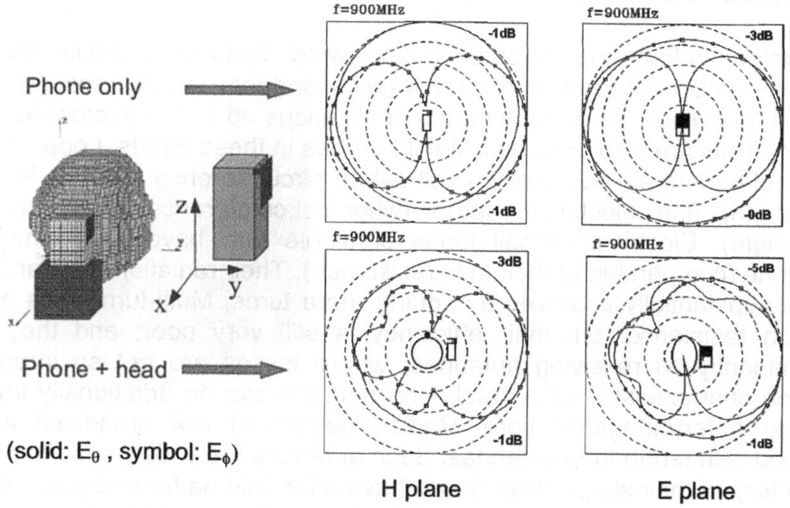

Figure 2.36: *Radiation patterns of a mobile phone*

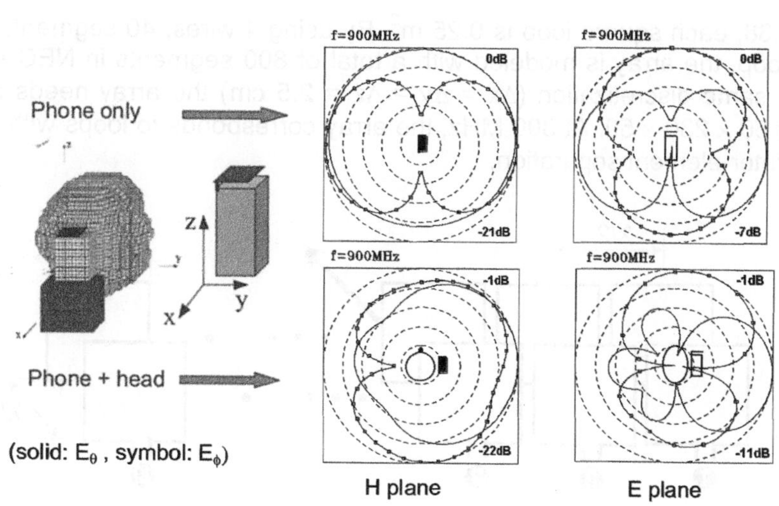

Figure 2.37: *Radiation patterns of a mobile phone*

2.6.5 MoM/FDTD Comparisons

MoM is an excellent tool in antenna engineering, when the complex radiating structure can be represented as a superposition of small segments. It may be a wire antenna, linear and/or planar array, antennas over PEC surfaces, etc. A very detailed ground screen layout is also possible with this tool. On the other hand, FDTD and TLM are also effective in antenna engineering. Having both FD and TD simulator tools also makes the comparison between them possible. To give comparisons between NEC and FDTD codes, a 2×10 square loop array is taken into account (see Fig. 2.38).

Loop antennas form another antenna type, which features simplicity, low cost, and versatility. Loop antennas can have various shapes: circular, triangular, square, elliptical, etc. They are widely used in applications up to the microwave bands (up to GHz). They are also used as EM field probes in these bands. Loop antennas are usually classified as electrically small (when circumference is much less than the wavelength) and electrically large (when circumference is in the order of wavelength). Electrically small loops of single turn have very small radiation resistance (comparable to their loss resistance). Their radiation resistance, though, can be substantially improved by adding more turns. Multi-turn loops have better radiation resistance, but their efficiency is still very poor; and they are used predominantly as receiving antennas, where losses are not so important. The radiation characteristics of a small loop antenna can be additionally improved by inserting a ferromagnetic core. Radio-receivers of AM broadcast are usually equipped with ferrite-loop antennas. Such antennas are widely used in pagers, too. Regardless of their shape, small loops have a far-field pattern very similar to that of a small dipole (normal to the plane of the loop). This should be expected because of the equivalence of a magnetic dipole and a small current loop. Of course, the field polarization is orthogonal to that of a dipole.

In Fig. 2.38, each square loop is 0.25 m^2. By using 4 wires, 40 segments for each square loop, the array is modeled with a total of 800 segments in NEC simulator. With the same discretization ($\Delta x = \Delta y = \Delta z = 2.5$ cm) the array needs an FDTD space of $50 \times 230 \times 50$. At 300 MHz, the array corresponds to loops with $\lambda/4$ sides and $\lambda/2$ inter-element separation.

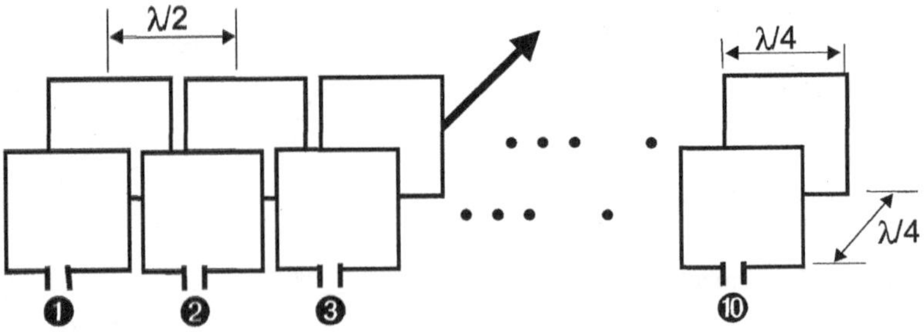

Figure 2.38: *A 2×10 square loop array*

The simulated radiation patterns and steered beams obtained via both NEC and FDTD simulators are given in Fig. 2.39. Here, vertical field components are plotted.

The top of the figure points array boresight. As shown in the figure, very good agreement between the two simulators is obtained.

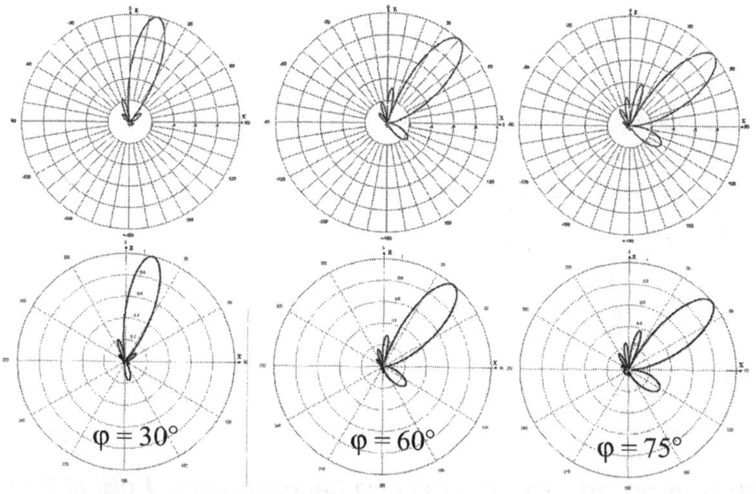

Figure 2.39: *Beam steering with the loop array NEC (top), and FDTD (bottom)*

The final structure used in MoM/FDTD comparisons is illustrated in Fig. 2.40. Here, a cellular hand-set with a quarter-wavelength monopole is taken into account.

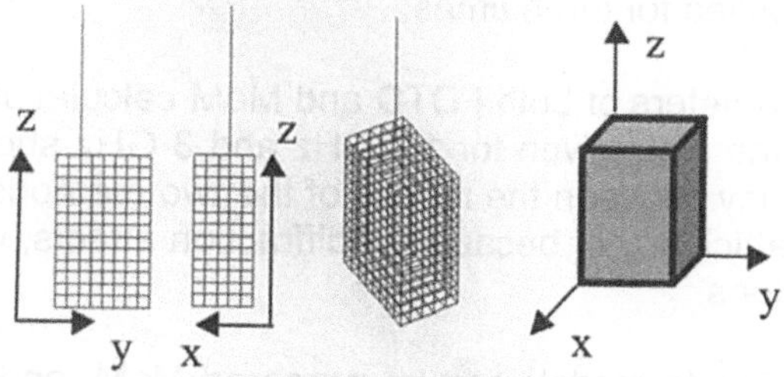

Figure 2.40: *A handset model for FDTD and NEC calculations*

Again a 5 cm × 8 cm × 13cm rectangular PEC box and 13 cm thin wire monopole are simulated both in FDTD and MoM calculations. In the NEC model, 845 segments are used with 1 × 1 square cm grids. The same structure is modeled by 10 × 16 × 26 PEC box in a 50 × 60 × 100 dimension FDTD volume with a $\Delta x = \Delta y = \Delta z$ = 0.5cm cell size. The monopole is excited with a Gaussian-type pulsed voltage source. Near fields are simulated via FDTD iterative equations and far fields are extrapolated at each time-step. A total of 600 time steps is found to be enough to let all transients be absorbed in the absorbing boundary blocks. Once the simulation is over, broad-band frequency behaviors are obtained via off-line analysis. Typical results are given in Fig. 2.41, where vertical radiation patterns are plotted at three different frequencies.

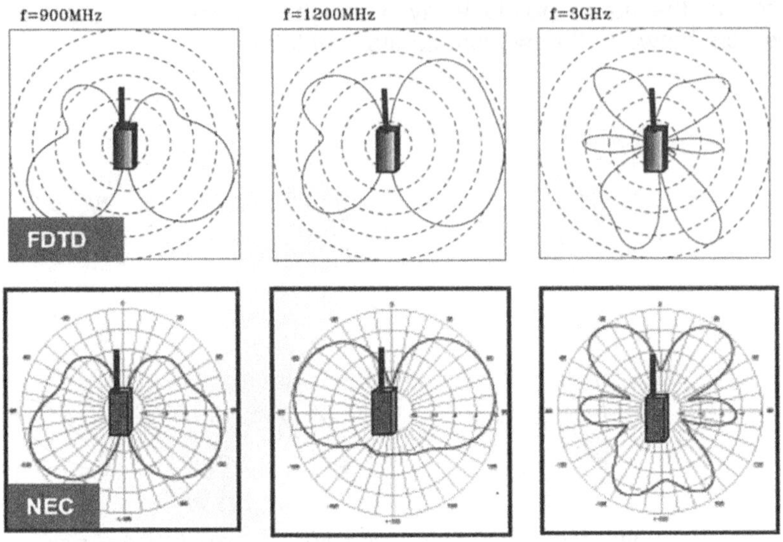

Figure 2.41: *Vertical radiation patterns of a handset (gain: 4 dBi at 900 MHz, 4.3 dBi at 1.2 GHz, 4.6 dBi at 1.2 GHz)*

Here, the 20-dB dynamic range is used with a 4 dB margins between the circles and the gains are normalized to 0 dB. To obtain these plots, an NEC package is run for the number of frequencies. But only one FDTD simulation is run, and off-line DFT is applied for three times.

Since the parameters of both FDTD and MoM calculations are optimized according to 900 MHz, the plots given for 1.2 GHz and 3 GHz should be evaluated carefully. The discrepancy between the results of the two methods may be either because of parameter inefficiency or because of diffraction effects, which cannot be handled in NEC calculations.

As long as discrete models can be prepared MoM, and FDTD simulations are run and results can be compared. They are both effective and efficient to use up to certain frequencies. For example, different HF, VHF, and UHF radiatiors and arrays can be tested on an aircraft, a ship or a military tank. One example is given in Fig. 2.42, where three radiators are located at the back of an airplane. Both NEC (wire mesh) and FDTD (staircase cubical) models can be prepared, either manually or by using powerful graphical tools. These examples (and more) are given in Chapter 3, where different RCS analysis tools are tested.

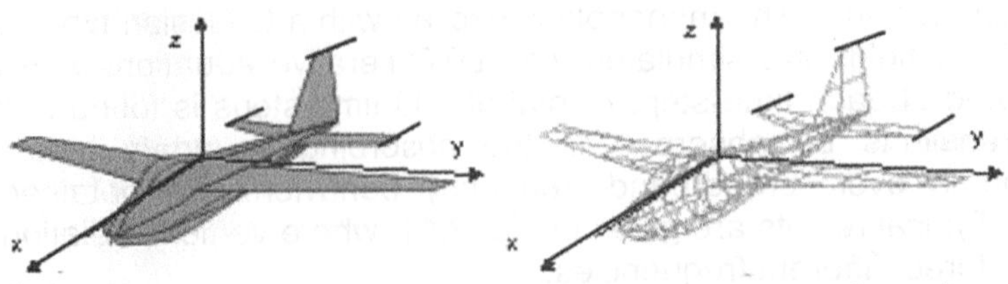

Figure 2.42: *Discrete (left)and FDTD (right) MoM models of an aircraft*

Both techniques are effectively used in antenna simulations. While FDTD is easy to use for geometries in free space, lossy ground effects and ground screens are better handled via NEC package. FDTD is better in simulating huge structures such as antenna systems over a ship or an aircraft since large amount of segments are required in NEC modeling. MoM technique is based on matrix inversion, where the size of the matrix is defined by number of segments. Computation time exponentially increases with the number of segmentations.

2.7 SAMPLE FDANT SIMULATOR

The dipole and monopole radiators are two most widely used antennas in communication systems, from very low frequencies to microwaves. For example, they are used in LF, MF, HF and VHF radio broadcasting, in VHF direction finding, in FM transmitting, in radio and TV receivers, etc. Arrays of dipoles are commonly used as base-station antennas in land-mobile systems. The monopole is perhaps the most common antenna for portable equipment, such as cellular telephones, cordless telephones, automobiles, trains, etc. It has attractive features such as simple construction, relatively broadband characteristics, and small dimensions at high frequencies.

Usually the radiation patterns of single-element antennas are relatively wide; i.e., they have relatively low directivity (gain). In long-distance communication, antennas with very high directivity are often required. This type of antenna is possible to construct by enlarging the dimensions of the radiating element (maximum size much larger than λ). This approach, however, may lead to the appearance of multiple side lobes and technologically inconvenient shapes and dimensions. Another way to increase the electrical size of an antenna is to construct it as an assembly of radiating elements in a proper electrical and geometrical configuration called *antenna array*. Usually the array elements are identical. This is not necessary but it is more practical, simple, and convenient for design and fabrication. The individual elements may be of any type (wire dipoles or loops, apertures, etc.)

The total field of an array is a vector superposition of the fields radiated by the individual elements. To provide very directive pattern, it is necessary that the partial fields (generated by the individual elements) interfere constructively in the desired direction and interfere destructively in the remaining space. Basic methods to control the overall antenna pattern may be listed as follows:

(a) Geometrical configuration of the array (linear, circular, rectangular, etc.)
(b) Relative displacement between elements
(c) Excitation amplitude of individual elements
(d) Excitation phase of each element
(e) Relative pattern of each element

In this section a sample antenna simulator is presented by which array characteristics of a 1D and 2D linear, center-fed dipole elements can be analyzed. The FDTD-based sample simulator is supplied for the reader to practice antenna simulations. First, the structure and usage of the code is given. Then, the simulator is run for different scenarios supplied with the simulator, and the results are discussed.

2.7.1 The FDANT Code

FDANT is a sample simulator based on the FDTD method. It can be used for antenna radiation pattern analysis. The structure of the FDANT is given in Fig. 2.43.

The simulator FDANT first reads the pre-selected parameters from the parameter file FDANT.PAR and the user-defined parameters from the input file ANT.INP and runs for short pulse excitation. Transient analysis is done within FDTD computation space, and time-domain output data are stored in three different files. SOURCE.dat contains time vs. source amplitude to allow the user to see the excitation signal in TD. AN-NEAR.dat has near field E_z values in two columns: time [nanosec] and E_z [V/m]. Finally, AN-FAR.dat have TD far field data along the number of chosen directions, obtained via NTFF transformation directly in TD.

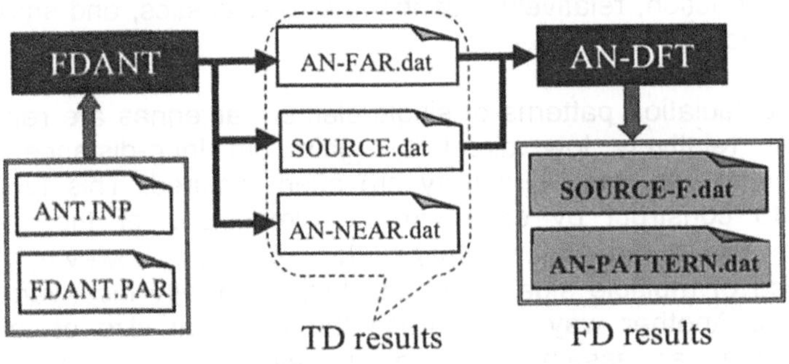

Figure 2.43: *The structure of the FDANT*

Off-line DFT analysis is required to obtain FD results. AN-DFT is prepared for this purpose. It receives TD data from AN-FAR.dat and SOURCE.dat files and yields the frequency spectrum of the source and antenna radiation pattern in output files, SOURCE-F.dat and AN-PATTERN.dat, respectively. A typical antenna array system which can be modeled in FDANT is depicted in Fig. 2.44.

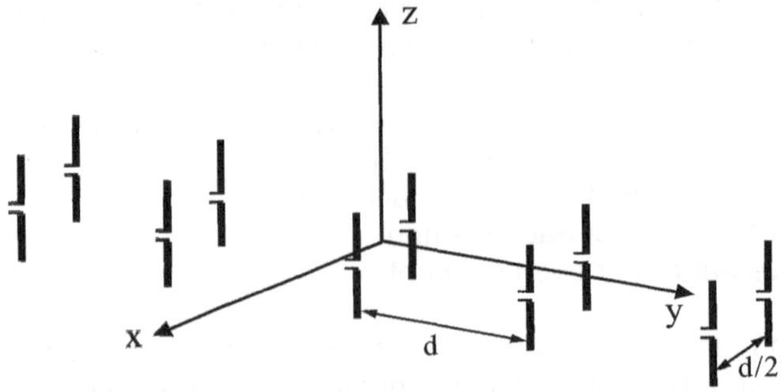

Figure 2.44: *The array structure that can be modeled in FDANT*

FDANT is prepared to investigate vertical or horizontal radiation patterns for a 2D array of maximum 2 by 6 elements. The array is located longitudinally along the *y*-direction. The elements are half-wavelength vertical center-fed dipoles. The inter-element distance along the *y*-direction is *d* and supplied by the user. The inter-element distance along the *x*-direction is automatically set as *d*/2. The lengths of the dipoles are set as *d*/4. The excitation is a short pulse of modulated Gaussian function. The modulation signal and bandwidth is determined automatically to include frequencies for the first few resonances. The user may decide which row to excite; both, front row or back row. These are all given by the user in the input file ANT.INP explained in Table 2.15.

Table 2.15: The ANT.INP file and required parameters

Row No:	Parameter
1	# of elements in X [1-2], # of elements in Y [1-6]
2	Inter-element distance along Y [m]
3	Excitation (0, Both rows; 1, front row; 2, back row)
4	Delay [sec]
5	Observation point for near fields Obsx, Obsy, Obsz
6	Pattern plane [1-0] (1, Vertical pattern; 2, horizontal pattern)
7	Angle [deg] PHI (for vertical pattern), Theta (for horizontal pattern)

In the ANT.INP file, the fourth line is reserved for time delay to allow the antenna array to steer its beam in azimuth. The calculation of this delay for an array to steer φ degrees from antenna bore-sight (from *x*-direction) is given in Fig. 2.45. Since the array is in free-space velocity of light *c* and inter-element distance *d* (supplied by the user) are used to calculate the delay τ. This is the delay of the second element with respect to the first one. The third, fourth, etc., elements have delays of 2τ, 3τ, etc., respectively.

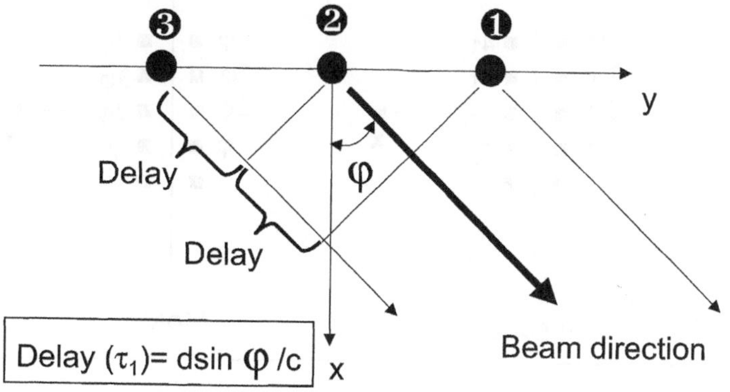

Figure 2.45: *Beam forming by giving time delays*

Line 5 in the ANT.INP file is for the near-field observation point. The user supplies three integer numbers to represent the (*x*,*y*,*z*) coordinate. The sample simulator uses 50 × 230 × 41 FDTD computation space; therefore these three integers may be any number between 3 and 47, 3 and 227, and 3 and 38, respectively. The last two lines of ANT.INP correspond to antenna pattern requirements. In line 6, the user gives either "1" or "0" for vertical or horizontal radiation patterns, respectively. If a vertical (horizontal) pattern is chosen, Phi = constant (Theta = constant) angle must be supplied in the last line.

For example, the ANT.INP file in Table 2.16 belongs to Fig. 2.44. Here, d = 60 cm ($d = \lambda$ at 500 MHz, $d = \lambda/2$ at 250 MHz, etc.), all dipoles are excited, no delay is given between elements, the observation point is (25,80,35), the horizontal pattern is chosen, and vertical angle is fixed at 90 degrees.

Table 2.16: ANT.INP input file of Fig. 2.44

Row No:	Parameter
1	2, 5
2	0.6
3	0
4	0.
5	25,80,35
6	0
7	90

2.7.2 Time-Domain Beam Steering (FDANT vs. NEC2)

The simulator FDANT is run with the parameters in Table 2.16 for different excitations, and inter-element delays and horizontal beams are plotted. The results are given in Figs. 2.47-2.51, together with MoM results computed via the NEC2 code. First, element locations and the delays given are pictured in Fig. 2.46.

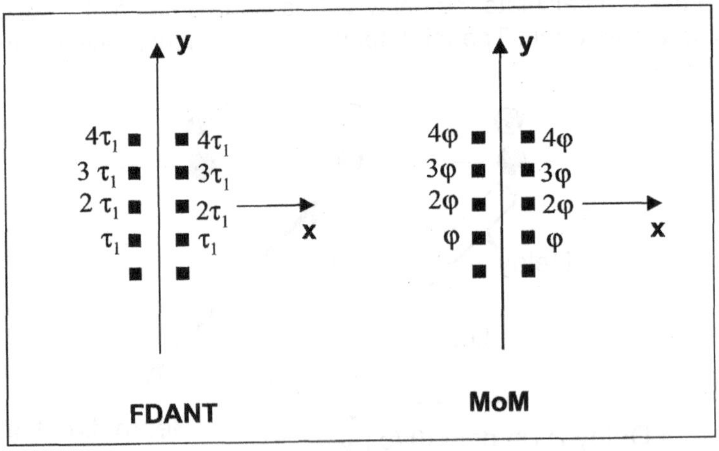

Figure 2.46: *Delays in FDANT and beam steering angles in MoM*

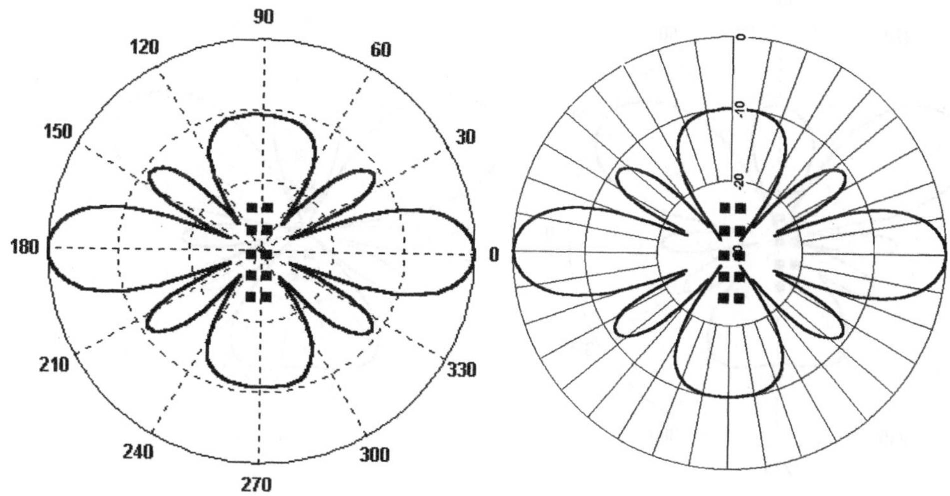

Figure 2.47: *Horizontal radiation pattern of 2×5 array given in Fig. 2.44. d = 60 cm., f = 250 MHz, zero delay. Left, FDANT; right, MoM*

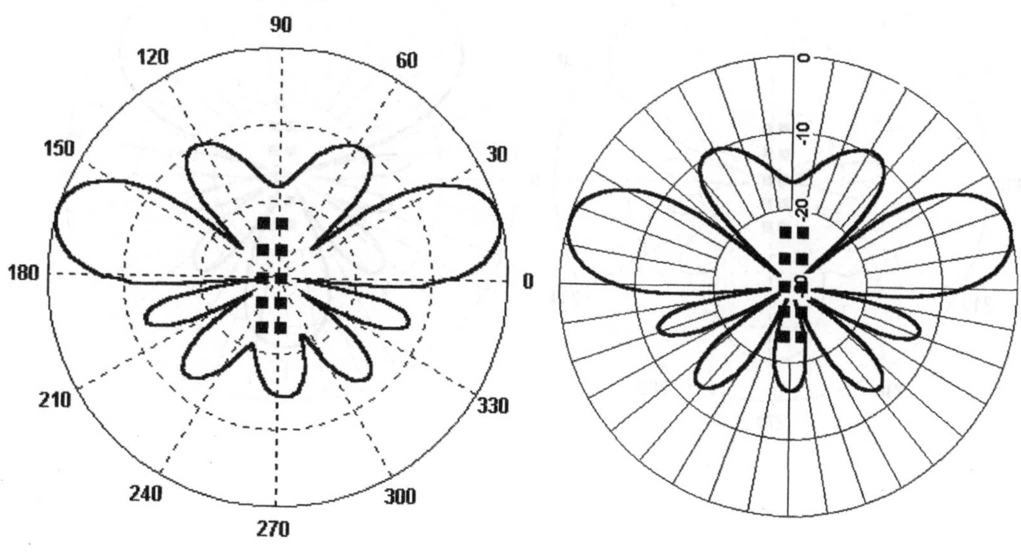

Figure 2.48: *Horizontal radiation pattern of 2×5 array given in Fig. 2.44. d = 60 cm., f = 250 MHz, 0.51764 nsec delay, $\varphi = 15°$. Left, FDANT; right, MoM*

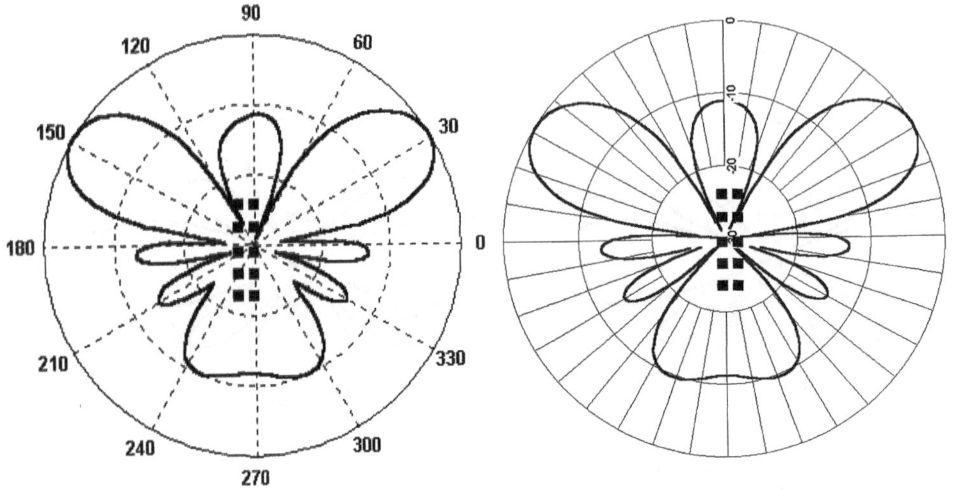

Figure 2.49: *Horizontal radiation pattern of 2×5 array given in Fig. 2.44. d = 60 cm., f = 250 MHz, 1.0 nsec delay, φ=30°. Left, FDANT; right, MoM*

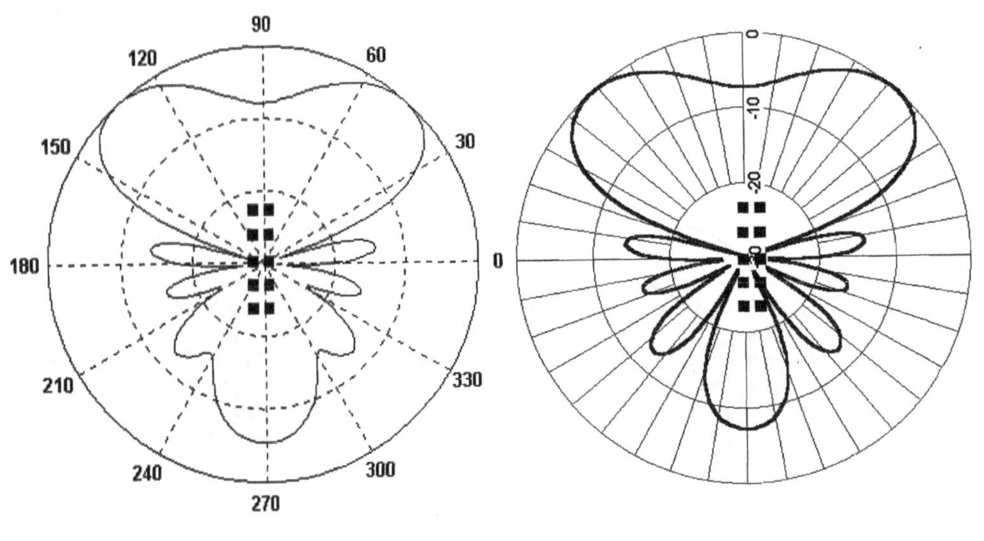

Figure 2.50: *Horizontal radiation pattern of 2×5 array given in Fig. 2.44. d = 60 cm., f = 250 MHz, 1.4142 nsec delay, φ=45°. Left, FDANT; right, MoM*

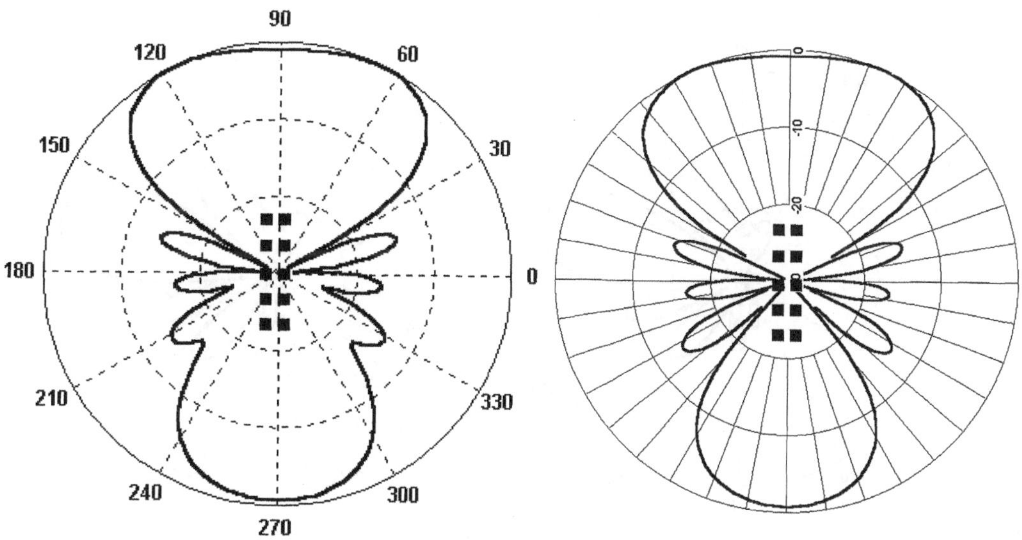

Figure 2.51: *Horizontal radiation pattern of 2×5 array given in Fig. 2.2. d = 60 cm., f = 250 MHz, 1.732 nsec delay, $\varphi = 60°$. Left, FDANT; right, MoM*

Note that the array cannot steer beams beyond φ = 30° as clearly observed in Figs. 2.50 and 2.51. The beams formed in Fig.s 2.47-2.51 are symmetric forward and backward with zero FTB ratio, since both rows have same excitation properties. To direct most of the energy forward the two corresponding elements in the *x*-direction must be an endfire property. At 250 MHz (i.e., inter-element distance along *x* is λ/4 and inter-array distance along *y* is λ/2) this may be satisfied by giving extra delay of d/2c in FDANT (90° phase lag in MoM). Figs. 2.52 shows the excitation, delays in FDANT and phase shifts in MoM. Figs. 2.53-2.56 belong to examples for this scenario with different delays.

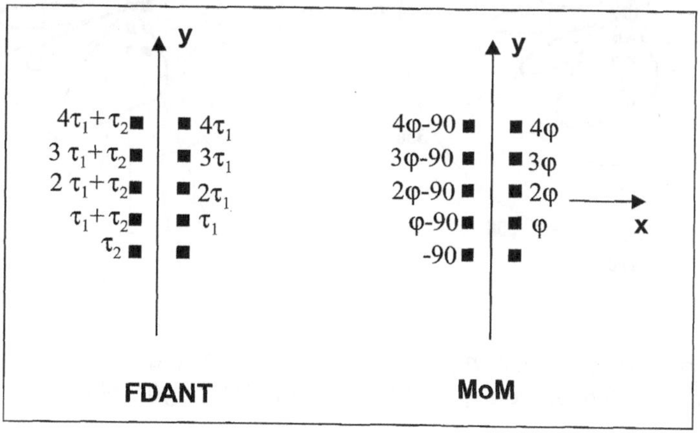

Figure 2.52: *Delays in FDANT and beam steering angles in MoM*

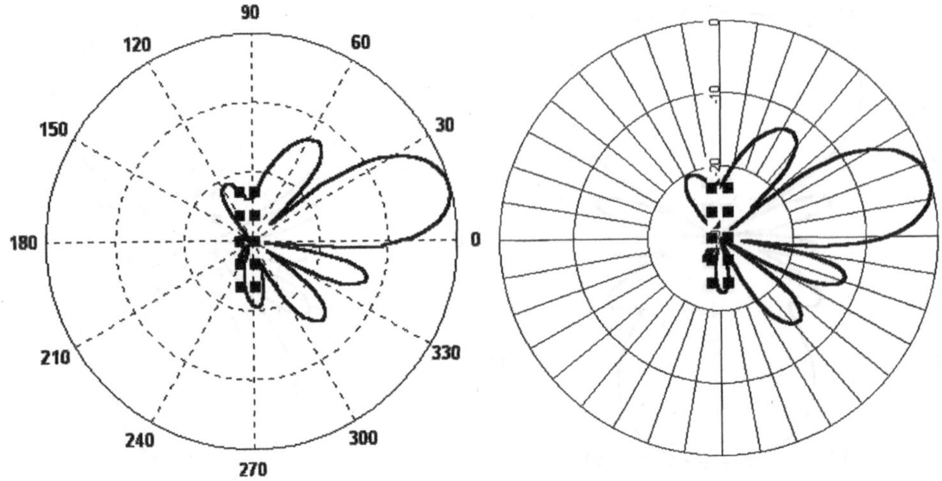

Figure 2.53: *Horizontal radiation pattern of 2×5 array given in Fig. 2.44. d = 60 cm., f = 250 MHz, 0.512764 nsec delay, φ=15°. Left, FDANT; right, MoM*

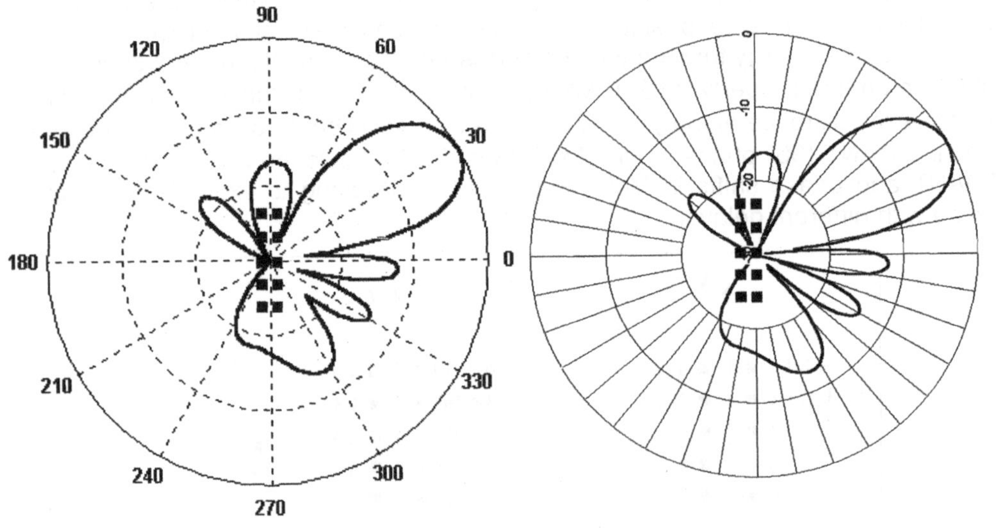

Figure 2.54: *Horizontal radiation pattern of 2×5 array given in Fig. 2.2. d = 60 cm., f = 250 MHz, 1.0 nsec delay, φ=30°. Left, FDANT; right, MoM*

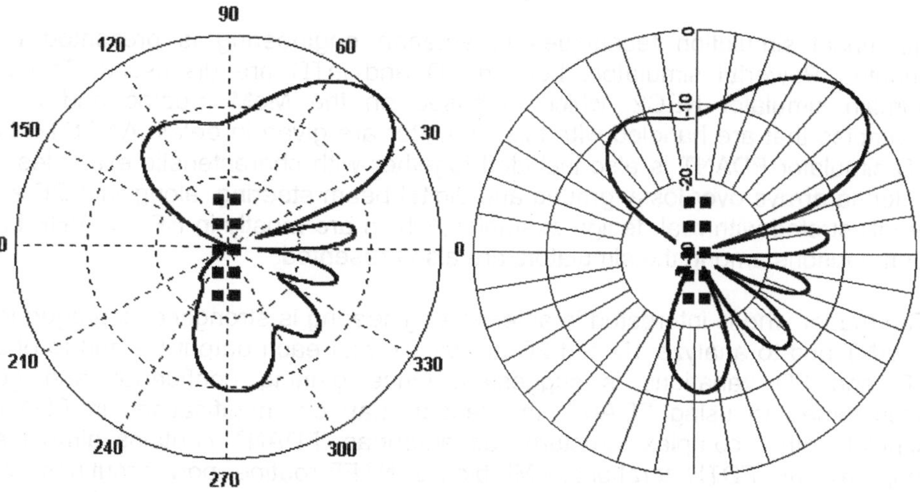

Figure 2.55: *Horizontal radiation pattern of 2×5 array given in Fig. 2.44. d = 60 cm., f = 250 MHz, 1.4142 nsec delay, φ=45°. Left, FDANT; right, MoM*

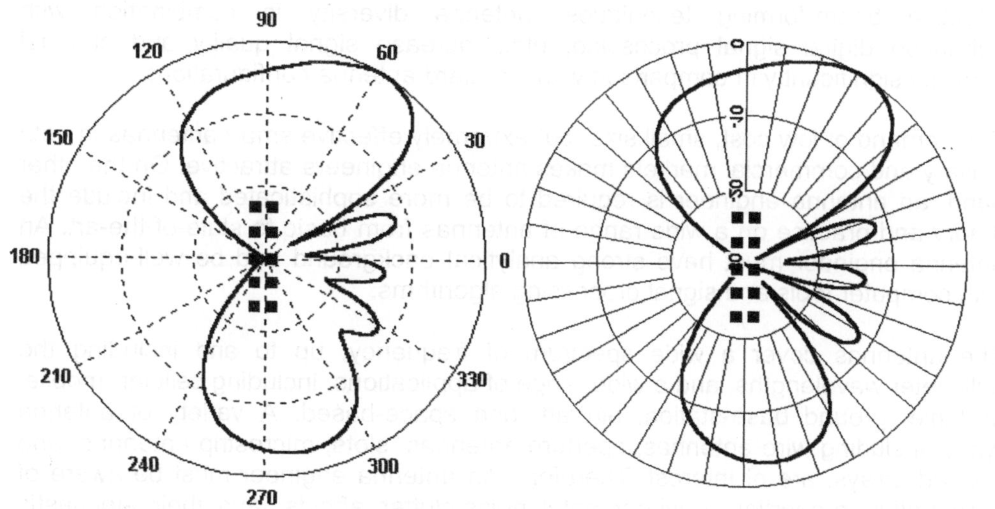

Figure 2.56: *Horizontal radiation pattern of 2×5 array given in Fig. 2.2. d = 60 cm., f = 250 MHz, 1.732 nsec delay, φ=60°. Left, FDANT; right, MoM*

It is interesting to note that although the array pattern is distorted, the endfire array may still steer horizontal beams up to 60°.

2.8 CONCLUSIONS AND DISCUSSION

Numerical simulation techniques in antenna engineering is presented in this chapter. Powerful simulators both in FD and inTD are discussed. The public domain simulator NEC2, which is based on the MoM method and complex examples that are handled with this simulator are given in detail. An FDTD-based TD simulator FDANT is also included together with characteristic examples. Wire antenna arrays over lossy ground and digital beam steering, along with DF arrays, are illustrated with real design examples, which are mostly in HF and VHF bands. Comparisons among the simulators are also presented.

The reader who is interested in antenna engineering is strongly encouraged to use FDANT and to analyze 1D and 2D arrays. To plot each output file and to observe TD and FD variations is suggested. Once gaining confidence and getting accustomed to using FDANT, the reader may do modifications in FDANT to simulate other complex structures as antennas. FDANT contains almost every major routine, FDTD iterations, PML blocks, NTFF routine, power routines, etc., to direct interested reader towards building his/her own antenna simulator, by just adding a specifically designed GUI.

Recent years have witnessed a lot of research activities focused on the topics of antennas and arrays. For example, smart (intelligent) antennas have started to be used widely to enhance the performance of cellular radio systems. Switched or adaptive beam-forming techniques, antenna diversity in combination with enhanced digital signal processing, etc., increase signal quality and channel capacity significantly in comparison with standard antenna configurations.

The demand of low cost, small size, but extremely effective smart antennas in both military and commercial markets makes antenna engineers attractive. On the other hand, an antenna engineer is required to be more sophisticated and include the theory and practice on a wide range of antennas from basic to state-of-the-art. An antenna engineer must have strong analytical background and be well-equipped with computer tools and signal processing algorithms.

The antennas cover a wide spectrum of frequency, up to and including the millimeter wavelengths, and a wide range of applications, including cellular, mobile, portable, ground base-station, aircraft, and space-based. A variety of antenna types, including wire antennas, aperture antennas, slots, microstrip antennas, and phased arrays, are of interest. Therefore, an antenna engineer must be aware of propagation properties, environmental noise/clutter effects, and their stochastic characteristics at different frequency ranges. But, first thing he/she must do is to keep in mind that time-domain/frequency-domain relations in electromagnetics is very similar to the relation between spatial and wave number domains. A simplest propagating plane wave is in the form of

$$\Phi(r,t) \propto e^{i(\omega t - kr)} \tag{2.9}$$

and, whatever characteristics $\exp(i\omega t)$ have are also the case for $\exp(ikr)$. Figure 2.57 is given for this illustration.

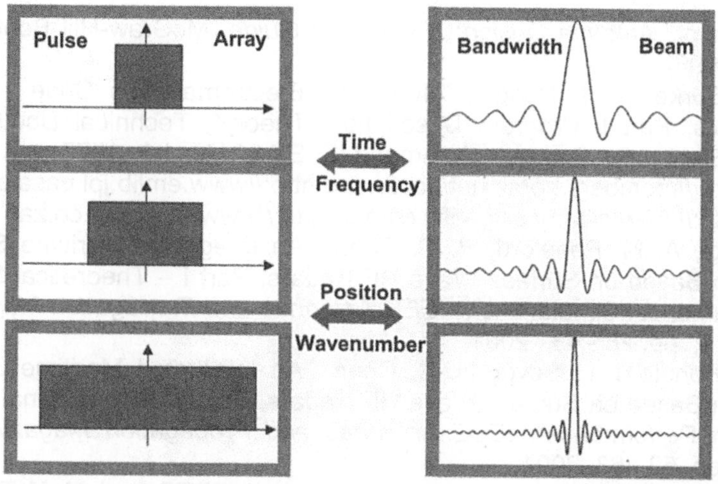

Figure 2.57: *Time-frequency and position - wavenumber relations*

Figure 2.57 illustrates that:

- A rectangular time (frequency) window corresponds to a beam-type (Sinc(.) function) variation in frequency (time) domain.
- Similarly, a rectangular aperture (array) in spatial domain corresponds to a beam-type (Sinc(.) function) variation in wavenumber domain.
- The wider the antenna aperture, the narrower the antenna beam; or, the narrower the pulse in time domain, the wider the frequency band.

References

[1] W. L. Shutzman, G. A. Thiele, *Antenna Theory and Design*, Second Edition, John Wiley & Sons, New York, 1998

[2] C. A. Balanis, *Antenna Theory*, John Wiley & Sons, New York, 1982, second edition, 1997

[3] J. D. Kraus, *Antennas*, second edition., McGraw-Hill Book Co., New York, 1988

[4] E. Jordan, K. Balmain, *Electromagnetic Waves and Radiating Systems*, Prentice-Hall, Englewood Cliffs, NJ, 1950, second edition. 1968

[5] R. E. Colin, *Antennas and Radiowave Propagation*, McGraw-Hill Book Co., New York, 1985

[6] Thomas A. Milligan, *Modern Antenna Design*, McGraw-Hill Book Co., New York, 1985

[7] R. C. Hansen, *Phased Array Antennas*, John Wiley & Sons, NY, 1997

[8] E. Brookner, *Practical Phased Array Systems*, Artech House, Norwood, MA, 1991

[9] R. P. W. King, *The Theory of Linear Antennas*, Harward University Press, Cambridge, MA, 1956

[10] R. S. Elliot, Antenna Theory and Design, Prentice-Hall, Englewood Cliffs, NJ, 1981

[11] G. Manser, *Antenna Design: A Practical Guide*, McGraw-Hill Book Co., New York, 1996

[12] G. J. Burke, A. J. Poggio, "Numerical Electromagnetic Code - Method of Moments, Part I: Program Description, Theory", Technical Document, 116, Naval Electronics System Command (ELEX 3041), July 1977

[13] Visit http://members.home.net/nec2/ and http://www.emlib.jpl.nasa.gov

[14] Visit http://www.nittany-scientific.com or http://www.poynting.co.za

[15] L. Sevgi, A. M. Ponsford, H. C. Chan, "An Integrated Maritime Surveillance System Based on Surface Wave HF Radars, Part I – Theoretical Background and Numerical Simulations", *IEEE Antennas and Propagation Magazine*, Vol. 43 No.4, pp. 28 - 43, 2001

[16] A. M. Ponsford, L. Sevgi, H. C. Chan, "An Integrated Maritime Surveillance System Based on Surface Wave HF Radars, Part II – Operational Status and System Performance", *IEEE Antennas and Propagation Magazine*, Vol. 43 No.5, pp. 52 - 63, 2001

[17] L. Sevgi, "HF Wire Antenna Array Design via FDTD and MoM Techniques", *Proc. of EEE Canadian Conference on Electrical and Computer Engineering*, pp. 7 -12, (ISBN: 0-7803-5579-2), May 9 - 12, 1999, Alberta, Canada

[18] Y. Bahadırlar, L. Sevgi, "Sensors and Signal Processing: Electronic Beam Steering and Direction Finding", *SIU2000, Proc. Of Turkish Signal Processing and Its Applications Symposium*, pp. 215 - 220, June 12 - 14, 2000, Antalya, Turkey

[19] L. Sevgi, S. Paker, "FDTD Based RCS Calculations and Antenna Simulations", *AEU, International J. of Electronics and Commun.*, Vol. 52 No.2, pp. 65 - 75, March 1998

[20] F. Akleman, M. O. Özyalçın, L. Sevgi, " Comparison of TLM and FDTD Techniques in RCS Simulation and Antenna Modelling", *Proc. Of AP 2000 (ICAP + JINA) Millenium Conference on Antennas and Propagation*, 9 - 14 April 2000, Davos, Switzerland

[21] M. N. O. Sadiku, *Numerical Techniques in Electromagnetics*, CRC Press, Boca Raton, FL, 1992

[22] F. Akleman, M. O. Özyalçın, L. Sevgi "Radar Cross Section Modeling and Stealth Target Design", *1999 Symposium on Aeronautics and Astronautics Defense Technologies*, Turkish Airforce Academy, 34807, April 19 - 20, 1999, Yesilyurt, Istanbul Turkey

3

RCS Prediction and Reduction

3.1 INTRODUCTION

Time-varying electric currents produce time-varying electric and magnetic fields in free space. These fields interact and induce time-varying electric currents in materials that they encountered. These currents ,in turn, generate electric and magnetic fields of their own, and these fields propagate in free space at the speed of light. EM wave-material interaction in radar technology is called target reflectivity and is specified in terms of radar cross section (RCS).

RCS is the area a target would have to occupy to produce the amount of reflected power (echo) that is detected back at the radar. RCS of a target is dependent upon several factors, such as

- Viewing direction
- Radar frequency
- Physical size of the target
- Geometry of the target
- Composition of the target
- Polarization of the radar signal, etc.

Radar cross section (RCS) plays an essential role in understanding EM phenomena and designing radars operating in different frequency regions. RCS modeling and reduction is crucial in designing "low visible" (stealth) weapon systems in defense technologies, such as combat aircraft and/or missiles.

Before discussing RCS, a kind of radar equation is summarized with explanations of the parameters. The received power of a radar from a target at distance *R*, where both transmitter and receiver are co-located, is

$$P_r = \frac{P_t G_t G_r \sigma \lambda^2}{(4\pi)^3 R^4} \tag{3.1}$$

where

P_r: Receive peak power,	P_t: Transmit peak power
G_t: Transmit antenna gain,	G_r: Receive antenna gain
σ: Target RCS	λ: Radar wavelength.

This is the simplest form of the radar equation and it is quite different in, for example, high-frequency surface wave radar (HFSWR) than in microwave radar

(MWR). As given in (3.1), the received power of a radar is directly proportional with the target RCS.

RCS is defined as

$$\text{RCS} = \sigma = \frac{\dfrac{\text{Power reflected to receiver}}{\text{Solid angle}}}{\dfrac{\text{Incident power density}}{4\pi}} \tag{3.2}$$

and given as [1]

$$\sigma = \lim_{R \to \infty} \left\{ 4\pi R^2 \frac{|E_s|^2}{|E_i|^2} \right\}; \quad \left(E_{i,s} = E_\theta \quad \text{or} \quad E_\varphi \right) \tag{3.3}$$

where E_s and E_i are the scattered and incident electric fields, respectively. By definition, range should be infinity, which means that far-field behaviors are of interest in practice, where plane wave representation is valid. Since both incident and scattered fields may be either E_θ or $E\varphi$, we understand that spherical coordinates must be used. Before preceding with target reflectivity and RCS reduction the definitions and physical meaning of far field, plane wave, and polarization must be well-understood.

Measurement or calculation of RCS is essentially a matter of finding the scattered electric fields from the target. RCS is expressed in terms of area. The square meter (m^2) is usually used as a reference to express RCS as a relative value using decibels (dB). For example, a 0.34-m^2 RCS corresponds to $10\text{Log}_{10}(0.34) = -4.7$ dB, or σ = 28 dB (or dBm^2 to illustrate that the value is with respect to 1m^2) is equal to $10^{28/10} = 631$ m^2.

Most of EM effects are categorized in terms of wavelength, such as near field, far field, RCS regimes, etc. A distance within a wavelength or two is generally referred to as near field. A distance of ten or more wavelengths is assumed to be far-field distance. EM interaction is very complex in the near field, where all three electric and three magnetic field components exist. Waves from a point source propagates outward with a spherical wavefront that is perpendicular to the direction of propagation. In the far field, wavefront is assumed to be planar. Therefore, when we mean far field, we mean a plane wave. A plane wave is a fictitious wave, where electric field, magnetic field, and propagation vector are all perpendicular to each other. The ratio of the magnitudes of electric and magnetic fields is the intrinsic (characteristic) impedance of the propagation medium and is equal to $377 = 120\pi$ Ω in free space.

In addition to direction of propagation EM waves are also characterized by *field direction* or *polarization.* Unless specifically stated otherwise, polarization refers to the spatial direction of the electric field vector with respect to time. If electric field is parallel to a specific plane the polarization is said to be linear. Moreover, if it is vertical (horizontal) with respect to earth's surface it is said to be a vertical (horizontal) polarization. If the tip of electric field vector traces an ellipse with respect to time, then it is *elliptic polarization*. A special case of the elliptic polarization is the *circular polarization*. RCS of a target is usually investigated for both vertical and horizontal polarization cases.

EM waves scatter when a change occur in EM boundary conditions. Some characteristic wave phenomena are listed as follows:

- *Specular reflection*: This is a mirror-like reflection, where Snell's law of reflection and refraction is valid. Lobes occur due to diffraction.
- *Diffraction*: Occur along discontinuities, where EM boundary conditions must be satisfied. Mainly, edge and tip diffractions are of interest.
- *Traveling wave*: A long thin body with near nose-on illumination may cause these waves. Along the body, EM scattering may occur due to surface discontinuity, and change in material (e.g., metal to plastic end of body).
- *Creeping wave*: Waves that propagate in the shadow region of smooth bodies are called creeping waves.
- *Ducting*: Also known as trapped (guided) waveguide modes. It occurs when a wave is trapped inside semi-open regions, such as an air inlet cavity of a jet.

RCS of a target is classified according to the types of radars; mono-static or bi-static radars. A mono-static radar is a radar whose transmitter and receiver are co-located. A bi-static radar has a transmitter and a receiver located at different sites. RCS of a target in a mono-static radar is for the back-scatter case, where angle of incidence and scatter are same (see Fig. 3.1). For a bi-static radar these angles are different.

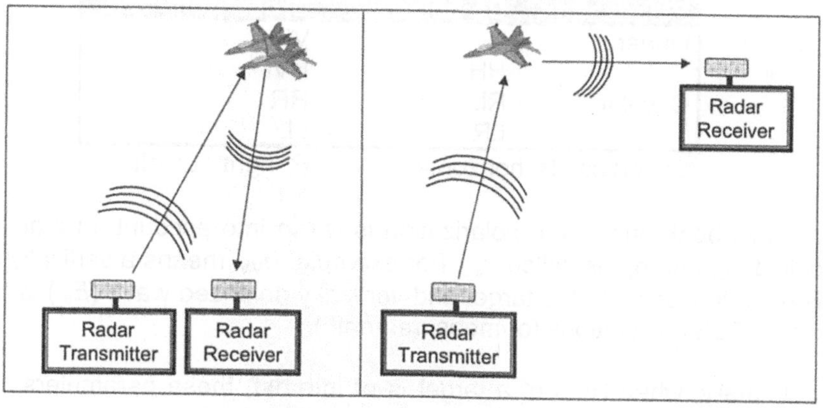

Figure 3.1: *Mono (left) and bi-static (right) RCS definitions*

The picture of RCS modeling is given in Fig. 3.2. The target is assumed to be located in the center of spherical coordinates, where both angle of illumination (incidence) and angle of observation (scatter) are determined with two angles: θ, vertical angle from z axis and φ, horizontal axis in xy plane, from x axis. These angles are defined according to scattering convention. Incidence angles are taken as angles the incident wave comes from (not the angles that the propagation vector makes with *z*- and *x*-axes). But the angles of the propagation vector are used for the scattering RCS direction. Thus, incident and scattered angles are equal if a back scatter RCS computation is the case.

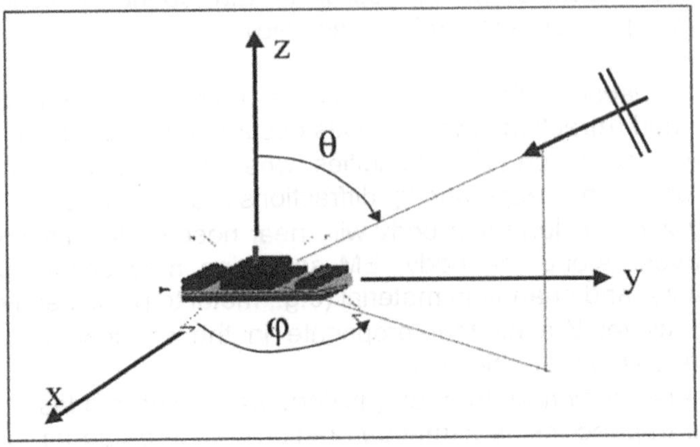

Figure 3.2: *Mono-static and bi-static RCS and definition of angles*

RCS of a target depends on the polarization of the transmitter and receiver. In most of practical cases, RCS of a target is mentioned for the four cases given in Table 3.1.

Table 3.1: Co- and cross-polarized RCS cases

Polarization	Co-polarized	Cross-polarized
Linear	VV	VH
	HH	HV
Circular	RL	RR
	LR	LL

V: vertical H: horizontal R: right L: left

Here, in this book only linear polarization is taken into account. In practice σ_{VV} is also called $\sigma_{\theta\theta}$ and σ_{HH} is called $\sigma_{\varphi\varphi}$. For example, σ_{VV} means a vertically polarized ($E_{\theta i}$) is used to illuminate the target and vertically polarized wave ($E_{\theta s}$) is taken into account in RCS calculations (or measurements).

As a summary, when RCS of a target is of interest, these parameters should be given:

- Angle of incidence (θ_i,φ_i)
- Angle of scatter (θ_s,φ_s)
- Incident field polarization $(E_{\theta i}, E_{\varphi i})$
- Scattered field polarization $(E_{\theta s}, E_{\varphi s})$
- Frequency and target geometry (usually as maximum length vs. wavelength relation)
- RCS value (in m^2 or dBm^2)

Some RCS values of radars are given in Fig. 3.3. As indicated, typical values of RCS range from -30 dB for insects to 40 - 45 dB for large tankers and bombers. Current radars are capable of detecting swarms of insects and flocks of birds, which means a radar computer has to eliminate them on the basis of their velocity or trajectory or RCS amplitude. This makes RCS design and/or prediction problem complex.

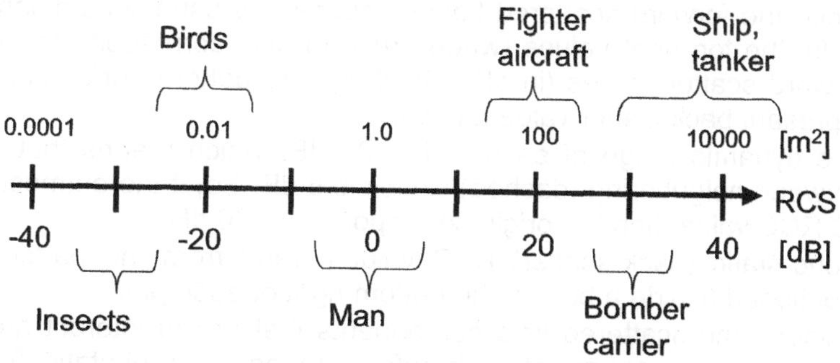

Figure 3.3: *Typical RCS values for some radar targets*

Plot Types

Different plots are presented in this chapter and three types of variations are given. They are:

1. *Back- and/or forward-scatter RCS vs. frequency*: The target is illuminated from a given direction and back-scatter or forward-scatter far fields are extrapolated. If the scattering angles are chosen same (different) with the angles of incidence back-scatter (forward-scatter), RCS is obtained. In FDTD and TLM, this is simulated directly in time domain and pulsed plane wave is used. Therefore a broadband RCS signature is obtained via a single simulation. Then, back-scatter or forward-scatter RCS vs. frequency plots are obtained via off-line DFT analysis. In MoM, the procedure must be repeated for a number of desired frequencies.
2. *Bi-static RCS patterns on vertical or horizontal planes*: These are obtained in a similar way in all (FDTD, TLM, and MoM) techniques. The target is illuminated from a given direction, and scattered far fields are obtained at vertical or horizontal planes. Far scattered fields are extrapolated in every 2°, which means that the extrapolation is carried out along 181 directions (for example, from $\varphi = 0°$ to $\varphi = 360°$ with $\Delta\varphi = 2°$ angular resolution, when θ_s is fixed at 90°). Each bi-static RCS pattern also contains back-scatter RCS information.
3. *Mono-static RCS patterns on vertical or horizontal planes*: Broad band TD simulations are run for back-scatter RCS (as in item 1) for each angle and TD far fields are extrapolated. Then, multi-frequency back-scatter RCS patterns can be obtained via off-line DFT.

Plot Readings

In all cases presented here, the polarization of EM waves are mentioned separately and other necessary explanations are arranged as follows:

- Each polar pattern is normalized to its maximum, and the maximum value is mentioned on top right of the plot. That is, the numbers on the top right of each plot is the value of largest circle.
- For bi-static patterns, angle of illumination is shown with an arrow. Two parallel thin lines show the equi-phase surface of the incident plane wave. Polar plots without arrows correspond to mono-static RCS variations.
- Since the forward-scattered field component is usually dominant in bi-static plots, the top right values (when present) also correspond to the maximum forward scatter values (in dB). Similarly, the bottom right values (if written) represent backscatter value (in dB).
- The dynamic range of each plot is 30 dB, which means that the distance between neighboring (dashed) circles is 6 dB. So, if the outmost circle has 0 dB RCS value then the origin corresponds to -30 dB.
- Mono-static (back-scatter) RCS values (read from the patterns) are also mentioned (inside a box on the bottom right of each plot).
- Incident and scattered field components that are of interest are mentioned for each case separately. Therefore, mono- and bi-static RCS patterns correspond to top and side views of the target RCS behavior.

3.2 TARGET REFLECTIVITY AND RCS REDUCTION

RCS of an object under a radar illumination is a fictitious area (in square meters, m^2) that describes electromagnetic reflectivity back to the radar [1]. RCS is the signature of the target under investigation, which is embedded in the signal environment that is externally limited either by noise/clutter or by other targets. This signature is an information due to the interaction between EM waves and target structure. EM wave - target interaction is classified in terms of structure dimensions (l) and radar wavelength (λ) relation. The three RCS frequency regions where qualitative as well as quantitative differences occur are [2];

- Low frequencies (*Rayleigh region*) where target dimensions (l) are much less than the radar wavelength ($l << \lambda$). In this region, RCS is proportional to the fourth power of frequency ($\sigma \approx f^4$).

- Medium frequencies (*resonance region*) where target dimensions and the radar wavelength are in the same order ($l \approx \lambda$).

- High frequencies (*optical region*) where target dimensions are very large compared to the radar wavelength ($l >> \lambda$). In this region, RCS is roughly the same size as the real area of target.

A PEC sphere is spherically symmetric: therefore it is a canonical test target in RCS modeling, where three characteristic RCS regimes are clearly illustrated. Frequency variation of RCS of a sphere is pictured in Fig. 3.4. As given in the figure, the RCS curve is almost linear for $ka = 0.5$ (where $k = 2\pi/\lambda$ is the free-space wavenumber and a is the radius), but oscillates for $0.5 < ka < 10$. The oscillations die out above ka = 10, and the curve is essentially a constant equal to πa^2 when $k \to \infty$.

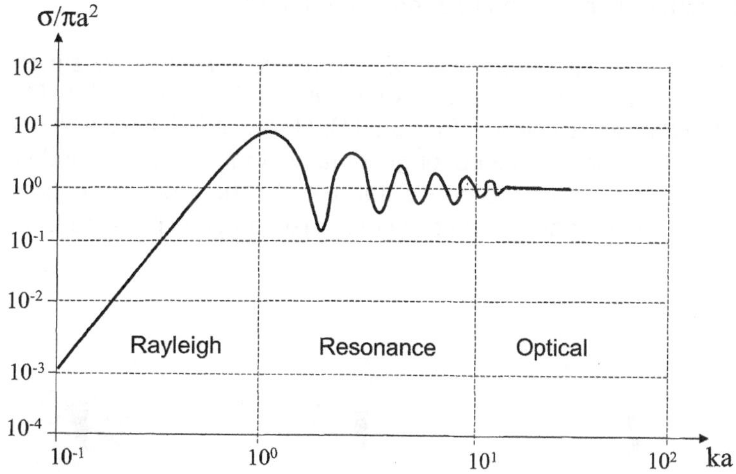

Figure 3.4: *RCS vs. frequency of a sphere with radius a*

Although RCS vs. frequency plots of realistic targets are not as typical as the PEC sphere presented in Fig. 3.4, three typical RCS regimes can still be observed. To illustrate this, a typical RCS vs. frequency plot is given in Fig. 3.5, where a simple prototype of a navy frigate (with 45 m length) is used as a target. An FDTD-based RCS simulator is used to simulate this target [3].

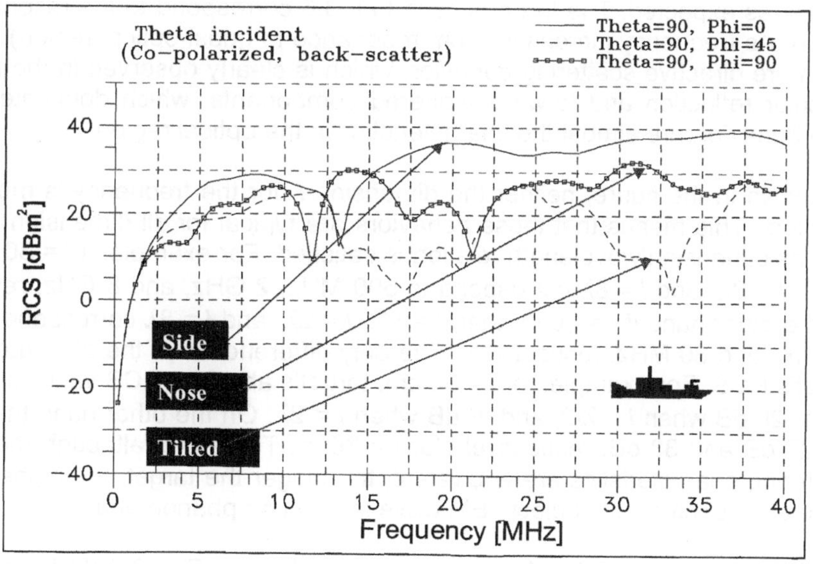

Figure 3.5: *RCS vs. frequency of a navy frigate (l=45 m)*

The frigate is located longitudinally along the *y*-axis on the horizontal (*xy*) plane and the plot corresponds to three theta incidence, co-polarized back-scatter case for three different illumination mentioned in the figure.

To better understand RCS behavior of a realistic target, first, simple canonical objects must be investigated in detail. A typical simple target and its RCS behavior at different frequencies is given in Fig. 3.6 [4]. Here, a PEC rectangular prism with dimensions of $l/3 \times l \times l/3$ is located longitudinally along the *y* axis and is illuminated from $\theta = 90°$ and $\varphi = 45°$ (in the horizontal plane). The plots correspond to σ_{VV} (or $\sigma_{\theta\theta}$), that is, both incident and scattered fields are vertically polarized.

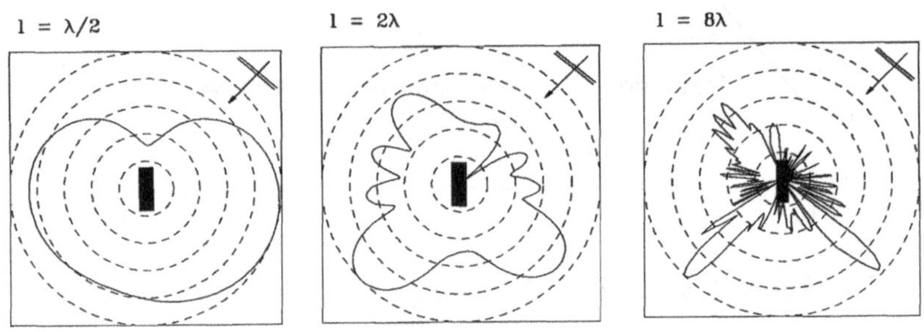

Figure 3.6: *Bi-static RCS vs. angle of a rectangular PEC prism*

Plots of lower frequencies (such as, $l = \lambda/10$ or $l = \lambda/5$ cases) are not included, since they are totally in the Rayleigh RCS region (where the target looks like a point source) and isotropic characteristics are observed with almost negligible RCS values, as expected. The three plots in Fig. 3.6 correspond to the resonance RCS region. As frequency increases (from resonance to quasi-optical region), the target has more directive scattering behavior, which is clearly observed in the plots. Two specular reflection and forward-scattered components, which dominate the polar RCS behavior, are almost the characteristics in the optical region.

As shown in the figure, neither the dimension *l* nor the frequency is mentioned in the plots. This means that these behaviors are typical for all dimension values, as long as dimension/wavelength relation is satisfied. For example, if $l = 30$ cm, then $l = \lambda/2$, $l = 2\lambda$, and $l = 8\lambda$ correspond to 500 MHz, 2 GHz, and 8 GHz, respectively. On the other hand if $l = 30$ m, then $l = \lambda/2$, $l = 2\lambda$, and $l = 8\lambda$ correspond to 5 MHz, 20 MHz and 80 MHz, respectively. The only difference is in the absolute values of target's RCS. For example, for $l = 30$ cm, target's absolute RCS value is calculated to be -20 dB when $l = \lambda/2$, and -9 dB when $l = 2\lambda$. On the other hand these values are 27 dB and 38 dB, respectively for $l = 30$ m. Therefore, although their bi-static scattering characteristics are equivalent, the bigger the target, the higher the RCS value as expected from object / EM wave interaction phenomena.

The last sample for RCS characteristics are given in Fig. 3.7 [5] for a triangular PEC prism. The top two plots are mono-static (back scatter) polar RCS behaviors, while the bottom two correspond to bi-static RCS polar behaviors for the illumination mentioned at two different frequencies. The maximum values are also mentioned in the plots. From these figures, the conclusions given below may be derived as general characteristics:

- Generally, specularly reflected and/or forward scattered components dominate RCS (with respect to secondary components, such as creeping waves, edge- and tip-diffracted waves, surface waves, etc.).
- RCS varies with frequency, polarization, angle of incidence, etc. In terms of frequency, three RCS regimes are important as mentioned above. In both Rayleigh and optical regions, RCS varies smoothly with changing wavelength. In the resonance region, RCS varies very rapidly with changing wavelength. On the other hand, the situation changes for changing aspect angle. That is, in both Rayleigh and resonance regions RCS varies smoothly with changing aspect angle, but RCS varies very rapidly with a slight change in aspect angle. For example, an X-band radar measured RCS of an aircraft is plotted in Fig. 3.8, where a 1-degree change in aspect angle results in a high RCS fluctuation. That is why statistical polar characteristics are used in the optical RCS region.

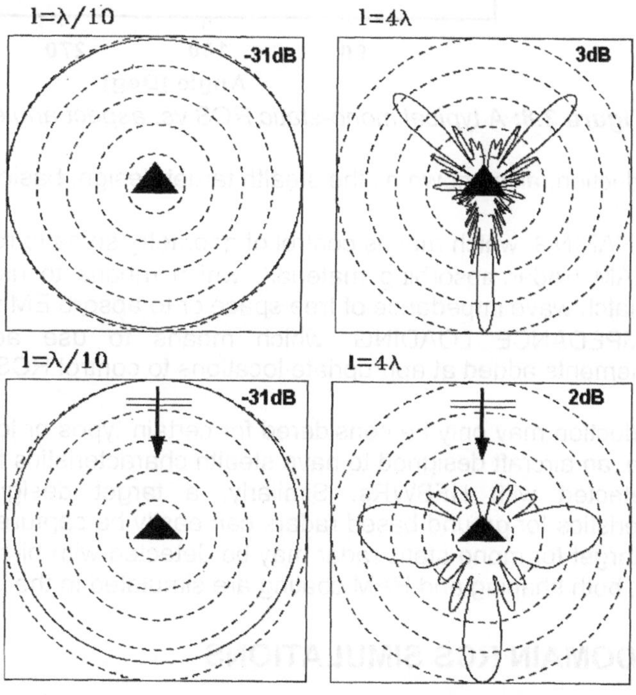

***Figure 3.7:** Mono (top) and bi (bottom) static polar RCS behaviors*

Radar wavelength of an X-band (i.e., operating in 8 - 12 GHz) radar is on the order of centimeters and it is the optical RCS regime, when aircraft or a ship is of interest as a target. Local characteristics dominate RCS behavior in this regime and well-known analytical approximate techniques, such as Geometric Optics (GO), Physical Optics (PO), Geometrical Theory of Diffraction (GTD), Physical Theory of

Diffraction (PTD), Uniform Theory of Diffraction (UTD), etc., are used in RCS modeling. On the other hand, radar wavelength is in between 50 m and 100 m for an HFSWR that usually operates in a 3 to 6 MHz band. The RCS region of interest is, therefore, *resonance regime*, where the whole structure contributes to target RCS. It is extremely difficult to apply these well-known analytical approximate techniques in RCS modeling. Except a few simple canonical structures, only pure numerical techniques can be used in RCS modeling in *resonance* regime [3 - 6].

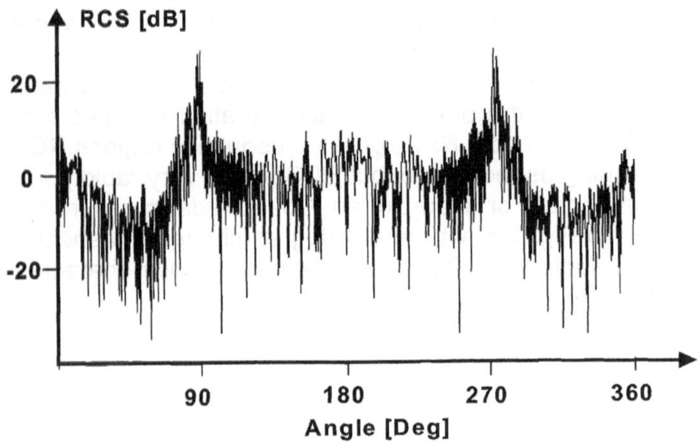

Figure 3.8: *A typical mono-static RCS vs. aspect angle in optical regime*

RCS reduction, also known as the stealth target design, basically relies on

- SHAPING, which means control of geometry so as to reduce RCS
- RAM (radar absorbing material), which means to use special material to match wave impedance of free space or to absorb EM wave energy
- IMPEDANCE LOADING, which means to use active and/or passive elements added at appropriate locations to control RCS

RCS reduction may only be considered for certain types or locations of radars. For example, an aircraft designed to have stealth characteristics for MW radars may be well detected with HFSWRs. Similarly, a target designed to have stealth characteristics for ground based radars can easily be captured by an aircraft radar; stealth target for mono-static radar may be detected with bi-static radar, etc. In this chapter, both shaping and RAM coating are simulated in the presented examples.

3.3 TIME-DOMAIN RCS SIMULATIONS

The RCS package is based on three important modules: the FDTD [10] or (SCN) TLM [12] module, where near scattered fields of a target are simulated; the NTFF transformation module, where far fields are extrapolated by accumulating the far zone TD vector potential due to the tangential electric and magnetic fields on a virtual, closed surface surrounding the target; and RCS-DFT, where off-line DFT analysis is performed together with Cartesian-polar transformations.

With today's personal computers, the TD (FDTD or TLM) computation volume may typically be chosen as 100 × 100 × 100, which means that a total of 1 million cells can be used. If cell sizes are chosen such that minimum wavelength component (highest frequency) is sampled with 10 samples (which is quite acceptable for numerical dispersion) the computation volume will then extend from the origin to 10λ × 10λ × 10λ in each axis. It was demonstrated that a space of 8 to 12 cells may be enough between the target and ABCs in the computation volume. Therefore, this space and ABC blocks occupy nearly 3λ/4λ of the 10λ computation volume at each coordinate axis, leaving a 6λ/7λ space for discrete target modeling (see Fig. 3.9). As a result, with these typical parameters, RCS analysis can be done up to frequencies where $\ell \approx 6\lambda$ is satisfied. Depending on the geometry, number of total cells in each axis ($N_x \times N_y \times N_z$) may differently be chosen to allow modeling higher frequencies. For example, halving the numbers in two axes enlarges the third axis four times, that is, 1 million cell FDTD volume may be built as 50 × 50 × 400. In that case, RCS analysis may be carried out up to $\ell \approx 24\lambda/28\lambda$, which is well beyond the resonance and quasi-optical regimes. Note that because of the more memory need in TLM method, it is not always possible to choose the volume and test structure as large as in FDTD method.

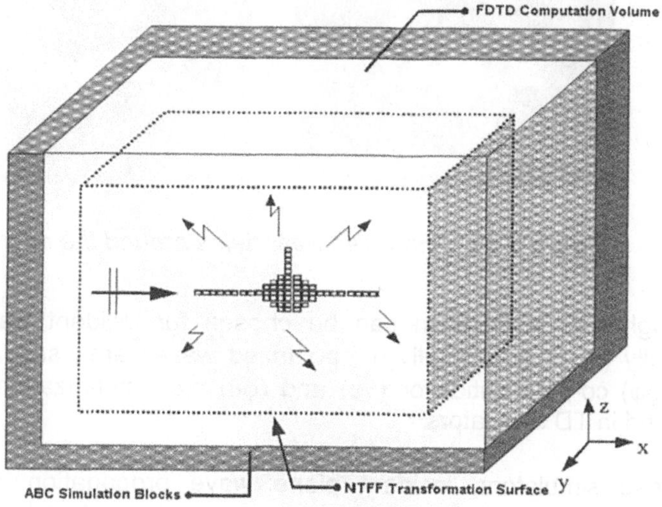

Figure 3.9: *FDTD and TLM computation volume, ABC blocks*

The simulation steps in FDTD- and TLM-based simulators are as follows:

- The target, located in computation space, is illuminated by a Gaussian (wide-band) plane wave from a given direction. The angle of incidence is defined by a horizontal angle φ (angle from *x*-axis) and a vertical angle θ (angle from *z*-axis). These angles are defined according to scattering convention. Incidence angles are taken as angles the incident wave comes from (not the angles that the propagation vector makes with *z*- and *x*-axes). But the angles of the propagation vector are used for the scattering RCS direction. Thus, incident and scattered angles are equal if back-scatter RCS computation is the case.

- Once the target is illuminated with this pulsed plane wave, the simulation starts and near scattered fields are simulated (for example, see Fig. 3.10) inside the computation volume until $t = T_{max}$. T_{max} must be long enough to let field values at each FDTD and TLM cell approach zero (all transients must dissipate out the computation volume).

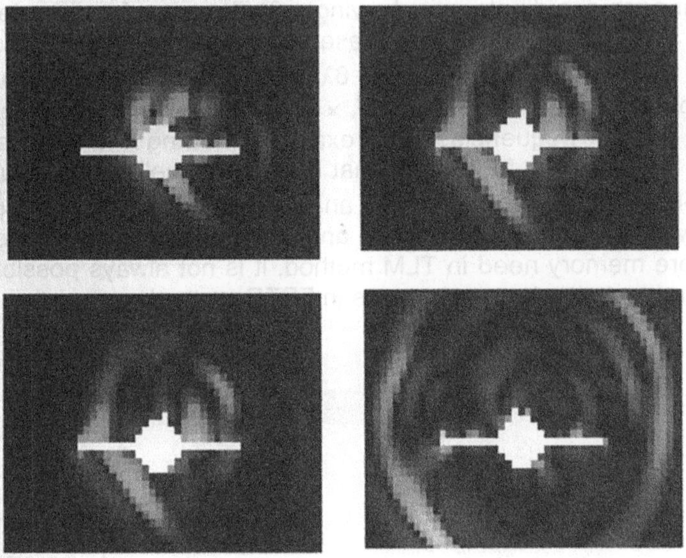

***Figure 3.10** : Simulated near fields around the aircraft*

- Although any polarization can be chosen for incident wave, in practice vertically (θ) or horizontally (φ) polarized waves are used. Therefore, (θθ) and (φφ) co-polarization or (θφ) and (φθ) cross-polarization cases can be explored in TD simulators.

- In these simulators, incident plane wave propagation is implemented analytically to reduce numerical errors, and scattered near fields are simulated directly in TD within computation volume.

- TD far fields for the desired directions are extrapolated by using equivalent currents via NTFF routine, over a closed, virtual surface surrounding the target (see for example, Fig. 3.11).

- While time history of the field components are simulated in Cartesian coordinates within the volume (i.e., E_x, E_y, E_z, H_x, H_y and H_z are used), two far field components (E_θ and E_φ) are extrapolated in spherical coordinates.

- During the simulation, scattered fields for desired directions, $E_{\theta s}(t)$ and $E_{\varphi s}(t)$, have been stored.
- Besides the simulated $E_{\theta s}(t)$ and $E_{\varphi s}(t)$, analytically calculated $E_{\theta i}(t)$ and $E_{\varphi i}(t)$ are also stored (actually, E_{xi}, E_{yi} and E_{zi} may be enough to use (3.3) since the amplitudes are used by definition).
- After the simulation, for either vertical or horizontal planes, co-polarized or cross-polarized RCS behaviors are computed via off-line DFT routine for a given frequency range.

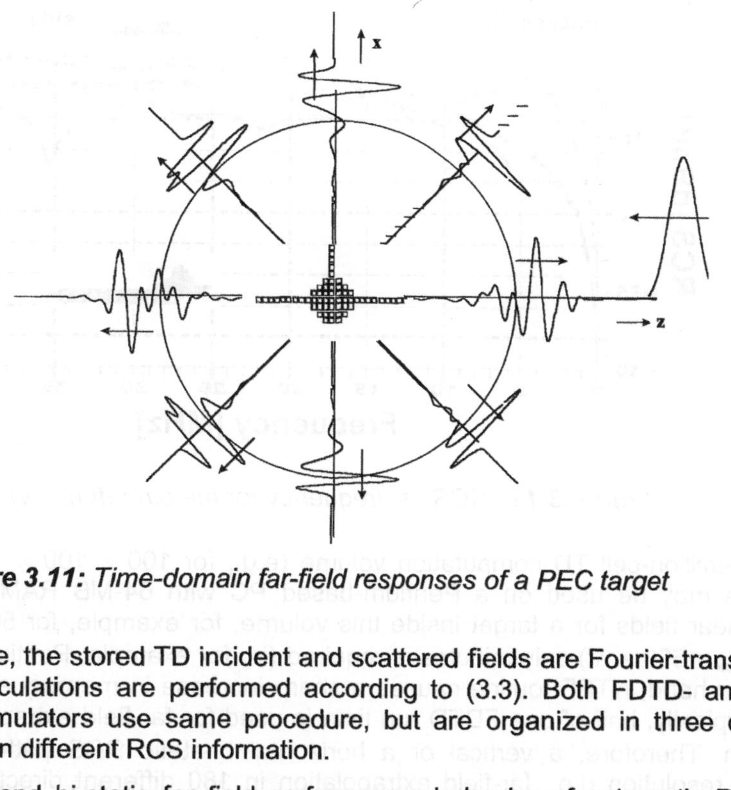

***Figure 3.11:** Time-domain far-field responses of a PEC target*

In DFT routine, the stored TD incident and scattered fields are Fourier-transformed and RCS calculations are performed according to (3.3). Both FDTD- and TLM-based TD simulators use same procedure, but are organized in three different forms to obtain different RCS information.

1. Broadband bi-static far field or frequency behavior of a target's RCS are computed along a single direction via **SN-RCS** packages. The program initializes with the given illumination direction (θ_i and φ_i) and the polarization of the incident wave. Then, scattered field direction (θ_s and φ_s) is given. If scattering and incidence angles are same (different), mono-static (bi-static) RCS results are computed.

2. Bi-static RCS behavior of the target for a chosen plane and for co- or cross-polarization case is computed with **BI-RCS** packages. First, direction of incidence (θ_i and φ_i angles) is given together with the polarization. Then, the plane (vertical or horizontal) of RCS pattern and the angular increments ($\Delta\theta$ and $\Delta\varphi$) are decided.

3. Finally, mono-static RCS patterns on vertical or horizontal planes can be simulated with **MN-RCS** packages. The simulators are run for each angle separately, and far-field TD values are calculated. The plane (vertical or horizontal) of RCS pattern and the angular increments ($\Delta\theta$ and $\Delta\varphi$) are given and patterns are computed.

An example output of the SN-RCS package is given in Fig. 3.12, where RCS vs. frequency is plotted for backscatter RCS case.

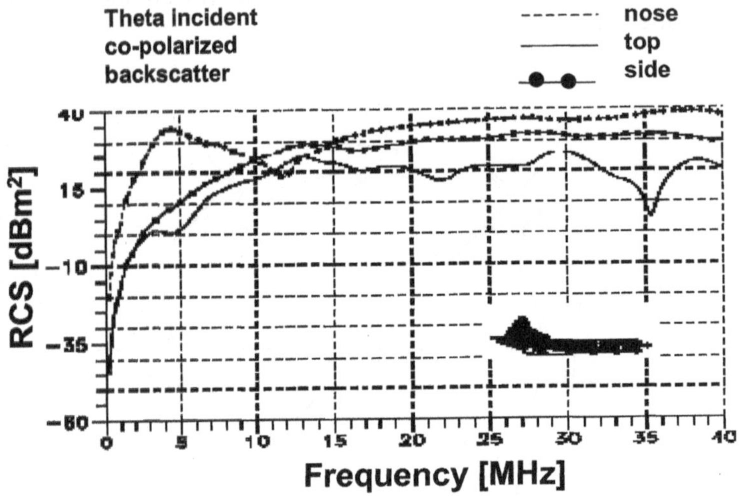

***Figure 3.12** : RCS vs. frequency for the aircraft model*

For a 1-million-cell TD computation volume (e.g., for 100 × 100 × 100) the RCS package may be used on a Pentium-based PC with 64-MB RAM memory. To obtain near fields for a target inside this volume, for example, for 500 time steps (i.e., $T_{max} = 500 \times \Delta t$) only minutes is required (in, for example, Pentium-III PC). On the other hand, NTFF routine causes a sharp increase in memory and simulation time. Typically, half of one FDTD run time is used for far-field extrapolation in one direction. Therefore, a vertical or a horizontal bi-static RCS pattern with a 2° angular resolution (i.e., far-field extrapolation in 180 different directions) may be obtained in 91 minutes (1.5 hours). Although the required memory and computation time is so intensive, the information which is obtained via this package is very valuable. The same calculations last longer with the TLM method. It has been shown that 128-MB RAM memory gives good results in TLM simulations.

3.4 SAMPLE RCS SIMULATORS

Three sample RCS packages are presented in this section. The first two are F-SNRCS and T-SNRCS packages, where mono- and bi-static RCS vs. frequency plots can be obtained. The third one is the F-BIRCS package, in which RCS patterns on vertical or horizontal packages can be simulated.

3.4.1 F-SNRCS Package

F-SNRCS is an FDTD-based sample simulator that can be used for mono- or bi-static RCS calculations of targets for a chosen angle of illumination and polarization, along a given direction of scattering. The structure of the F-SNRCS is given in Fig. 3.13. The simulator uses scattered field FDTD iterations in the main routine. An incident plane wave pulsed field is supplied analytically from a given direction [12]. The computation space is terminated by second-order MUR-type absorbing boundary blocks to simulate free space. NTFF routine is used for TD far field accumulation.

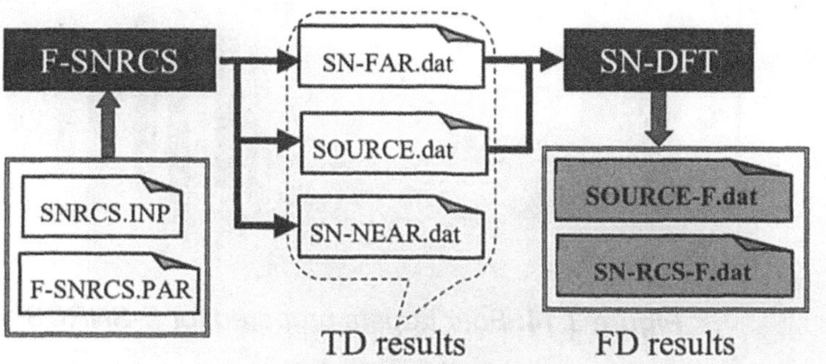

Figure 3.13: *The structure of the F-SNRCS*

The simulator F-SNRCS first reads the pre-selected parameters from the parameter file F-SNRCS.PAR, reads the user-defined parameters from the input file SNRCS.INP, and runs TD simulation for a short (once-differentiated) Gaussian pulse excitation. The parameters in input file SNRCS.INP are given in Table 3.2.

Table 3.2: The SN-RCS.INP file and required parameters

Row No:	Parameter
1	Type of the target [1-4]
2	Longest size of the target [m]
3	Observation point for near fields Obsx, Obsy, Obsz (three integer numbers)
4	Angle of incidence (θ_{inc}, φ_{inc})
5	Polarization (1: θ-pol, 0: φ-pol)
6	Angle of scattering (θ_{scat}, φ_{scat})

In SNRCS.INP file, line 1 is reserved for pre-prepared PEC targets, which are shown in Fig. 3.14. The longest size (length) of the chosen target is given in line 2. The user supplies an observation point of TD near field variations in line 3. The angle of incidence is determined in line 4, by giving two angles in degrees. The polarization of the once-differentiated Gaussian plane wave (incident field) is chosen in line 5. Finally, line 6 is for angle of scattering in which TD far fields will be extrapolated.

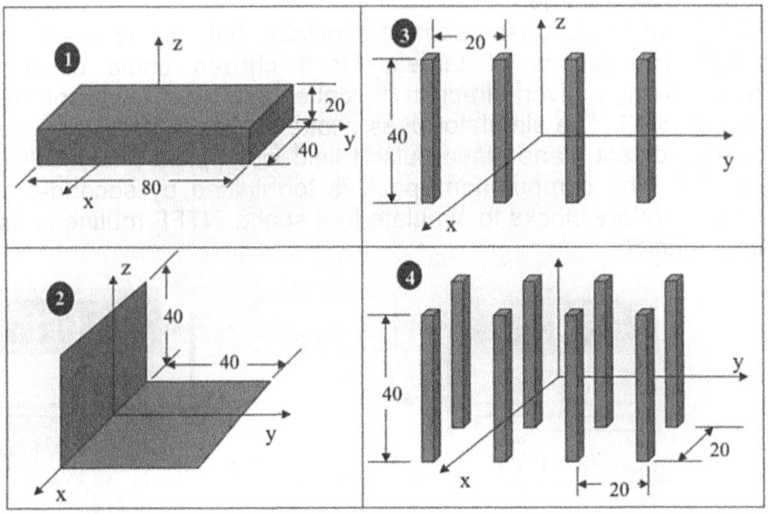

Figure 3.14: *Four targets prepared for F-SNRCS*

The computation space is 70 × 110 × 70. The target is located at the origin of the computation space, longitudinally along the *y*-direction. The near-field observation point better be in the first or last 10 cells of each axis (e.g., (9,9,9,) or (8,102,8) etc.), which guarantees the observation point not to fall in or on the target.

Transient analysis is done within the FDTD computation space, and TD data are stored in three different files. SOURCE.dat contains four columns. The first column is reserved for time [nsec], and the other three columns are for E_x, E_y, and E_z in [V/m], respectively. This allows the user to see the excitation signal in TD. SN-NEAR.dat has near time in [nsec] in the first column and has E_x, E_y, and E_z in [V/m] in the following three columns, respectively. Finally, SN-FAR.dat have TD far-field data along the chosen direction, obtained via NTFF transformation directly in TD. Off-line DFT analysis is required to obtain FD results. SN-DFT is prepared for this purpose. It receives TD data from SN-FAR.dat and SOURCE.dat files and yields the frequency spectrum of the source and RCS vs. frequency for the chosen parameter set. A typical input file and corresponding scenario is depicted in Fig. 3.15.

For the targets included in this package the number of time steps is chosen as 600, which is enough for all transients fade away from the computational volume. All other parameters are set automatically when the longest dimension of the target is specified by the user. This longest size is sampled with 80 cells for the first target and 40 cells for the rest. Therefore, cell sizes are $l/80$ for the first scenario and $l/40$ for the rest. Time step Δt is chosen accordingly as $c/2\Delta x$, which satisfies stability criteria. The incident field is a plane wave short (once-differentiated) Gaussian pulse. Its bandwidth is specified automatically inside the algorithm in a way, which allows RCS calculations up to $l \approx 4\lambda$, with a minimum numerical dispersion ratio of 1/10.

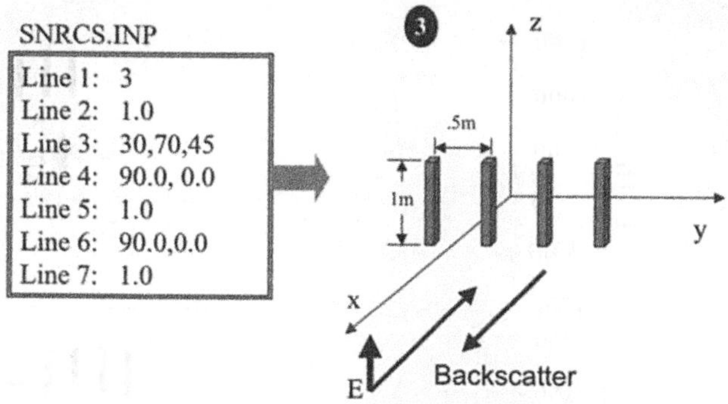

***Figure 3.15:** An SNRCS.INP file and corresponding RCS scenario.*

An F-SNRCS simulator is run with the parameters given in Fig. 3.15. Here the heights of the elements are 1 m, and they are located 0.5 m apart along *y*-direction. The plots of the three output files are given in Figs. 3.16-3.18.

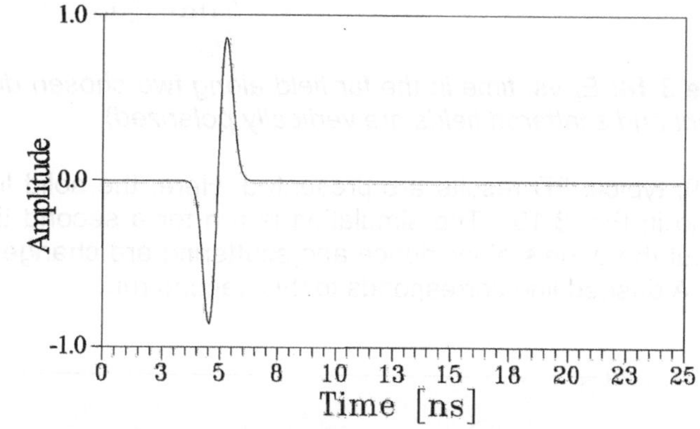

***Figure 3.16:** TD variation of once-differentiated Gaussian plane wave*

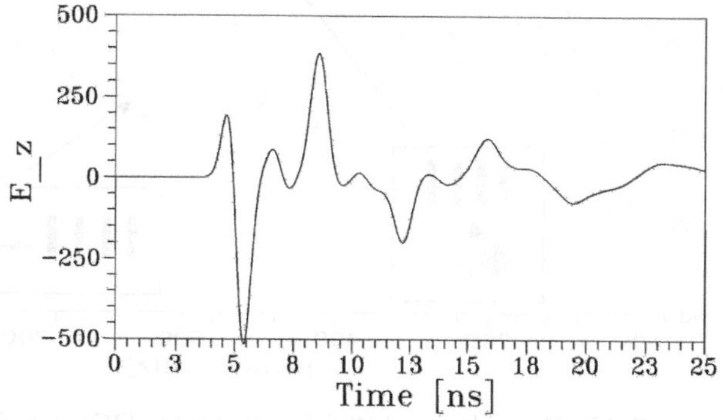

***Figure 3.17:** E_z vs. time in the vicinity of the target*

Time variation of the source pulse is given in Fig. 3.16. In Fig. 3.17, near E_z field at cell (30,70,45) vs. time, and in Fig. 3.18, far field vs. time, is plotted for two different illuminations.

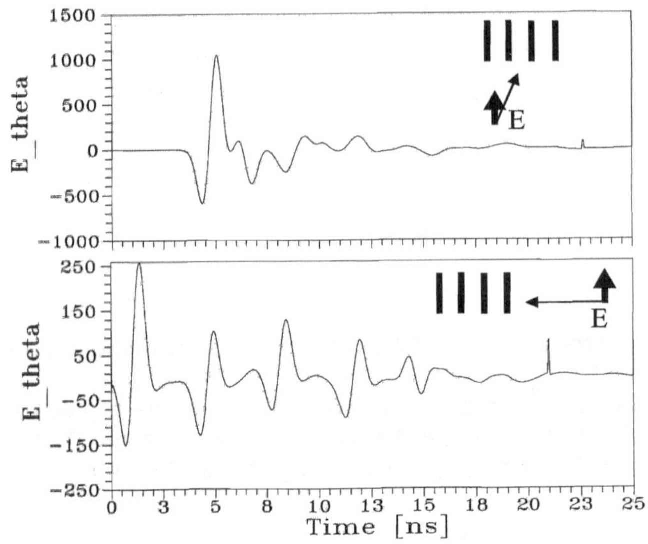

Figure 3.18: *E_θ vs. time in the far field along two chosen directions (both incident and scattered fields are vertically polarized)*

In Fig. 3.19, typical FD results are presented. Here, the solid line corresponds to the scenario in Fig. 3.15. The simulation is run for a second time with the same scenario, but the angels of incidence and scattering are changed from 90.0, 0.0 to 90.0,90.0. A dashed line corresponds to this second run.

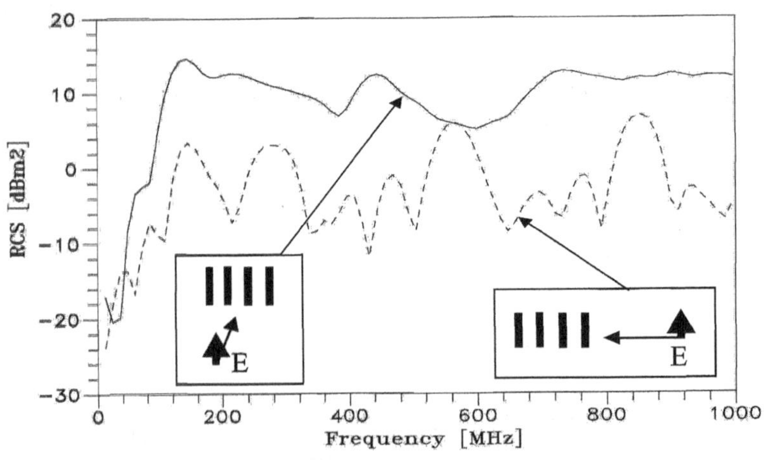

Figure 3.19: *Co-polarized ($\theta\theta$) backscatter RCS vs. frequency*

In other words, solid and dashed lines in the figure correspond to backscatter RCS simulations for broadside and endfire illuminations, respectively. In the figure, 300 MHz represents $l = \lambda$. Resonances at 150 MHz, 300 MHz and 600 MHz, etc.

belongs to cases, $l = \lambda/2$, $l = \lambda$, $l = 3\lambda/2$, etc., respectively (resonances of a single wire).

3.4.2 T-SNRCS Package

In RCS packages the total field is split into the incident and scattered fields, and only scattered fields are simulated in TD. In TLM algorithm, calculation of scattered field components is different, because this method uses voltage pulses. Each unit cell in TLM is represented with a scattering matrix, S (12×12 for free space), and reflected voltage pulses are equal to multiplication of S with incident voltage pulses. The scattering matrix for PEC bodies is diagonal with the elements equal to −1. So, assigning the negative of the incident voltage to the reflecting voltage on PEC boundaries is enough to obtain scattered fields. Since electric and magnetic fields are defined in terms of 12 voltage pulses in TLM, scattered fields are automatically obtained in the observation space when analytical incident plane wave illuminates PEC geometries.

The TLM-based T-SNRCS package has exactly the same structure as the F-SNRCS simulator. Its structure and names of the input/output files are pictured in Fig. 3.20. They both use SN-DFT for TD to FD data conversion. T-SNRCS and F-SNRCS packages locate the target under investigation into 3D computation volume, illuminate the target by a short pulse plane wave from a given direction, simulate near scattered fields and extrapolate TD far fields via the same NTFF routine. The only difference between these two simulators is the absorbing blocks they use: F-SNRCS uses second-order MUR approach, T-SNRCS uses 6-cell PML termination. The two packages are applied to the same scenario (given in Table 3.2) and the results are plotted together in Figs. 3.21 and 3.22, for two different illuminations.

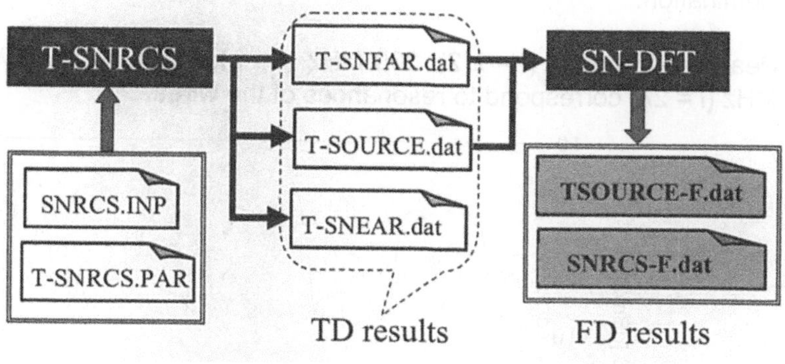

Figure 3.20: *The structure of the TSN-RCS simulator*

Results presented in Fig. 3.21 and 3.22 illustrate quite good agreement between FDTD- and TLM-based simulators. It should be noted that much better agreement can easily be obtained when smaller cell sizes are used and parameters are optimized accordingly. Regarding these figures, some remarks may be listed as follows:

- Height of four wires in this scenario are l = 1m and are located 0.5 m apart, therefore maximum frequency 1200 MHz in the figures correspond to $l = 4\lambda$.

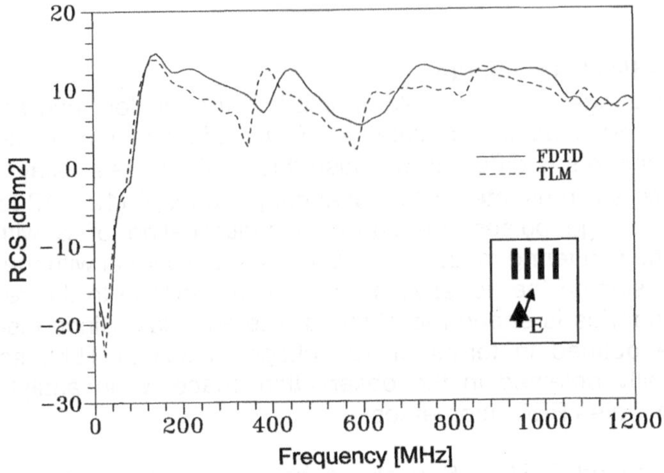

Figure 3.21: *Co-polarized (θθ) backscatter RCS vs. frequency FDTD (solid) and TLM (dashed), (broadside illumination)*

- Frequencies below 100 MHz correspond to Rayleigh scattering, where RCS drops rapidly from 5/10 dB to -20 dB.

- RCS fluctuation for broadside illumination is much less than that for endfire illumination. For broadside illumination an average value of 10 dB is obtained for this frequency range. This drops approximately 10 dB for endfire illumination.

- Peaks at 150 MHz ($l = \lambda/2$), 300 MHz ($l = \lambda$), 600 MHz ($l = 3\lambda/2$), and 900 MHz ($l = 2\lambda$) correspond to resonances of the wires.

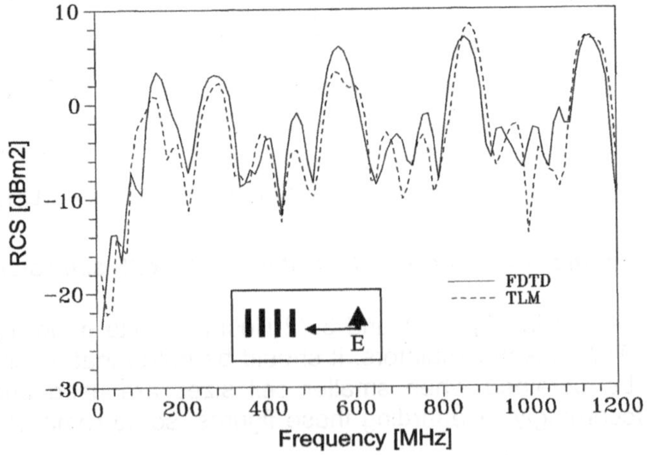

Figure 3.22: *Co-polarized (θθ) back-scatter RCS vs. frequency FDTD (solid), and TLM (dashed), (endfire illumination)*

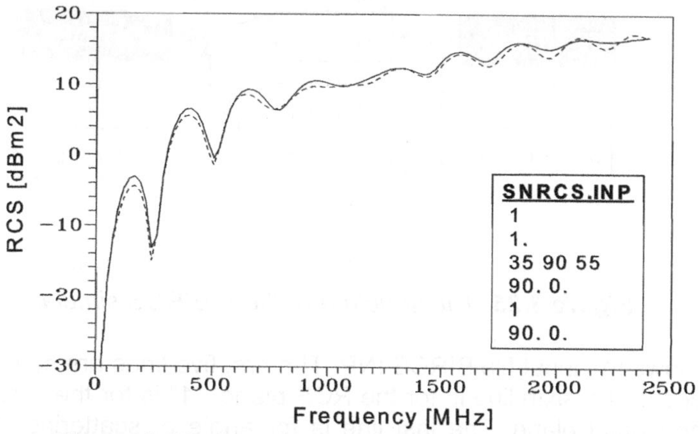

Figure 3.23: *A scenario and RCS vs. frequency*

Two more examples are plotted in Figs. 3.23 and 3.24, together with the input files. Determination of the scenario, the dimensions and analysis of the RCS results are left to the reader.

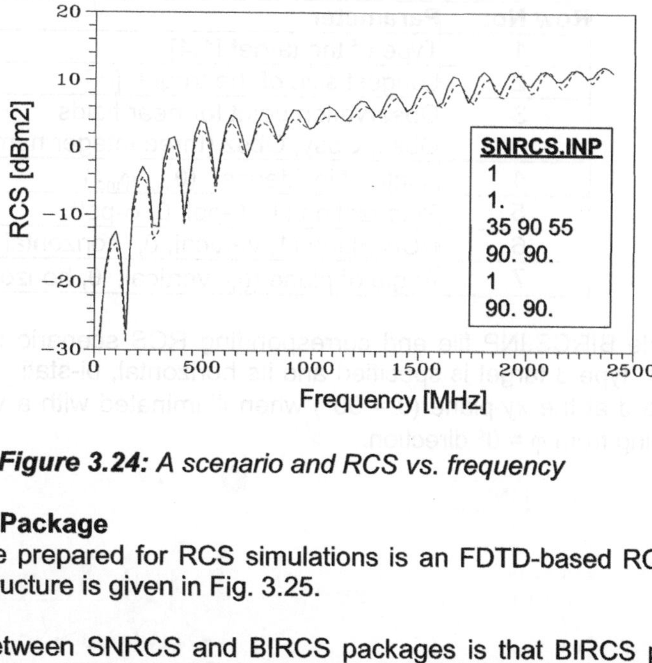

Figure 3.24: *A scenario and RCS vs. frequency*

3.4.3 F-BIRCS Package

The third package prepared for RCS simulations is an FDTD-based RCS pattern calculator. The structure is given in Fig. 3.25.

The difference between SNRCS and BIRCS packages is that BIRCS plots RCS patterns on vertical or horizontal planes. Again, simulation time is set 600. NTFF needs to accumulate early and late time responses, therefore the total time step used for storage is 800. During FDTD simulation, NTFF transformation is applied 145 times on a chosen (vertical or horizontal) plane and are all written into the output file F-BIFAR.dat. This file is therefore multi-MB in size, since it stores 800 by 145 lines. After the TD simulation, an off-line DFT is applied and the patterns are obtained on the chosen plane with every 2.5°. F-BIDFT is prepared for this purpose.

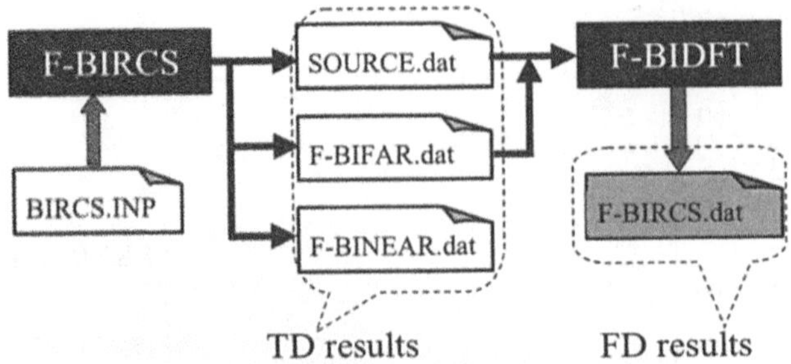

Figure 3.25: *The structure of the F-BIRCS simulator*

Table 3.3 shows the input file BIRCS.INP. The first five lines are same with the SN-RCS packages. The sixth line is for the RCS plane, "1" is for the vertical plane, "0" is for the horizontal plane. The last line is for angle of scattering. If a horizontal (vertical) plane is chosen by the user, then angle of scattering is θ = constant (φ = constant).

Table 3.3: The BIRCS.INP file and required parameters

Row No:	Parameter
1	Type of the target [1,4]
2	Longest size of the target [m]
3	Observation point for near fields Obsx, Obsy, Obsz (three integer numbers)
4	Angle of incidence (θ_{inc}, φ_{inc})
5	Polarization (1, θ-pol; 0, φ-pol)
6	RCS Plane (1, vertical; 0, horizontal
7	Angle of plane ((φ, vertical; θ, horizontal)

An example BIRCS.INP file and corresponding RCS scenario is depicted in Fig. 3.26. Here, type 3 target is specified and its horizontal, bi-static RCS behavior will be simulated at the *xy*-plane (θ = 90°) when illuminated with a vertically polarized wave coming from φ = 0° direction.

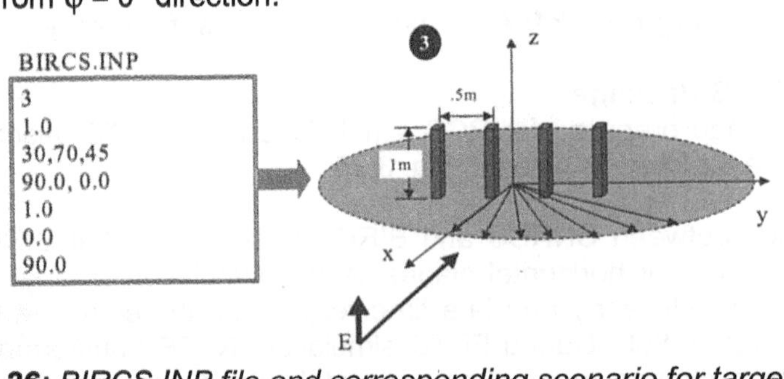

Figure 3.26: *BIRCS.INP file and corresponding scenario for target type 3*

Two results obtained via F-BIRCS simulator are plotted in Figs. 3.27 and 3.28. In Fig. 3.27, a horizontal bi-static RCS pattern is shown, when a type 3 target is illuminated from a positive x-axis. The arrows in the figure correspond to the angle of illumination. Both incident and scattered fields are vertically polarized (E_θ is of interest); therefore the figures present co-polarization results. A type 3 target has four vertical PEC wires, which have lengths of *l* and are located along the *y*-axis *l*/2 apart. In this scenario *l* is chosen as 1 m. Therefore $l = \lambda/4$ and $l = \lambda/2$ cases in Figs. 3.27 and 3.28 correspond to $\lambda = 2$ m (or $f = 150$ MHz) and $\lambda = 1$ m ($f = 300$ MHz), respectively. The plots are normalized to their maxima, so the outer dashed circles correspond to 0 dB. Real maximum RCS values are mentioned on the top right of each plot. Everything in this scenario is kept the same and the RCS plane is changes to vertical. The results are plotted in Fig. 3.28.

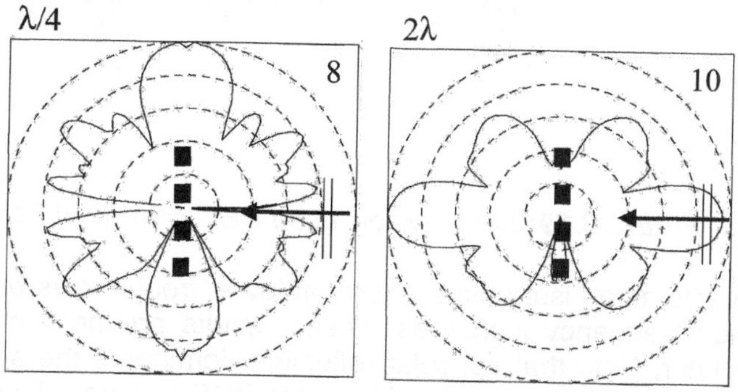

Figure 3.27: *Bi-static RCS patterns at two frequencies (horizontal)*

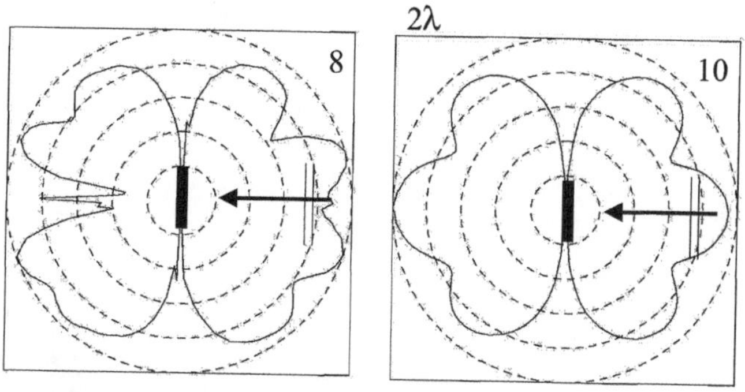

Figure 3.28: *Bi-static RCS patterns at two frequencies (vertical)*

3.5 APPLICATIONS

3.5.1 RCS Analysis with Canonical Objects

EM wave / object interaction is a complex process. To reduce complexity and to better understand this interaction, targets with simple geometries have long been studied. Although simple in geometry, these objects are canonical since their RCS values can analytically be computed in asymptotic sense and serve as reference solutions. Some of them are illustrated in Fig. 3.29, together with analytical RCS equations that are used in optical regime (for the maximum scatter directions mentioned there).

RCS tools presented in this Chapter may be used to obtain the examples given below. The first one is mono-static RCS patterns of a 1-m^2 PEC plate. Figure 3.30 shows horizontal pattern of this plate at six different frequencies mentioned by target length / radar wavelength relations.

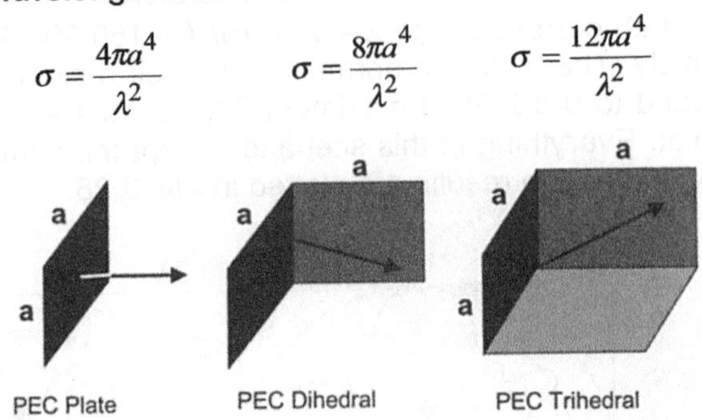

***Figure 3.29:** Some widely used PEC canonical RCS targets*

The plate acts as an isotropic scatterer at lower frequencies but becomes directive target when frequency increases. These results are produced via FDTD based RCS simulator. Note that specular reflection dominates the scattering mechanism at high frequencies (as expected in optical RCS regime). The difference between RCS values of $l = \lambda/10$ and $l = 4\lambda$ cases is nearly 40 dB.

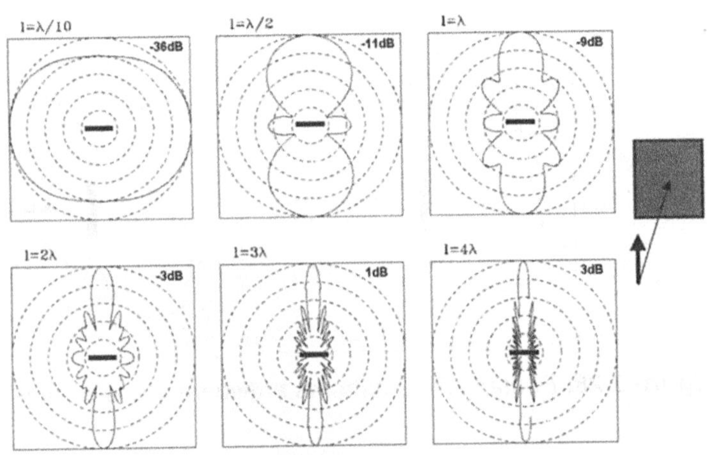

***Figure 3.30:** mono-static RCS patterns on horizontal plane (θθ polarization)*

A second example (also produced with FDTD based RCS simulator) is given in Fig. 3.31, where RCS vs. frequency of four canonical objects are plotted. A 10 cm × 10 cm × 30 cm PEC rectangular prism, 10-cm^3 PEC cube, 10-cm^2 PEC plate and 30-cm-long PEC wire are used as targets in this figure. The plots correspond to θθ-case backscatter RCS behaviors. Angle of illumination and object locations are also given in the figures. Some remarks can be listed as follows:

- PEC prism and PEC cubes are illuminated perpendicular to their sides. As seen in the figure, the difference between RCS values of two targets approaches to 10 dB as the frequency increase. This is consistent with the analytical results given for optical regime. The ratio of the facet areas of PEC prism and PEC cube is 3. Since RCS is proportional to the square of the area for flat PEC surfaces (i.e., $10 \log_{10} 3^2 = 9.5$ dB), expected difference between their RCS values is nearly 10 dB (as seen from the figure).

- The third object (plate) is illuminated with an incident field that has an electric field perpendicular to its one edge. Comparison of the cube and plate RCS curves may give clues for differences in RCS between specular reflection and edge diffraction. There is more than 20-dB difference for the frequencies above 4 - 5 GHz. Since $l = 30$ cm in this example, 9 GHz corresponds to $l = 9\lambda$.

- The edge-illuminated plate and vertical wire has nearly the same RCS amplitude at frequencies given in the figure.

- Note that the RCS regime for these curves is the resonance regime; therefore objects under investigation contribute as whole.

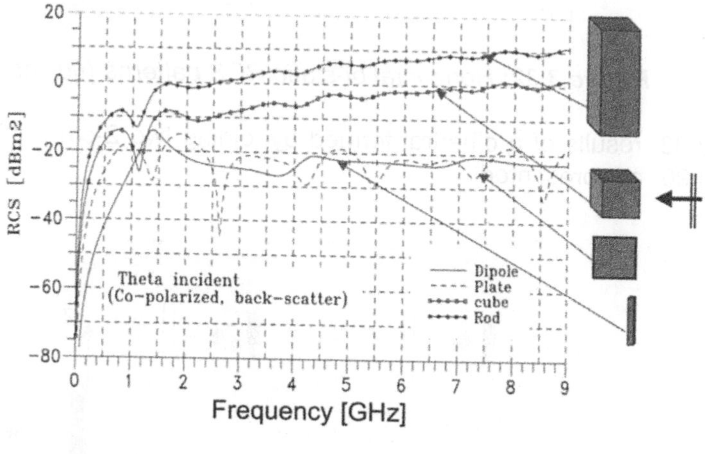

Figure 3.31: *Mono-static RCS vs. frequency of four objects*

Other examples included in this section belong to bi-static RCS patterns of a PEC plate and dihedral, which are produced both with FDTD and MoM packages. In Fig. 3.32, a 2-cm × 100-cm × 100-cm PEC plate is taken into account. Its horizontal RCS pattern is obtained for both perpendicular and oblique incidences for different frequencies. A 50 × 80 × 80 cell FDTD space is used with 2-cm cell sizes. The patterns at 1200 MHz correspond to $l = 4\lambda$. The same plate is modeled with 840 segments in a MoM package with 5 cm segment lengths. Wire radii are taken as 2 cm to correspond the same structure with FDTD (plate thickness is taken as 2 cm in FDTD simulations). In the figure, only 1.2-GHz results are plotted. A forward scattered component dominates in both cases and maximum RCS values are approximately 23 dB and 8 dB for perpendicular and 45° oblique illuminations, respectively. A very good agreement among the patterns are clearly observed in the figure.

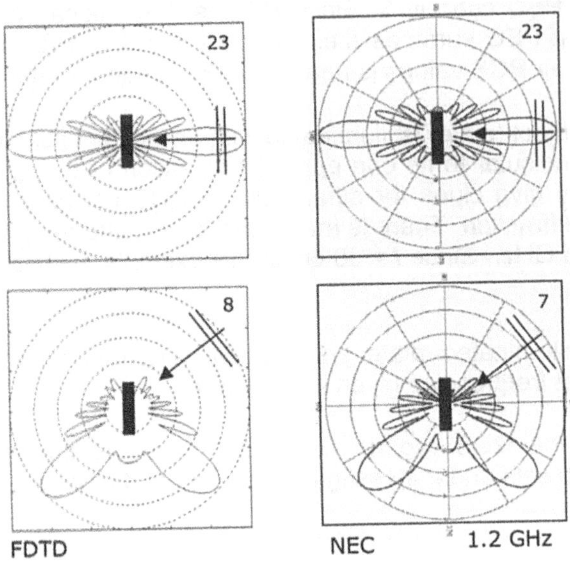

Figure 3.32: *Horizontal bi-static RCS patterns (θθ case)*

In Fig. 3.33, results of a dihedral formed by joining two 2-cm × 100-cm × 100-cm plates in 90° are presented.

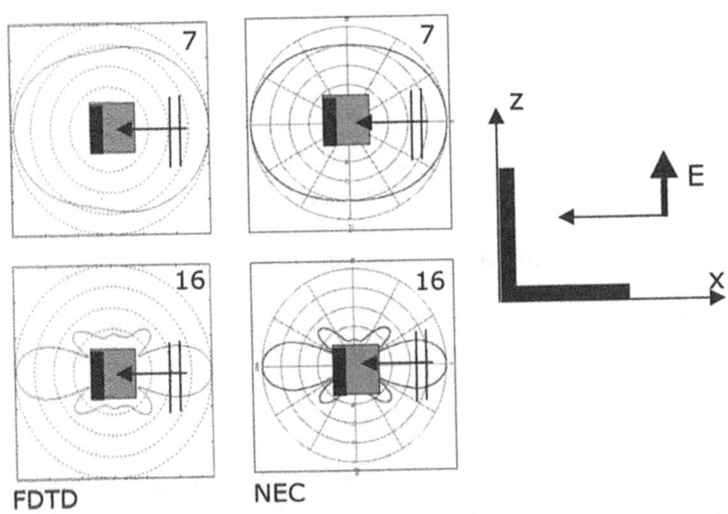

Figure 3.33: *Horizontal bi-static RCS patterns of the 90° dihedral (θθ case)*

The discretization parameters are kept the same as in the previous example. Angle of illumination and polarization are as mentioned in the figure inset. The top and bottom row are for 150 MHz ($l = \lambda/2$) and 600 MHz ($l = 2\lambda$), respectively. At 150 MHz dihedral still acts as a point scatterer, but becomes directive when frequency is increased to 600 MHz. Angle of illumination is changed to 45° on horizontal plane and simulations are repeated for the dihedral. All other parameters are kept same. The results are given in Fig. 3.34. Top and bottom rows correspond to 150 MHz ($l = \lambda/2$) and 600 MHz ($l = 2\lambda$), respectively.

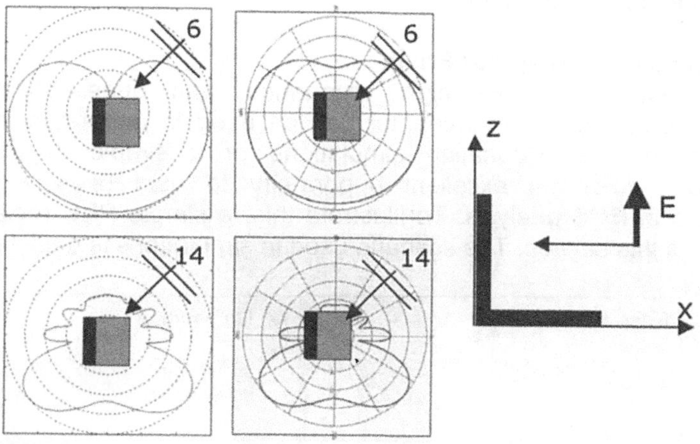

FDTD NEC

Figure 3.34: *Horizontal bi-static RCS patterns of the 90° dihedral ($\theta\theta$ case) for a 45° angle of incidence on a horizontal plane*

Finally, a bi-static RCS of a dihedral with 120° joint angle is simulated. Typical results are depicted in Fig. 3.35.

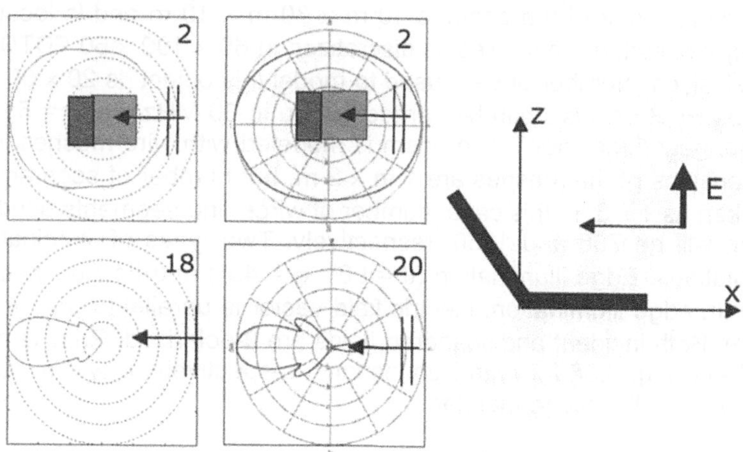

FDTD NEC

Figure 3.35: *Horizontal bi-static RCS patterns of the 120° dihedral ($\theta\theta$ case)*

Again, top and bottom rows correspond to 150 MHz ($l = \lambda/2$) and 600 MHz ($l = 2\lambda$), respectively. The FDTD volumes are taken as 80 × 80 × 80 and 130 × 80 × 80 for 90° and 120° dihedrals, respectively. This example demonstrates that avoiding a perpendicular surface along the radar direction reduces RCS significantly. In this example using a 120° dihedral, instead of a 90°, one reduces backscatter RCS more than 20 dB (note that the values mentioned in the plots correspond to maximum values, and they belong to a forward-scatter lobe. Back-scatter lobe values may be read directly from the plots, by taking into account that each circle corresponds to a 6 dB difference).

3.5.2 A PEC Rectangular Prism

Although limited with Rayleigh, resonance, and quasi-optical regimes, the packages presented in this chapter can be used in understanding reflectivity of variety of objects. Repeating simulations of a simple object with different parameters yields an excellent opportunity to gain physical insight and to experience in RCS analysis. To illustrate this, a simple PEC rectangular prism is analyzed in this section. The scenario used in simulations is depicted in Fig. 3.36.

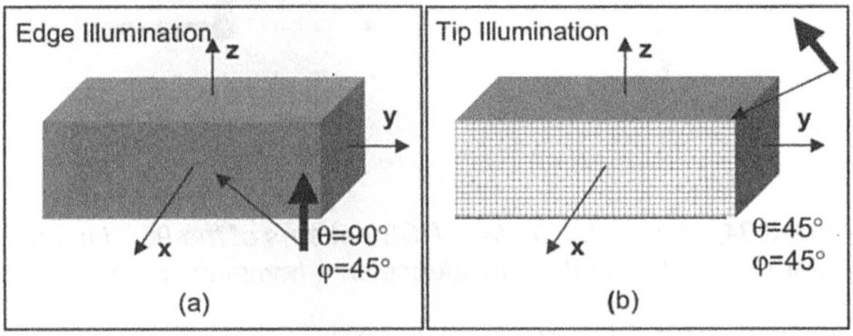

Figure 3.36: *A PEC rectangular prism, (a) edge, (b) tip illuminations*

The dimensions of the prism is 10 m × 30 m × 10 m and is located longitudinally along y-direction. This object is discretized in 60 × 100 × 60 FDTD volume with 0.5 m cell sizes. Number of cells used to model this object is 20 × 60 × 20. With these values RCS analysis can be carried out up to 60 MHz (i.e. λ = 5m, with $l = 6\lambda$, l is the longest dimension). The prism is modeled with wire meshes in MoM package. Dimensions of the meshes are 1 m × 2 m, but number of segments for each mesh is taken as 1 × 3. In this case, number of wires and segments used in modeling this prism will be 700 and 1260, respectively. Two types of illumination is chosen in simulations. Edge illumination (θ = 90°, φ = 45°) and tip illumination (θ = 45°, φ = 45°). In edge illumination, electric field vector is parallel to one of the edges of the prism. Both incident and scattered fields are θ-polarized. Results are plotted in Fig. 3.37 and Fig. 3.38 for both illuminations, respectively. Physical observations listed below is worthwhile to mention:

- FDTD and MoM results agree very well for the range of parameters used in simulations. Both can handle simulations up to l = 10 - 20λ with today's PCs.

- In the Rayleigh region the object acts as an isotropic point scatterer and its RCS is low (as compared high frequencies) for edge illumination. This is also true for perpendicular illumination, which is not given here (the reader may do these simulations with the F-BIRCS package included in this chapter. But, this is not true for tip illumination. RCS of the object decreases as frequency increases (see Fig. 3.38).

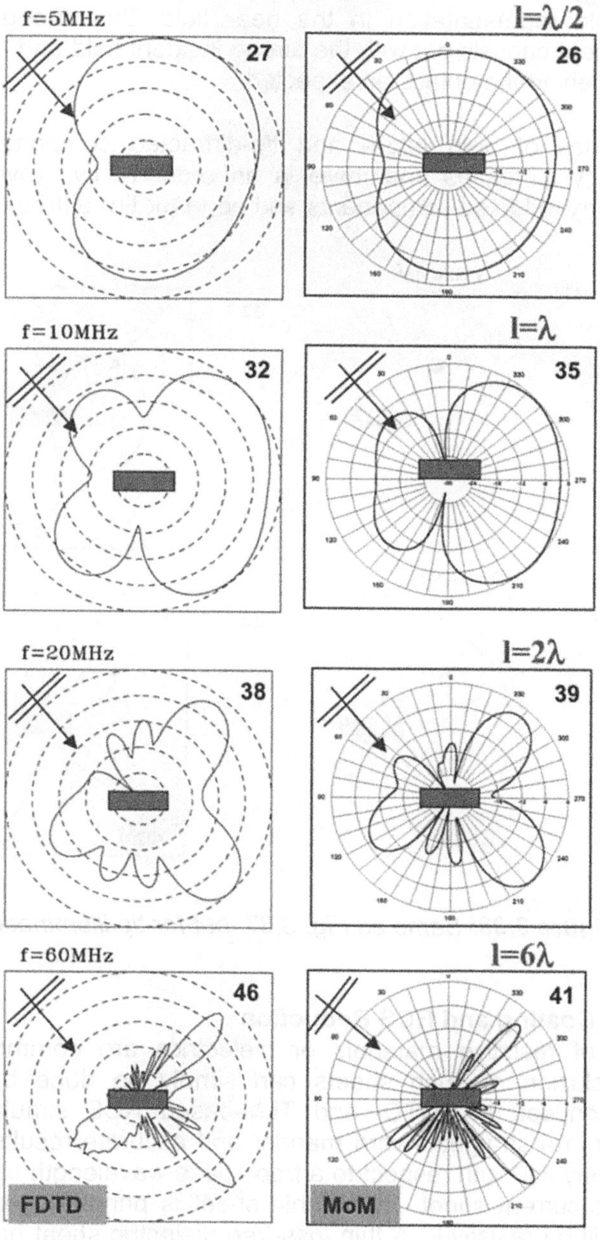

Figure 3.37: *Horizontal plane bi-static RCS patterns of the prism (θθ case)*

- Starting from approximately $l = \lambda$, scattering properties of the object become directive. For edge illumination, three scattering components dominate: specular reflections from planes on both sides of the edge, and forward scatterer. Except for a few resonating structures, forward-scattered components always dominate in RCS patterns.

- Note that RCS relates scattered EM field components to incident field. In total field representation in the near field, this strong forward-scattered component cancels out with the strong incident field and results in a shadow region behind the object, as expected.

- Amplitudes of both edge- and tip-diffracted components decreases as frequency increases and there is an order of two (even three) between specularly reflected components and edge (or tip)-diffracted components.

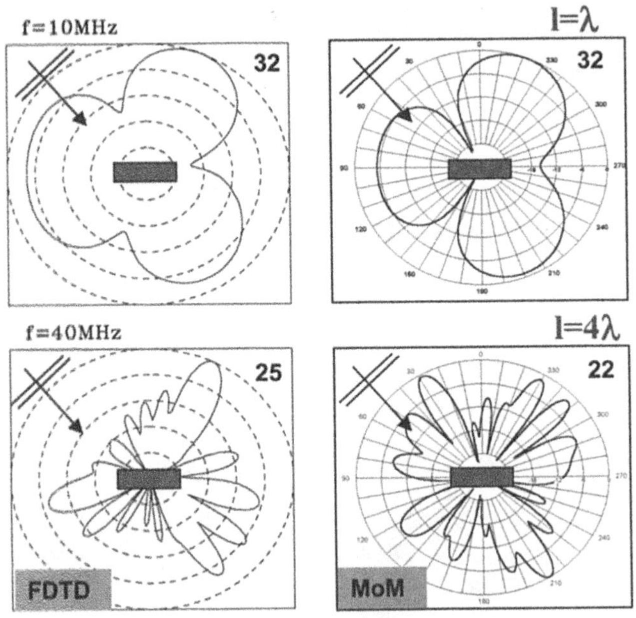

Figure 3.38: *Same as Fig. 3.37, but for tip illumination (θθ case)*

3.5.3 RAM Coating and RCS Reduction

Thin sheets of resistive materials or dielectrics are commonly used in RCS modeling. Inclusion of such sheets can simply be done by using the sheet impedance concept. In FDTD- and TLM-based RCS simulators, this can be accomplished in a straightforward manner and accurate results may be obtained. Sheets relatively thin with respect to a free-space wavelength can be approximated by an electric current sheet. If the thin sheet is primarily conductive, the sheet impedance will be resistive. A thin loss-free dielectric sheet has a purely reactive sheet impedance, while in general the sheet impedance is complex. These sheets are characterized by a discontinuity in the tangential magnetic fields on either side of the sheet. The sheet impedance can be defined in several ways. Here the sheet impedance Z_{sh} is modeled as

$$Z_{sh} = \frac{1}{\sigma d + j\omega\varepsilon_0(\varepsilon_r - 1)d} \tag{3.4}$$

where d is the thickness, and σ and ε_r are the electrical parameters of the sheet. Sheet impedance reduces to resistive sheet when relative permittivity of the material is 1. Two examples for resistive sheet coating is given below in Figs. 3.39 and 3.40.

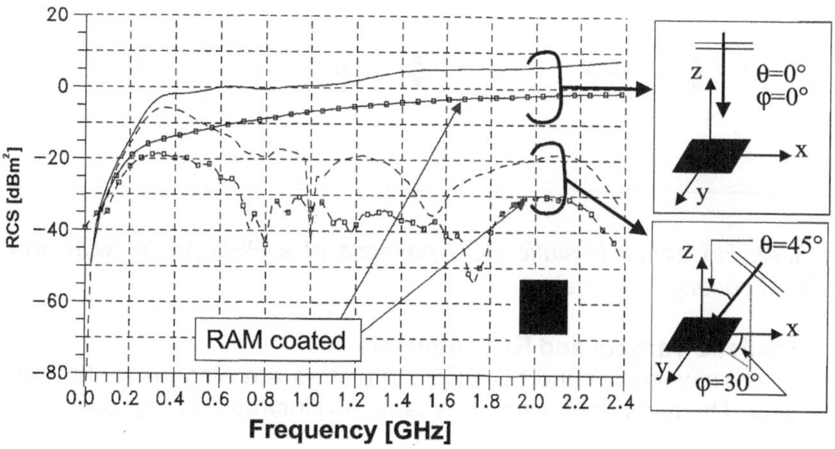

Figure 3.39: *RCS vs. frequency of a square plate for two different illuminations (with and without 500-Ω RAM coating, θθ case)*

In Fig. 3.39, RCS vs. frequency of a 0.3-m^2 square plate, with perpendicular and oblique incidences, is shown. Also in the figure, simulation results with RAM coated material is shown. Average reduction of 10 dB in RCS values is clearly observed from the figure. Note that, there are certain frequencies where RAM coating for these illuminations do not work. As much as 40 dB difference between perpendicular incidence and oblique incidence is observed at certain frequencies. The average difference is nearly 30 dB.

In Fig. 3.40, co-polarized bi-static RCS patterns of a PEC prism, without (top row) and with (bottom row) 500-Ω RAM coating, are plotted. As observed in the plots RAM coating becomes effective at high frequencies. For example, backscatter RCS values without and with 500-Ω RAM coating differ more than 10 dB in $l = 2\lambda$ case. It should be noted that RAM coating may decrease or increase backscatter RCS in resonance region; therefore one should be very careful while coating targets in this regime.

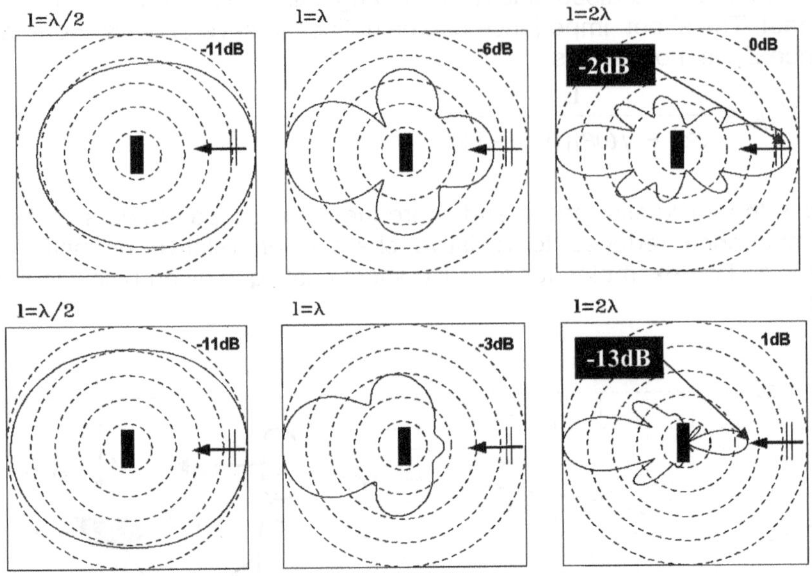

Figure 3.40: *Horizontal bi-static RCS patterns of a PEC prism (with and without 500-Ω RAM coating, θθ case)*

3.5.4 Realistic Targets and RCS Analysis

Comparisons between TD and FD RCS simulators are also presented for realistic radar targets. The first target is an army tank, as illustrated in Fig. 3.41.

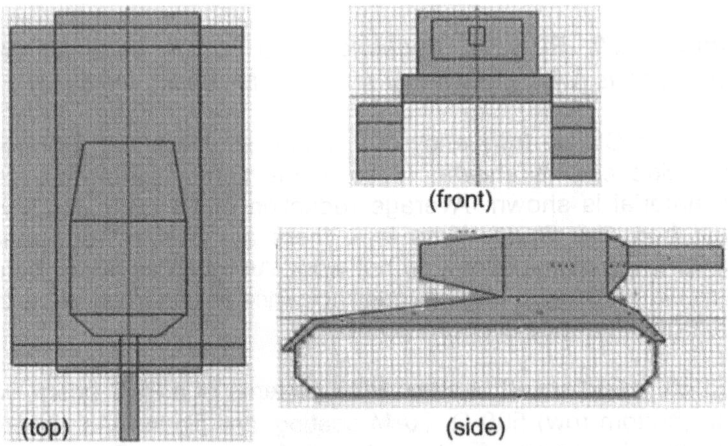

Figure 3.41: *Discrete tank model for TD simulations*

The dimension of the tank is 8.2 m × 3.8 m × 2.8 m and is located longitudinally along the *x*-axis as shown in the figure. Although difficult, FDTD discrete model of this tank is built [9] in a 100 × 70 × 60 space with 0.12-m cell sizes. The tank occupies 68 × 32 × 23 cells. These values permit RCS analysis up to 250 MHz, with a $\lambda_{min}/10$ numerical dispersion ratio. The segment length is chosen as 0.15 m in its MoM model. The number of segments is therefore 269. The tank is illuminated from its nose (θ = 90°, φ = 0°) by a vertical (θ) polarized wave and its bi-static RCS patterns at horizontal plane are plotted in Fig. 3.42.

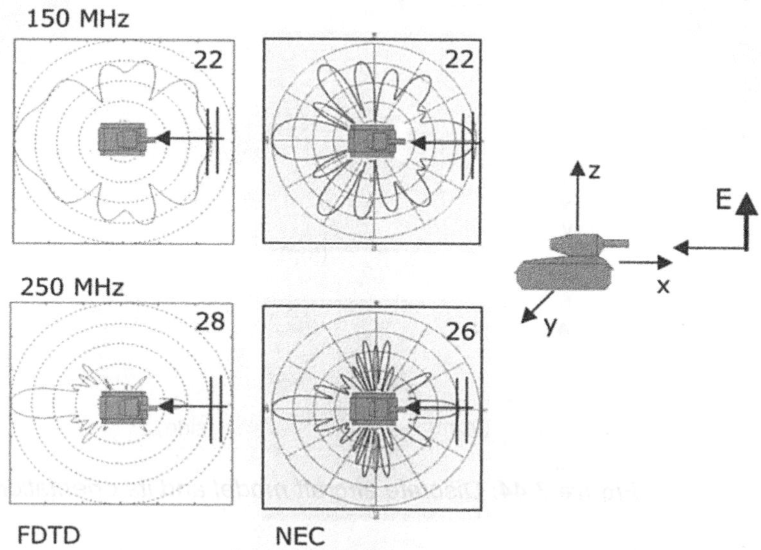

Figure 3.42: *Bi-static RCS patterns of the tank at the xz-plane ($\theta\theta$ case)*

The tank is then illuminated from the top (θ = 0°, φ = 0°) by a vertical (θ) polarized wave and its bi-static RCS patterns at horizontal plane are plotted in Fig.3.43.

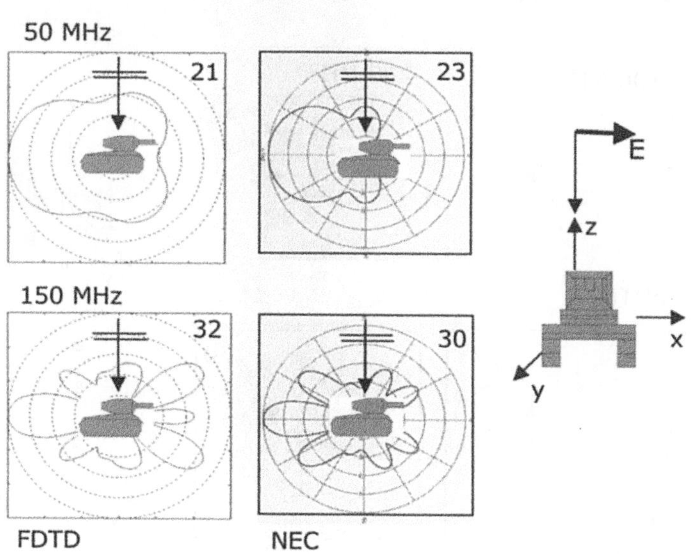

Figure 3.43: *Bi-static RCS patterns of the tank at the xz-plane ($\theta\theta$ case)*

Another model used in RCS simulations is the aircraft model given in Fig. 3.40, which has a size of 7.2 m × 6.8 m × 1.8 m. It is discretized in 110 × 100 × 50 FDTD volume with 0.12 m cell sizes, therefore occupies a space of 72 × 68 × 18.

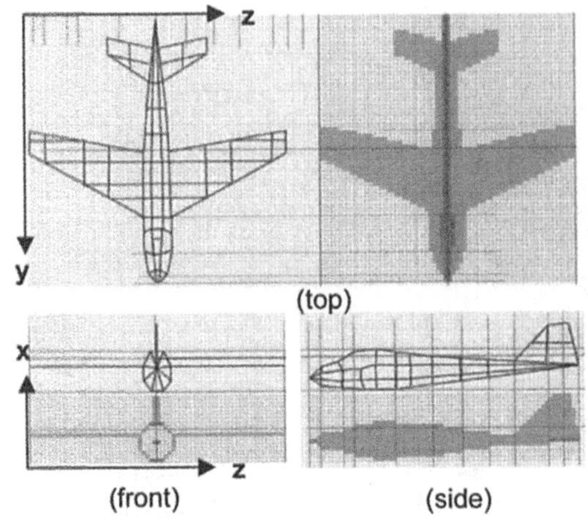

Figure 3.44: *Discrete aircraft model and its orientation*

The NEC model contains 255 wires (segments) with 0.05-m segment length. Two typical results are plotted in Figs. 3.45 and 3.46. Angle of illuminations and pattern planes are also mentioned in the figures. FDTD and MoM results are given on top and bottom rows, respectively. The plots correspond to approximately $l = 2\lambda$ (for f=100 MHz) and $l = 5\lambda$ (for $f = 250$ MHz).

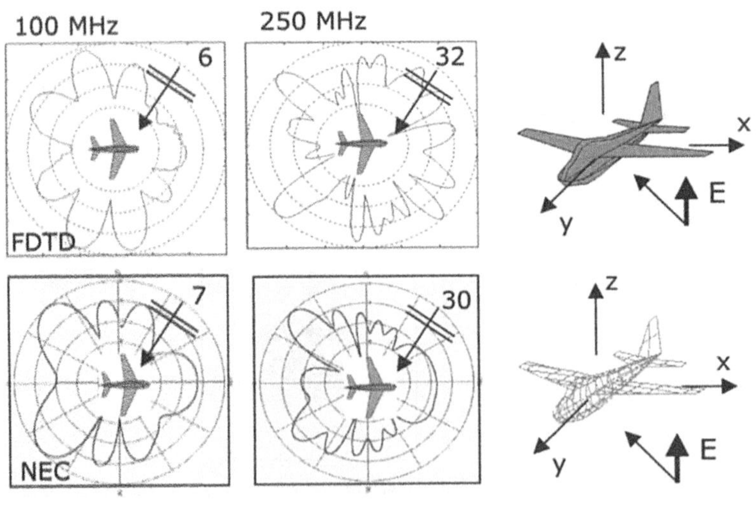

Figure 3.45: *Bi-static RCS patterns of the aircraft at the horizontal plane (θθ case)*

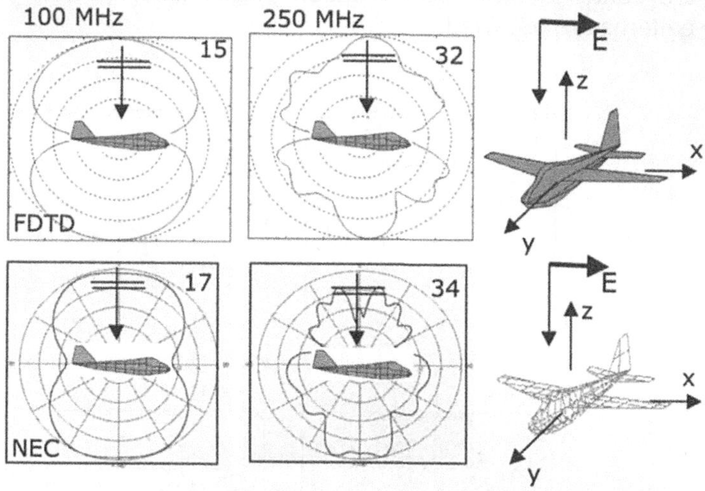

Figure 3.46: *Bi-static RCS patterns of the aircraft at the horizontal plane (θθ case)*

3.5.5 RCS / Antenna Patterns of Wire Arrays

The last examples in this chapter are reserved for the analogy between radiation patterns of wire arrays (when used as antenna elements) and their RCS behaviors (when illuminated by a plane wave). Take a four-element wire array (scenario 3) as shown in Fig. 3.47.

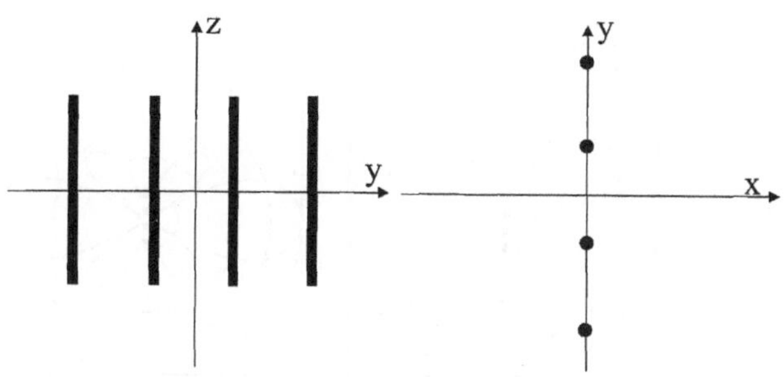

Figure 3.47: *A four-element array and its location in the Cartesian system*

A center-fed vertical wire antenna has a vertical electric field component when radiating, caused by its current distribution. The same current distribution is obtained when the wire is illuminated by a vertically polarized EM wave. This is also true for an array of vertical wires. The only requirement is that there must not be any other sources (except wire feeds) when radiating. This also valid for an array of vertical wires. Typical examples are given in Figs. 3.48 and 3.49.

A 20 × 80 × 60 FDTD space with 0.01-m cell sizes is reserved for the simulations. Four thin wires (1 cm × 1 cm × 40 cm) with length l = 40 cm are located $d = l/2$ apart from each other. RCS vs. angle on the horizontal plane are plotted in Fig. 3.48, and horizontal plane radiation patterns are plotted in Fig. 3.49, when all elements are center-fed with no excitation phase difference. Exactly the same horizontal patterns are obtained.

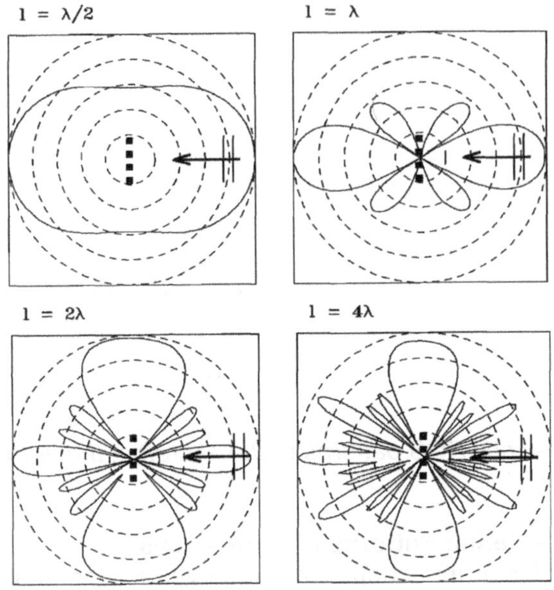

Figure 3.48: *Horizontal RCS patterns of a four-element vertical array , broadside illumination (θθ case)*

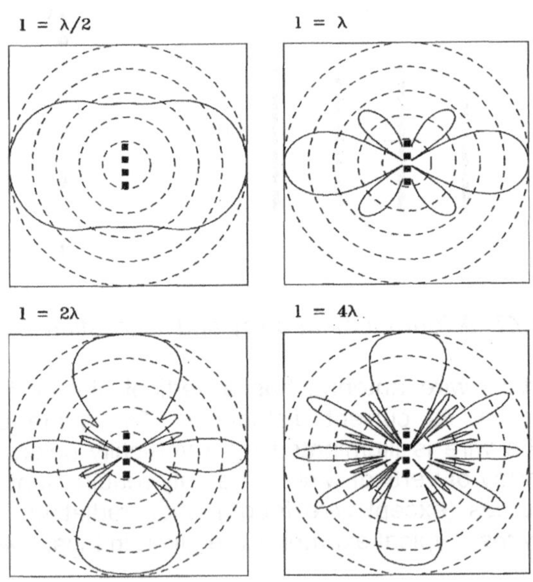

Figure 3.49: *Horizontal radiation patterns of a four-element vertical array. All elements are center-fed (θθ- case)*

Finally, the same comparison is given for opposite phase feedings in Fig. 3.50. When used as radiating elements, the four wires are center-fed with phases 0°, 180°, 0°, and 180°, respectively. The corresponding illumination angles are θ_i = 90°, φ_i = 90°, and θ_i = 90° (when $d = \lambda/2$), φ_i = 45° (when $d = \lambda$), respectively.

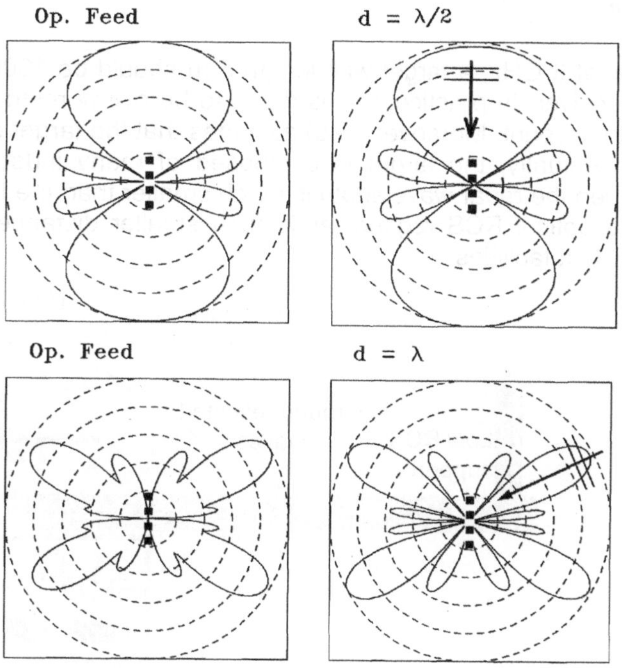

Figure 3.50: *(Left) Radiation patterns of the array when the elements are all fed, but with 180° phase difference. (Right) horizontal bi-static RCS patterns at the same operating frequencies for two different illuminations*

3.6 RCS MEASUREMENTS

RCS measurements play an important role in EM (radar) engineering. Measurement RCS data can serve different purposes:

- To establish target's RCS to design a low RCS military aircraft or ship
- To make comparisons with the simulated data
- To diagnose or troubleshoot

RCS measurement is as difficult as RCS simulations. Measurement facilities can be grouped into two types: outdoor and indoor test ranges. A typical picture of the outdoor test range is given in Fig. 3.51. Here, the target is elevated on top of a low RCS support. A radar transmitter / receiver is located across the target on a leveled ground, far enough to meet far-field conditions. The required distance in between the radar transmitter/receiver and the target depends on the frequency and target length. Based on limiting the EM round-trip phase error to $\pi/8$, the minimum distance can be obtained from

$$R_{\min} = \frac{4L^2}{\lambda} \tag{3.5}$$

For example, at 3 GHz, a target with length 5 m should be 1000 m from the radar transmitter/receiver. In practice, this is found to be overly restrictive and is usually not taken into account. Moreover, (3.5) assumes that the target contributes its RCS as a whole. Actually, the target is composed of many independent scatterers, especially when the radar wavelength is much smaller than the target's length. This is met in the optical RCS regime, and much smaller distances may be used in outdoor RCS test facilities.

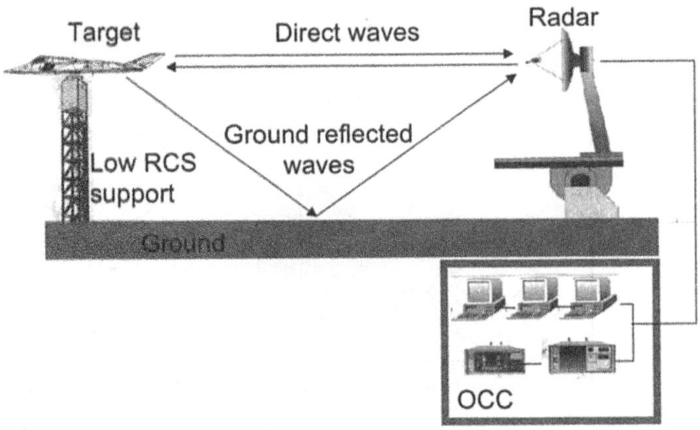

Figure 3.51: *Open-field RCS measurement system*

Outdoor RCS facilities suffer from environmental conditions. The majority of RCS measurement facilities are indoor test ranges, which provides isolation from the outside environment. Indoor test ranges are also desirable from technical and security point of views. On the other hand, the obvious limitation of indoor test ranges is the size of the building required for large targets. One solution to the demand of accurate and practical measurements with small-size indoor test facilities is the compact range. A typical layout of a compact range is depicted in Fig. 3.52.

The facility is an anechoic chamber, with a low RCS target support (that can be turned, preferably, 360° around its axis). Two reflectors are used to supply a plane-wave illumination for the target. Indoor RCS test and measurement facilities must be well-designed, maintained, and carefully calibrated. Signal coupling, leakage, noise and unwanted reflections degrade RCS measurement accuracy. Signal coupling may be too strong if a common antenna is used for both transmit and receive. Unwanted reflections and coupling may be mitigated by repeating RCS measurements without and with the target. This is called *background subtraction*.

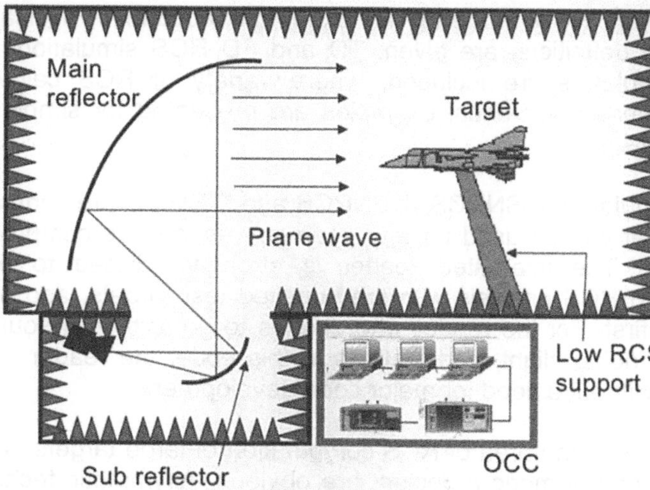

Figure 3.52: *Indoor RCS test and measurement system inside an anechoic chamber*

3.7 CONCLUSIONS AND DISCUSSION

Today's radar targets effectively reduce their observability features applying shaping and new material technologies, and employing cover, concealment, and deception to deny acquisition. Consequently, modern radar systems employing sophisticated signal processing must keep up with the new scenarios through exploitation of target features revealed only by precision simulation and/or measurements. The majority of advanced radar processors perform surveillance, detection, discrimination, and non-cooperative target identification functions on these bases. This necessitates, first, a basic understanding of radar and signal processing technologies used for target RCS evaluation, and then requires interpretation of radar signatures data for extraction of maximum information from radar returns.

An RCS engineer should enhance his/her understanding of the

- EM wave / target interaction phenomena
- Radar technology and target discrimination principles
- Interpretation target radar signatures with emphasis on signal processing
- Correlation of radar signatures with target features and comparison of data sets processed by different methods
- Evaluation of radar target engagement scenarios
- Effects of radar system parameters, signal processing parameters, and of the environment on observed target characteristics
- Practical techniques to exploit target features for discrimination and identification

In this chapter, RCS modeling is taken into account. Physical meaning and mathematical definitions are given. TD and FD RCS simulations are discussed. Powerful simulators are included, where variety of RCS calculations can be performed. Finally, numerical examples are presented for simple, canonical and realistic targets.

The three simulators F-SNRCS, T-SNRCS and F-BIRCS are compact, easy to use simulators that can be used for any geometry as long as computer requirements are satisfied. The interested reader is strongly advised to play with these simulators, do RCS analysis with the supplied test targets, and understand RCS phenomena first. For the reader who desires to go further, a source code is also supplied. By doing slight modifications in this code, the reader may extend their applicability without a need for major code development.

The challenges in the area of RCS computation of large targets, such as aircrafts, ships, and tanks of modern design, are obvious. Asymptotic techniques, such as the GTD and PTD, have been successfully applied, but emphasis has been given to finding more and more efficient and effective computer simulation during the last couple of decades. MoM, FDTD, and TLM are among the few powerful numerical RCS computation techniques. Today's computers (with GB RAMs and a few-GHz CPU speeds) make it possible to numerically simulate complex targets successfully in Rayleigh and resonance (and quasi-optical) regimes, but all fail in optical RCS regime. For example, realistic discretization of an aircraft or a tank for an X-band radar may require hundreds of GB RAM memories, which is quite far from today's computer capacities. Therefore, simulations in optical RCS regime still rely on asymptotic techniques or their hybridized forms.

One promising technique that can be used in RCS simulations in the optical regime is the Fast Multipole Method [13 - 15], which gives more than a million degrees of freedom in solving large MoM problems. FMM and its hybridization with asymptotic techniques, such as the Physical Optics or ray-based methods, has also been proposed. Unfortunately, these strategies have limited success until now, because of the lack of methodological merge between numerically rigorous and asymptotic techniques. The question is still on for finding ways by which one can enhance the FMM. The FMM has still several fundamental limitations when applied to large targets at higher frequencies that are difficult to overcome in practice.

In conclusion, RCS modeling and simulation has a variety of challenging characteristics, and analytical, numerical, and hybrid methods are open to further investigations.

References

[1] U. F. Knott, *Radar Cross Section Measurements,* Van Nostrand Reinhold, New York, 1993

[2] J. F. Shaeffer, M. T. Tuley, E. F. Knot, *Radar Cross Section*, Artech House, Norwood, MA, 1985

[3] L. Sevgi, S. Paker, "FDTD Based RCS Calculations and Antenna Simulations", *AEU, International J. of Electronics and Communication*, Vol. 52 No.2, pp. 65 - 75, March 1998

[4] F. Akleman, L. Sevgi, "Radar Cross Section and Antenna Modeling with FDTD Method", *Proc. of JINA'98, 10th Int. Symposium on Antennas*, Nov 14 - 16, Nice, France, 1998

[5] F. Akleman, M. O. Özyalçın, L. Sevgi, "Radar Cross Section Modeling and Stealth Target Design", *Symposium on Aeronautics and Astronautics Defense Technologies*, April 19 - 20, 1999 , Yesilyurt, Istanbul, Turkey

[6] L. Sevgi, A. M. Ponsford, H. C. Chan, "An Integrated Maritime Surveillance System Based on Surface Wave HF Radars, Part I – Theoretical Background and Numerical Simulations", *IEEE Antennas and Propagation Magazine*, Vol. 43 No.4, pp. 28 - 43, 2001

[7] A. M. Ponsford, L. Sevgi, H. C. Chan, "An Integrated Maritime Surveillance System Based on Surface Wave HF Radars, Part II – Operational Status and System Performance", *IEEE Antennas and Propagation Magazine*, Vol.43 No.5, pp. 52 - 63, 2001

[8] L. Sevgi, "Target Reflectivity and RCS Interaction in Integrated Maritime Surveillance Systems Based on Surface Wave HF Radar Radars", *IEEE Antennas and Propagation Magazine*, Vol. 43 No. 2, pp. 36 - 51, 2001

[9] B. Çiber, *RCS Modeling and Reductions techniques with FDTD and MoM Numerical Methods, M.Sci. Thesis*, Istanbul Technical University, November 2001, Istanbul, Turkey

[10] K.S. Yee, "Numerical Solution of Initial Boundary Value Problems Involving Maxwell' s Equations", *IEEE Transactions on Antennas and Propagat.* Vol. 14 No.3, pp. 302 - 307, May 1966

[11] K. S. Kunz, R. J. Luebbers. *The Finite-Difference Time-Domain Method for Electromagnetics*, CRC Press, Boca Raton, FL, 1993

[12] C. Christopoulos, *The Transmission-line Modeling Method*, Oxford University Press, Oxford, 1995

[13] N. Engheta, W. D. Murphy, V. Rokhlin, M. S. Vassiliou, "The Fast Multipole Method (FMM) for Electromagnetic Scattering Problems", *IEEE Transactions on Antennas and Propagat.*, Vol. 40, pp. 634 - 641, June 1992.

[14] J. Song, C. C. Lu, W. C. Chew, "Multilevel Fast Multipole Algorithm for Electromagnetic Scattering by Large Complex Objects", *IEEE Transactions on Antennas and Propagat.*, Vol. 45, pp. 1488 - 1493, Oct. 1997.

[15] N. Geng, A. Sullivan and L. Carin, "Multilevel Fast Multipole Algorithm for Scattering from Conducting Targets Above or Embedded in a Lossy Half Space", *IEEE Trans. Geosci. Remote Sensing*, Vol. 38 No. 4, pp. 1561 - 1573, 2000

4

Microwave Network Design

4.1 INTRODUCTION

Commercial wireless communication has escalated the use of RF and microwave technologies in commercial products the since early 1990s, before which military applications were dominant. Commercial products have been developed under significantly different circumstances than military systems. Cost, design and product period, performance criteria, and market size are totally different in commercial products. Because of the highly demanding design process, engineers in commercial companies are often called upon to work in technical areas where they may not have a strong background. Most products in commercial wireless technology require some sort of antenna, transmission line, and microwave network theories. Although first-generation products have been able to use simple antennas and networks, today's market demands smarter and smaller products, which necessitates having a knowledge of intelligent signal processing and programming besides RF and microwave engineering background.

The majority of these products are based on printed circuit technology, and understanding of the microstrip transmission line is a must for engineers in this area. Microstrip antennas, multi-port microstrip networks, such as impedance matchers, filters, couplers, etc., based on multi-level printed circuit technology are inexpensive and are small in size, and production in large quantities is easy.

The purpose of this chapter is to describe the operation of microstrip lines, networks first, and to introduce powerful simulators that can be used in designing microstrip networks.

4.1 MICROSTRIP NETWORKS

When first introduced in the early 1950s, microstrip transmission lines did not make a strong impact, because of (i) striplines which support TEM (transverse electromagnetic) waves, and (ii) lack of active semiconductor components [1-3]. Microstrip lines and networks became popular with the advent of microwave transistors, and other active devices and complete circuits such as amplifiers, phase shifters, and mixers are then possible to realize in hybrid form.

Designing active and passive microwave networks in hybrid form requires an understanding of both transmission line theory and microstrip structure. Although it is assumed that the reader has these backgrounds, the following sections outline these topics briefly.

4.2.1 Transmission Line Theory

A transmission line is more than a set of long, parallel lines; it is a distributed-parameter physical system. Transmission lines are used to transmit electric energy and communication signals from one point to the other. The basic transmission line connects a source to a load. This may be a transmitter and an antenna, a television or a radio antenna and a receiver, one port of a coupler and a power meter, etc. A typical two-wire transmission line is sketched in Fig. 4.1.

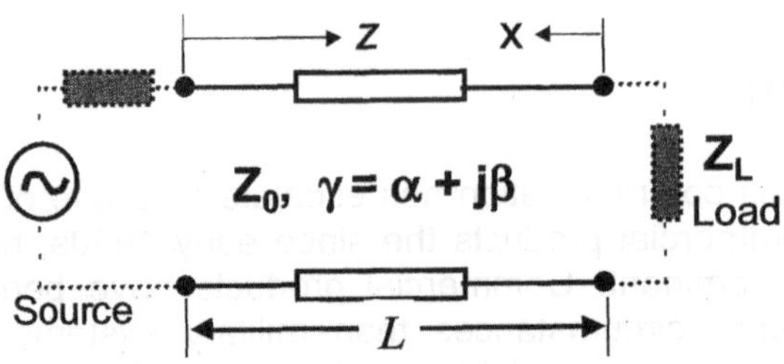

***Figure 4.1:** A two-wire transmission line*

A transmission line has two sets of constitutive parameters. The primary parameters are listed in Table 4.1. The secondary parameters are the characteristic impedance Z_0 and the complex propagation constant γ, which are mentioned in Fig. 4.1. The real part of the propagation constant is loss per unit length (α [dB/km]) and phase constant (β [radian/km]). For a loss-free transmission line $\alpha = 0$ and only phase shift occurs along the line.

Table 4.1: Primary parameters of a transmission line

Parameter	Symbol	Unit
Unit length resistance	R	Ω/m
Unit length conductivity	G	S/m
Unit length inductance	L	H/m
Unit length capacitance	C	F/m

There is a direct analogy between a uniform transmission line and the uniform plane wave. The field distributions for the uniform plane wave and for the uniform transmission line are both known as transverse electromagnetic (TEM) waves, where both electric and magnetic fields are perpendicular to the direction of propagation. That is, they both lie in the transverse plane. Because of this analogy the TEM wave approach may be used in analyzing transmission lines, which is the field approach. Another approach is the circuit (or network) approach where voltages and currents in terms of two independent variables, time and distance, can be used to establish sets of differential equations given as

$$-\frac{\partial}{\partial z}v(z,t) = Ri(z,t) + L\frac{\partial}{\partial t}i(z,t) \tag{4.1}$$

$$-\frac{\partial}{\partial z}i(z,t) = Gv(z,t) + C\frac{\partial}{\partial t}v(z,t) \tag{4.2}$$

which are called coupled transmission lines. The steady state sinusoidal form of these equations are

$$\frac{d}{dz}V(z) = -(R + j\omega L)I(z) \tag{4.3}$$

$$\frac{d}{dz}I(z) = -(G + j\omega C)V(z) \tag{4.4}$$

where $\omega = 2\pi f$ is the angular frequency in [rd/s]. By using these first-order coupled equations, second-order de-coupled transmission line equations can be obtained as

$$\frac{d^2}{dz^2}V(z) = \gamma^2 I(z) \tag{4.5}$$

$$\frac{d^2}{dz^2}I(z) = \gamma^2 V(z) \tag{4.6}$$

where

$$\gamma = \sqrt{(R + j\omega L)(G + j\omega C)} = \alpha + j\beta \tag{4.7}$$

is the complex propagation constant. In a lossless line (i.e., $R = G = 0$), complex propagation constant reduces to $\gamma = j\beta$. In this case, the solution can be written as

$$V(z) = V_1 e^{-j\beta z} + V_2 e^{j\beta z} \tag{4.8}$$

$$I(z) = \frac{1}{Z_0}\left(V_1 e^{-j\beta z} - V_2 e^{j\beta z}\right) \tag{4.9}$$

where Z_0 is the characteristic impedance given as

$$Z_0 = \sqrt{\frac{R + j\omega L}{G + j\omega C}} \rightarrow \sqrt{\frac{L}{C}} \text{ (when R=G=0).} \tag{4.10}$$

and V_1 and V_2 are complex constants that are going to be determined from the boundary conditions at the source end ($z = 0$) and load end ($z = L$). It is practical to measure distances from the load, therefore (4.8) and (4.9) reduce to

$$V(x) = V^+ e^{j\beta x} + V^- e^{-j\beta x} \tag{4.11}$$

$$I(x) = \frac{1}{Z_0}\left(V^+ e^{j\beta x} - V^- e^{-j\beta x}\right) \tag{4.12}$$

by replacing $z = L\text{-}x$ and using $V^+ = V_1 e^{-j\beta L}$ and $V^- = V_2 e^{j\beta L}$. Equations (4.11) and (4.12) express voltage and current at any junction in terms of traveling voltage and current waves towards the load (i.e., $V^+e^{j\beta x}$) and towards the source (i.e., $V^-e^{-j\beta x}$). These two terms are also called incident and reflected voltages, respectively.

Two important parameters in transmission line theory are the voltage reflection coefficient (Γ) and the standing wave ratio (SWR or s). The voltage reflection coefficient at the load is defined as

$$\Gamma(x=0) = \Gamma_0 = \frac{V^-}{V^+} = \frac{Z_L - Z_0}{Z_L + Z_0} \tag{4.13}$$

and can be found at any point along the transmission line from

$$\Gamma(x) = \Gamma_0 e^{-j2\beta x} \tag{4.14}$$

SWR is defined as a ratio of maximum voltage to the minimum voltage along the line as

$$SWR = s = \frac{V_{\text{max}}}{V_{\text{min}}} = \frac{1+|\Gamma_0|}{1-|\Gamma_0|} \Rightarrow |\Gamma_0| = \frac{s-1}{s+1}\,. \tag{4.15}$$

Finally, impedance of a lossless transmission line at any distance x, with characteristic impedance Z_0 and propagation constant β, terminated by a load impedance of Z_L is given as

$$Z_{in} = Z_0 \frac{Z_L + jZ_0 \tan(\beta x)}{Z_0 + jZ_L \tan(\beta x)}\,. \tag{4.16}$$

Understanding the physical meanings of these parameters are very important in dealing with numerical simulation techniques; therefore, they are given below.

4.2.2 Characteristic Impedance

Characteristic impedance is an important parameter in transmission line analysis. As given in (4.10), it is a parameter related to the primary line constants and is real for an ideal (loss-free) transmission line. Its definition can be stated as:

- The ratio of voltage divided by current at any point along an infinite-length homogeneous transmission line
- The ratio of voltage divided by current at any point along a reflection-free homogeneous transmission line

As understood from (4.16), the impedance seen at a distance x of a transmission line terminated with Z_L depends on the characteristic impedance, load impedance, and the distance. It may take any complex value even when characteristic and load impedances are real. It is important to note that the impedance of a line at any point reduces to the characteristic impedance when the load impedance is real and equal to the characteristic impedance.

4.2.3 Voltage Reflection Coefficient

A discontinuity occurs when impedance Z (different from Z_0) is inserted at any point along a transmission line. A discontinuity means an occurrence of an impedance value different from characteristic impedance of the line. Voltage reflection coefficient, given in (4.13) and (4.14), state that voltage reflection occurs at these discontinuities, and is directly determined from the load and characteristic impedances. A voltage at any point is the sum of incident and reflected voltages. When incident and reflected voltages are in (opposite) phase, a voltage maximum (minimum) occurs. There is no reflected voltage when the line is terminated by a load impedance equal to its characteristic impedance.

Reflection coefficient is a complex number with a modulus in the range of [0,1]. The extreme cases are as follows:

- When $Z_L = Z_0$, reflection coefficient is zero ($\Gamma = 0$). The power along the line totally transferred to the load. This is called *matched transmission line.*

- When $Z_L = 0$ (SC, short circuited), reflection coefficient is minus one ($\Gamma = -1$). In this case, the power along the line totally reflects back (with a phase shift of 180°).

- When $Z_L \to \infty$ (OC, open circuited), reflection coefficient is one ($\Gamma = 1$). In this case, the power along the line totally reflects back (without phase shift).

4.2.4 Standing Wave Ratio

Having a discontinuity along a line means voltage waves traveling in both directions. These voltage waves interfere, and voltage maxima and minima occur. These waves are called standing waves. The definition of the SWR is given in (4.15). SWR is a real number and is in the range $[1, \infty)$.

- SWR = 1 when there is no impedance discontinuity along the line (matched line)

- SWR $\to \infty$ when the line is terminated by OC or SC. Voltage fluctuation along the line is maximum and all power is reflected back to the source.

The input impedance of a SC or OC transmission line is pure imaginery, either inductive or capacitive, and depends on the length of the line. This is illustrated in Fig. 4.2. Some remarks could be listed as follows:

- Any reactance and/or capacitance can be obtained from an OC or SC transmission line by just selecting suitable length.

- The effects (voltage, current and impedance) along a transmission line are periodic with a period of half wavelength.
- Any reactance and/or capacitance obtained via a SC line with length l can also be obtained via OC line with length $l + \lambda/4$.

These properties of SC and OC terminated transmission lines are used in impedance matching, which will be explained in detail in the next section.

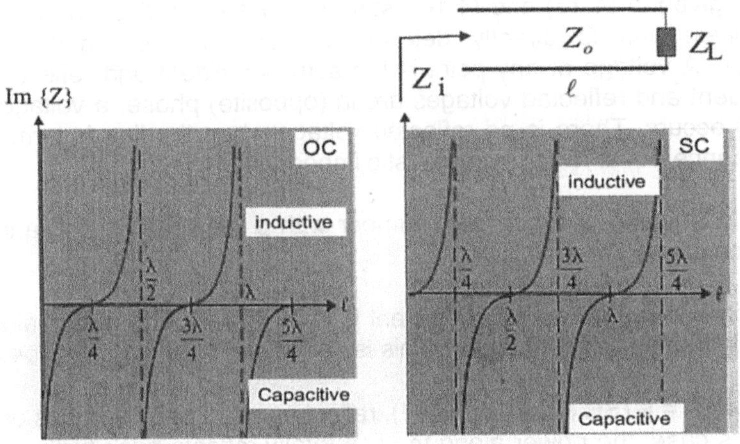

***Figure 4.2:** OC and SC terminated lines and impedance vs. length*

An important feature of a transmission line could be observed from (4.16) when the length of the line is equal to $\lambda/4$. In this case, it is reduced to

$$Z_{in} = \frac{Z_0^{\,2}}{Z_L} \tag{4.17}$$

which is also used in impedance matching circuits. Suppose Z_L is real and is different from the source impedance Z_{in}. The characteristic impedance of the quarter-wavelength line is then chosen as $Z_0 = \sqrt{Z_{in} \times Z_L}$ and impedance matching is automatically satisfied. If Z_L is not real, the distance from the load, which makes it real, is found first and then a quarter-wavelength line is added, which satisfies $Z_0 = \sqrt{Z_{in} \times Z_L}$. These are explained via characteristic examples in the following section.

4.2.5 Smith Chart and Impedance Matching

As seen from the equations above, transmission lines often involve manipulations with complex numbers. Although it is worthless to mention this as a problem with today's computers, it used to be very time consuming before widespread usage of numerical calculations. A means of reducing the labor without seriously affecting the accuracy of the results was found to be using transmission line charts. The most widely used one is the Smith chart [4].

The basic relationship upon which the Smith chart is constructed is (4.13). Since both Z_L and Γ are complex in general (i.e., $Z_L = x + jy$, $\Gamma = u + jv$), this bilinear equation can be thought as a mapping from polar to rectangular coordinates. The Smith chart is presented in Fig. 4.3.

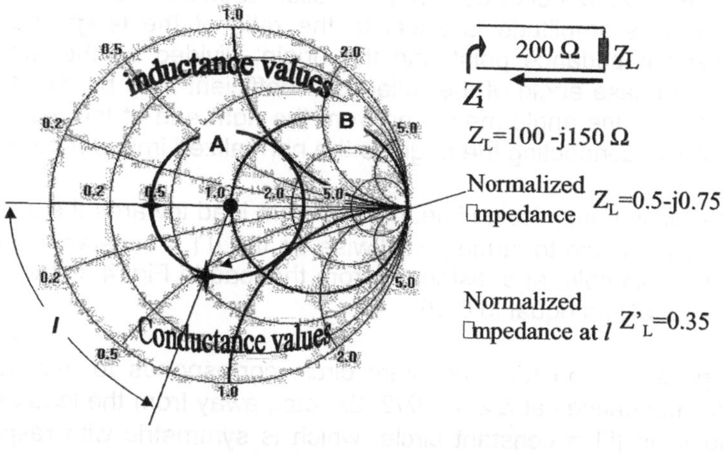

Figure 4.3: *The Smith chart*

The polar coordinates of the Smith chart are the amplitude and phase angle of the reflection coefficient. The Cartesian coordinates are the real and imaginary parts of the reflection coefficient. The entire chart lies within the unit circle $|\Gamma| = 1$. The chart also shows curves of normalized constant resistances and constant reactances. Normalized impedance is defined as the ratio of impedance over characteristic impedance of the line used. Some practical considerations about Smith chart could be listed as follows:

- The horizontal axis, which divides the chart into top and bottom semi-circles, is the normalized resistance axis. It starts from zero at left (SC point) and goes to infinity at right (OC point). The circles which are tangential to right boundary are normalized resistance = constant circles. The center of the chart is 1, where $Z_L = Z_0$. On the left (right) $Z_L < Z_0$ ($Z_L > Z_0$). The circle which is tangential to both center point and right boundary is labeled with the letter B in Fig. 4.3. On this loci, real part of the normalized impedance is equal to 1.

- The arcs above the resistance axis (initiated from right) are normalized inductance values (i.e., $X_L/Z_0 = \omega L/Z_0$). The arcs below the resistance axis correspond to normalized capacitances (i.e., $Y_L/Z_0 = 1/\omega CZ_0$). Therefore, normalized complex impedances with (positive) negative imaginary part are located at the top (bottom) semi-circle.

- Any point on the Smith chart corresponds to a normalized complex impedance. To locate an impedance on the chart, first it is normalized to the characteristic impedance of the transmission line. For example, the load impedance Z_L = 100 - j150 Ω is normalized with respect to the line impedance Z_0 = 200 Ω, and z_L = 0.5 - j0.75 is obtained. Then, the intersection of the real 0.5 circle and negative 0.75 arc are marked on the chart. This point is marked with a star in Fig. 4.3.

- The circle A is the loci of constant reflection coefficient amplitude (i.e., $|\Gamma|$ = constant. This amplitude is equal to the ratio of the length between the normalized impedance point and the origin, divided by the radius of the chart. The phase angle of the reflection coefficient may be directly read on the chart. It is the angle measured from the right end of the resistance axis up to the line connecting the origin to the normalized impedance point.

- For a lossless transmission line, moving from load towards the source along the line correspond to turning clockwise on the $|\Gamma|$ = constant circle, or vice versa. For example, at a distance *l* from the load in Fig. 4.3, the impedance becomes real and equal to 0.35.

- A complete tour on $|\Gamma|$ = constant circle corresponds to half-wavelength. That is, impedances at $\lambda/2$, λ, $3\lambda/2$, 2λ, etc., away from the load are also Z_L. The point on $|\Gamma|$ = constant circle, which is symmetric with respect to the origin is the normalized admittance at the same point, or the normalized impedance at $\lambda/4$ away.

- At the center of the chart, reflection coefficient is zero (SWR is equal to 1). The philosophy of impedance matching is to bring any point on the chart to the center. This is illustrated in Fig. 4.4. The normalized impedance at the load point (point A in the figure) is first moved to point B along $|\Gamma|$ = constant circle and the distance *d* is determined (step one). At this point on the transmission line a certain capacitance is added and impedance is matched at this point (step two). That is, the point B on the chart is transferred to point C (to the center).

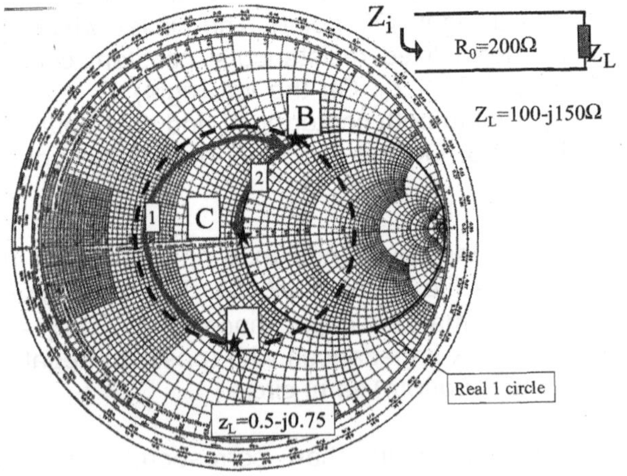

Figure 4.4: *Impedance matching on the Smith chart*

The procedure and steps of impedance matching with a single series OC stub on the Smith chart is given in Fig. 4.5.

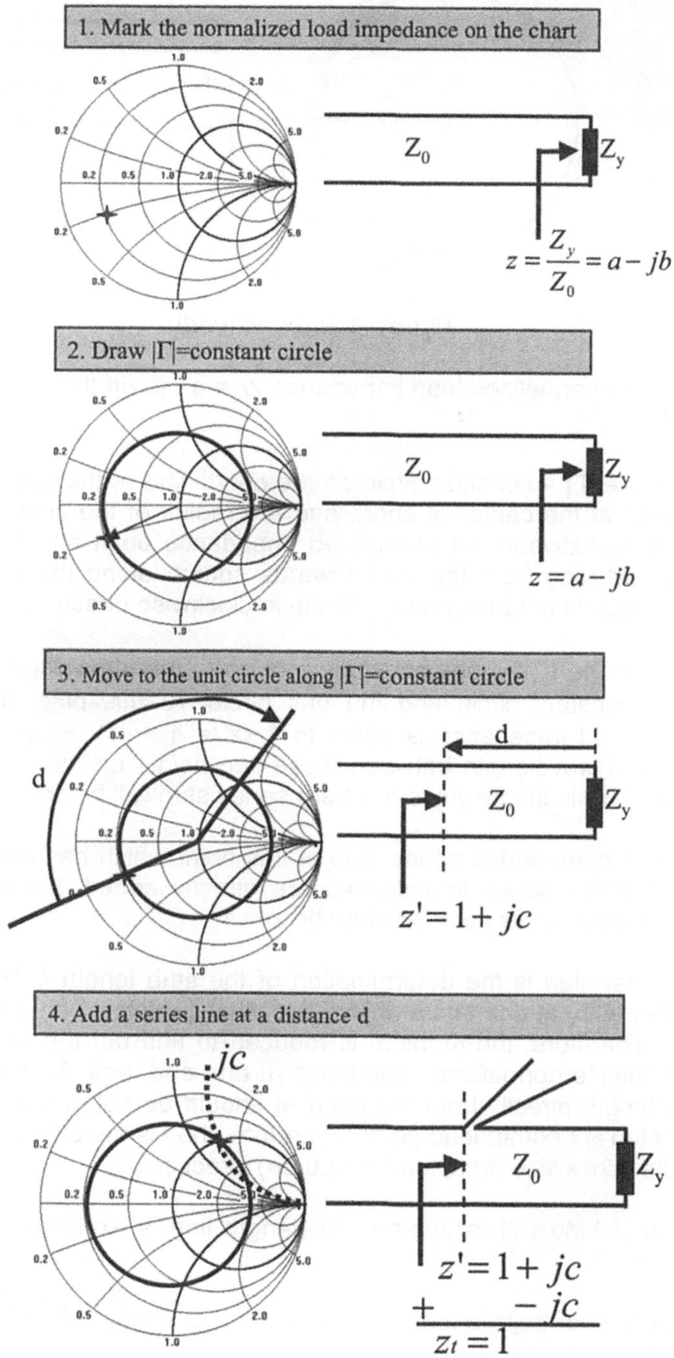

Figure 4.5: *Impedance matching with series stub*

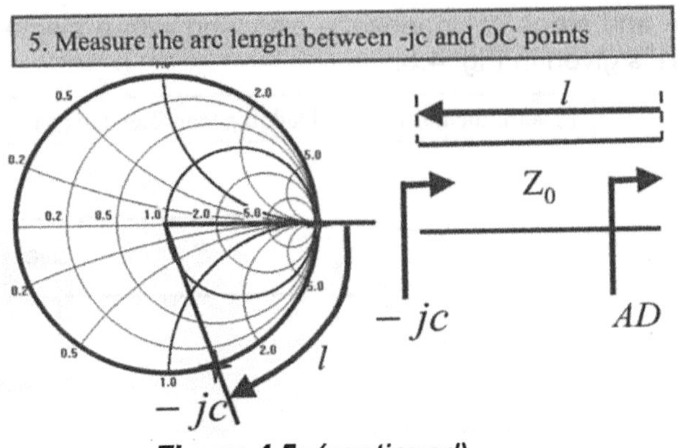

***Figure 4.5:** (continued)*

Step 1: Mark the normalized load impedance $z_L = a - jb$ (in this case) on the Smith chart.

Step 2: Draw the $|\Gamma|$ = constant circle on the chart. This is the circle with its center located at the center of chart, and the radius of the circle is equal to the distance between the normalized impedance point and the center of the chart. Moving from the load towards source along the transmission line corresponds to turning on this circle in clockwise direction.

Step 3: Turn on the $|\Gamma|$ = constant circle and go to the intersection point between $|\Gamma|$ = constant circle and the unit circle. At this point the real part of normalized impedance is equal to 1 ($z' = 1 + jc$). Read the distance in terms of wavelength between these two points on the outer scale of the chart. This is the distance d, where series stub will be connected.

Step 4: Add an open-ended series stub at this point which has a capacitive effect of jc. Since series impedances are directly added, the total normalized impedance at this point will then be $z_t = 1$.

Step 5: The last step is the determination of the stub length l. The stub is open ended (OC) at one end and jc at the other (which is connected to the main line. Therefore, the problem is reduced to find out the length of the stub with infinite normalized impedance at one end, and jc at the other. Read this length directly from the chart. It should be noted that, the OC end of the stub is another load point. Therefore, the l is the distance from OC to jc in the clockwise (from load to source) direction.

Impedance matching with a quarter-wavelength line is explained step by step in Fig. 4.6:

Step 1: Mark the normalized load impedance $z_L = a - jb$ (in this case) on the Smith chart.

Step 2: Draw the $|\Gamma|$ = constant circle on the chart.

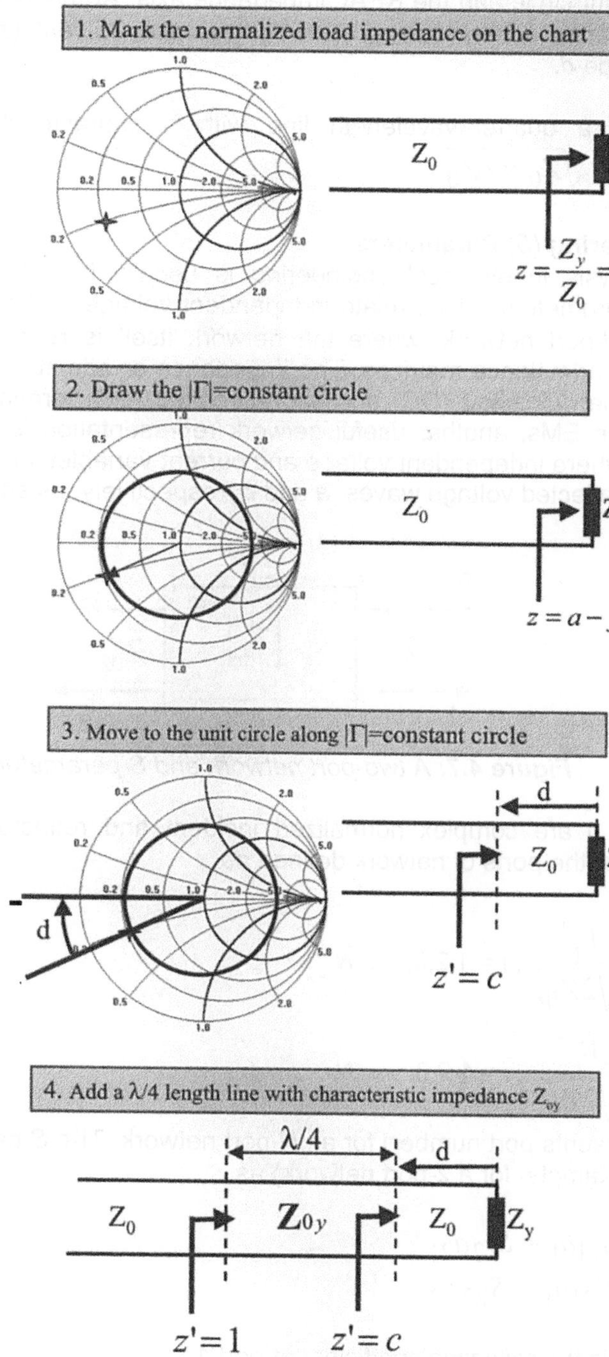

Figure 4.6: *Impedance matching with a quarter-wavelength line*

Step 3: Turn on the $|\Gamma|$ = constant circle until the intersection point between $|\Gamma|$ = constant circle and the REAL impedance axis. At this point the normalized impedance (when looking towards the load) is real ($z' = c$). Read the distance d.

Step 4: Add a quarter-wavelength line with a characteristic impedance of $Z_0 y = \sqrt{Z_0 \times cZ_0}$.

4.2.6 Scattering (*S*) Parameters

Network analysis in electrical engineering is usually based on impedance or admittance parameters, which relate independent voltage and current variables at ports of an *N*-port network, where the network itself is represented as $N \times N$ impedance or admittance matrices. The impedance or admittance parameters are also called open circuited (OC) and short-circuited (SC) termination parameters, respectively. In EMs, another useful network representation is the scattering (S) parameters, where independent voltage and current variables are replaced with the incident and reflected voltage waves, *a* and *b*, respectively, as shown in Fig. 4.7.

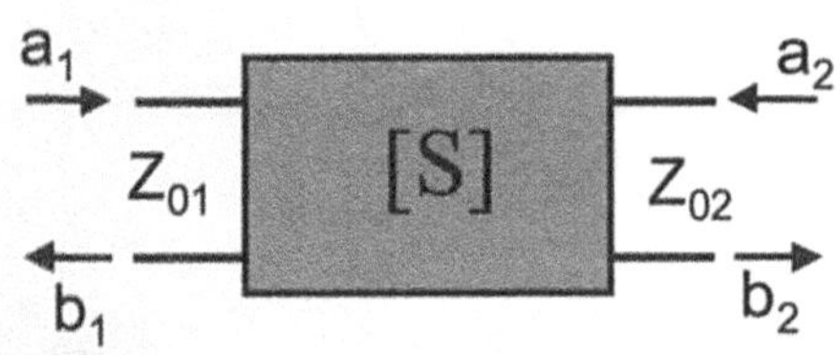

Figure 4.7: *A two-port network and S parameters*

Here, *a* and *b* are complex normalized incident and reflected voltage waves, respectively, at the ports of network defined as

$$a_i = \frac{V_i^+}{\sqrt{2Z_{0i}}}, \; i = 1,2,3, \ldots, N \tag{4.18}$$

$$b_i = \frac{V_i^-}{\sqrt{2Z_{0i}}}, \; i = 1,2,3, \ldots, N \tag{4.19}$$

(where *i* represents port number) for an *N*-port network. The *S*-parameters are then defined (for example, for a 2-port network) as

$$b_1 = S_{11}a_1 + S_{12}a_2 \tag{4.20}$$

$$b2 = S_{21}a_1 + S_{22}a_2 \tag{4.21}$$

where

S_{11} : is the reflection coefficient at port 1
S_{11} : is the reflection coefficient at port 2
S_{21} : is the forward transfer coefficient (from port 1 to port 2)
S_{12} : is the backward transfer coefficient (from port 2 to port 1)

under a special condition called *matched termination*. The reflection coefficient at port 1 is obtained as

$$S_{11} = \frac{b_1}{a_1} \Big|_{a_2=0} \tag{4.22}$$

Here, $a_2 = 0$ means that there is no incident voltage wave at port 2 and total power is transferred to the load connected to this port. Similarly, forward transfer coefficient S_{21} is defined as

$$S_{21} = \frac{b_2}{a_1} \Big|_{a_2=0} . \tag{4.23}$$

This is illustrated in Fig. 4.8. One of nice things about S parameters is that they are intuitive. This is not true for impedance or admittance parameters at high frequencies. S_{21} is simply the forward gain or loss of the two-port device, expressed in linear form.

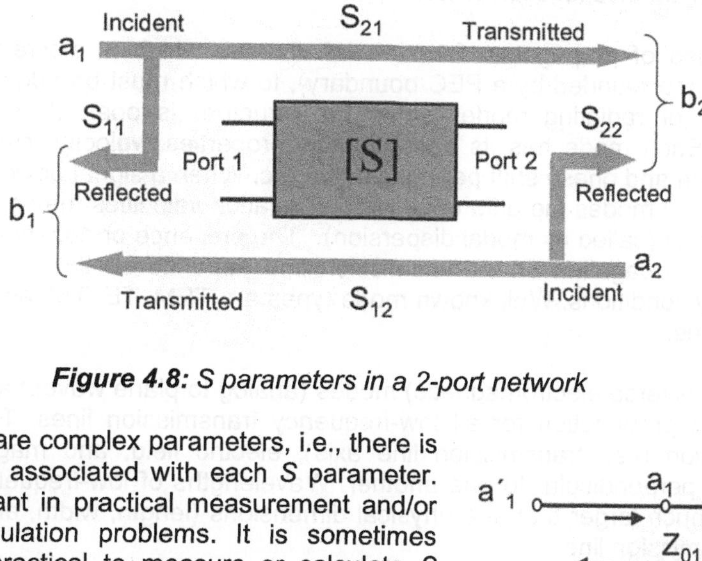

Figure 4.8: *S parameters in a 2-port network*

S parameters are complex parameters, i.e., there is a phase angle associated with each S parameter. This is important in practical measurement and/or numerical simulation problems. It is sometimes difficult or impractical to measure or calculate S parameters right at the junctions (or port entrance); therefore a connecting line is used for this purpose as illustrated in Fig. 4.9.

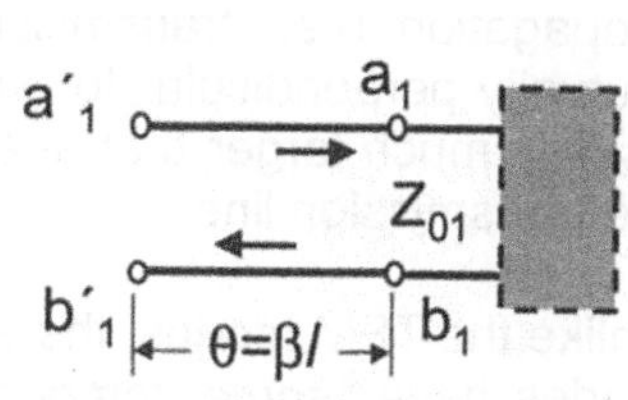

Figure 4.9: *Phase shift in S parameters*

In this case, incident and reflected voltage waves at the other end of the connecting line having length l may be written as

$$a_1' = a_1 e^{-j\beta l} \tag{4.24}$$

$$b_1' = b_1 e^{j\beta l} \tag{4.25}$$

which will yield, for example, reflection coefficient at port 1, as

$$S'_{11} = \frac{b'_1}{a'_1}\Big|_{a_2=0} = S_{11}e^{2j\beta l}\,. \tag{4.26}$$

Since S_{11} is of interest, it can directly be calculated from (4.26).

4.2.7 Modes

Electromagnetic theory is well established with the Maxwell equations. The second-order decoupled wave equation derived from Maxwell equation can be reduced to two sets of equations in waveguiding environments; the transverse wave equations, which is nothing but the eigenvalue problem, and the longitudinal wave equations, which are the transmission line equations. Eigenfuntions and eigenvalues are derived from the transverse wave equations. Eigensolutions are the solutions of homogeneous (source-free) Helmholtz wave equation and are called the modes of the structure. They are distinguished with amplitude distribution of the fields along the transmission line and their propagation constants. Modal excitations are derived from the longitudinal wave equations, which is a Green's function problem.

The modes of propagation form an infinite set, which is discrete for a closed structure (surrounded by a PEC boundary), to which must be added a continuous spectrum of radiating modes when the structure is open (fields extending to infinity). Each mode has its characteristic properties; velocity, cut-off frequency, attenuation and phase shift per unit length, etc. When a signal power is propagated over several modes the differences in propagation velocities may cause distortions of the signal (called as modal dispersion). The presence or absence of a particular mode of propagation on a transmission line depends on the excitation and the boundary conditions. Well-known mode types are TEM, TE, TM, and hybrid (EH or HE) modes.

TEM (transverse electromagnetic) modes (analog to plane waves) are the principal modes of propagation for all low-frequency transmission lines. The direction of propagation (i.e., transmission line axis), electric field, and magnetic field are mutually perpendicular to one another. Wavelengths of low-frequency signals are usually much larger than the physical dimensions (length, width, and/or height) of the transmission line.

Unlike the TEM modes, the electric field *E*, or the magnetic field *H*, or both, of other modes have some components along the effective direction of propagation. TE (transverse electric) types are the modes with no electric field component along the transmission line (propagation direction). Similarly, TM (transverse magnetic) are the modes which do not have magnetic field components along the propagation direction. Hybrid modes have both electric and magnetic field components along the transmission line.

Figure 4.10 illustrates some types of transmission lines and electric field vector distribution. Here, coaxial, microstrip, triple strip, and suspended strip lines are shown, all of which support propagation of TEM type modes. As shown in the figure, all but microstrip lines support trapped, propagating modes. Microstrip lines also support leaky modes as easily understood from the electric field vector distribution.

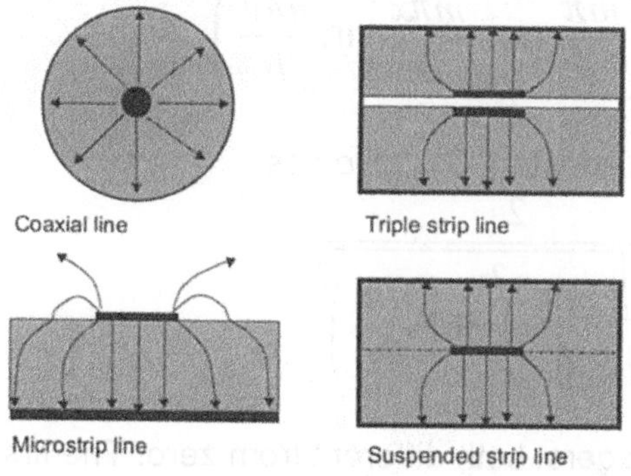

Figure 4.10: *Basic transmission lines and electric field vectors*

4.2.8 Waveguides

A waveguide (with either rectangular cross section as pictured in Fig. 4.10, or circular) has different characteristics among the other transmission lines. All transmission lines given in Fig. 4.10 are two-conductor lines, which supports TEM and/or quasi-TEM, and they have low-pass characteristics with different upper cut-off frequencies. On the other hand, the waveguide in Fig. 4.11 has a high-pass characteristic with a lower cut-off frequency.

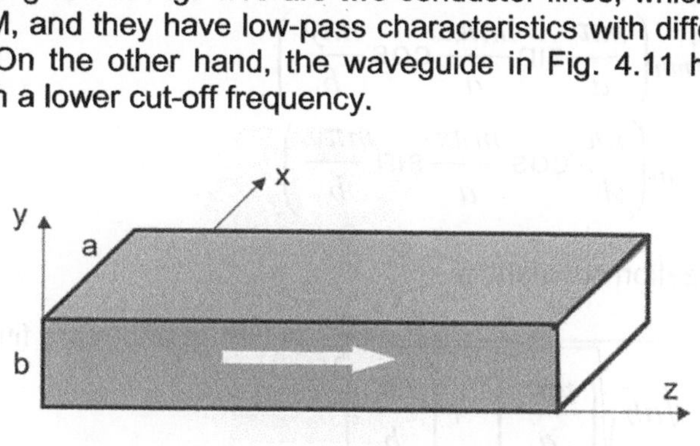

Figure 4.11: *Rectangular waveguide*

The waveguide with width *a* and height *b* (*a* > *b*) supports TM and TE modes and the solutions can be found analytically, since the eigenvalue equation is separable in the Cartesian coordinate system. The TM mode functions are

$$E_x = -C_{mn}\left(\frac{m\pi}{a}\cos\frac{m\pi x}{a}\sin\frac{n\pi y}{b}\right) \tag{4.27}$$

$$E_y = -C_{mn}\left(\frac{n\pi}{b}\sin\frac{m\pi x}{a}\cos\frac{n\pi y}{b}\right) \tag{4.28}$$

$$H_x = C_{mn}\left(\frac{n\pi}{b}\sin\frac{m\pi x}{a}\cos\frac{n\pi y}{b}\right) \tag{4.29}$$

$$H_y = -C_{mn}\left(\frac{m\pi}{a}\cos\frac{m\pi x}{a}\sin\frac{n\pi y}{b}\right) \tag{4.30}$$

with the normalization constant C_{mn} given as

$$C_{mn} = \frac{2}{\sqrt{ab}\sqrt{\left(\frac{m\pi}{a}\right)^2 + \left(\frac{n\pi}{b}\right)^2}}. \tag{4.31}$$

Here, m and n are integers, both different from zero. The first TM mode is TM_{11} and is called *dominant mode*. Similarly, TE mode functions are

$$E_x = C_{mn}\left(\frac{n\pi}{b}\cos\frac{m\pi x}{a}\sin\frac{n\pi y}{b}\right) \tag{4.32}$$

$$E_y = -C_{mn}\left(\frac{m\pi}{a}\sin\frac{m\pi x}{a}\cos\frac{n\pi y}{b}\right) \tag{4.33}$$

$$H_x = C_{mn}\left(\frac{m\pi}{a}\sin\frac{m\pi x}{a}\cos\frac{n\pi y}{b}\right) \tag{4.34}$$

$$H_y = C_{mn}\left(\frac{n\pi}{b}\cos\frac{m\pi x}{a}\sin\frac{n\pi y}{b}\right) \tag{4.35}$$

and the normalization constant is

$$C_{mn} = \frac{2}{\sqrt{ab}\sqrt{\left(\frac{m\pi}{a}\right)^2 + \left(\frac{n\pi}{b}\right)^2}} \quad \text{(when m and n are finite)} \tag{4.36a}$$

$$C_{mn} = \frac{\sqrt{2}}{\sqrt{ab}\sqrt{\left(\frac{m\pi}{a}\right)^2 + \left(\frac{n\pi}{b}\right)^2}} \quad \text{(when m or n vanishes).} \tag{4.36b}$$

The cut-off frequencies of both TM and TE modes are

$$f_{mn} = \frac{c}{2}\sqrt{\left(\frac{m}{a}\right)^2 + \left(\frac{n}{b}\right)^2} \tag{4.37}$$

When both m and n are non-zero the TE_{mn} and TM_{mn} modes are *degenerate*, i.e., modes with different transverse variations have the same transverse wavenumber,

$$k_x = \sqrt{\left(\frac{m\pi}{a}\right)^2 + \left(\frac{n\pi}{b}\right)^2}. \tag{4.38}$$

The longitudinal wavenumber (i.e., the eigenvalue) is

$$\beta = \sqrt{k^2 - k_x^2} = \sqrt{\left(\frac{2\pi}{\lambda}\right)^2 - k_x^2} \tag{4.39}$$

Phase and group velocities are

$$v_p = \frac{\omega}{\sqrt{k^2 - k_x^2}}, \quad v_g = \frac{c^2\sqrt{k^2 - k_x^2}}{\omega} \tag{4.40}$$

As mentioned above, waveguides act as high-pass filters. The eigenvalue of the lowest (dominant) mode is the cut-off frequency below which no transmission occurs. Between the cut-off and next eigenvalue, single-mode propagation occurs. When frequency is higher than the next mode, multi-modes propagate; and since their velocities and attenuation are different, signal distortion (dispersion) occur.

4.2.9 Microstrip Lines

Planar transmission lines are inhomogeneous, since they consist of two propagation media, like the dielectric substrate and air. Inhomogeneous lines do not support TEM modes. This is because the propagation constant of a TEM wave depends only upon frequency and the material properties of the medium in which it propagates. On the other hand, the dominant mode of many inhomogeneous transmission lines, such as an insulated two-wire pair, loaded coaxial and microstrip lines, etc., still behaves like a TEM mode at low frequencies. This is therefore called quasi-TEM. The longitudinal components of the quasi-TEM mode, while non-zero, have negligible amplitudes. The propagation of this mode can be determined approximately from electrostatics. The inhomogeneous structure is replaced by an equivalent homogeneous one, which is characterized by the effective permittivity obtained from electrostatic study of the structure. This simplified model is valid only at low frequencies.

The simple microstrip line is depicted in Fig. 4.12. It is a structure having a narrow PEC ribbon on one face of the dielectric substrate; the other face is entirely PEC.

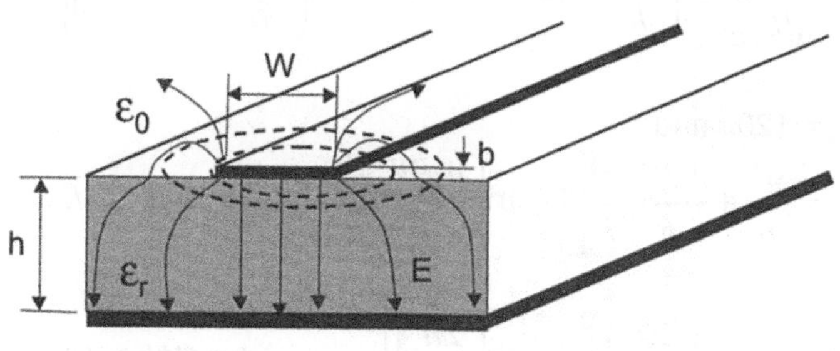

Figure 4.12: *Simple microstrip line*

The characteristic physical parameters of the microstrip line are:

- Relative permittivity ε_r of the substrate: The higher the permittivity and the more concentrated the fields within the substrate, the lower (undesired) the radiated power.
- The substrate thickness *h*: Generally is a fraction of a millimeter.
- The width *W* of the PEC ribbon: This width is of the same order as the substrate thickness, *h* ($0.1 \le W/h \le 3.0$). The characteristic impedance of the microstrip line is controlled by varying *W*.
- The thickness *b* of the PEC ribbon: Generally is negligible ($b/h << 1$).

As shown in Fig. 4.12, electric field lines go from the ribbon to the ground plane, most of which are concentrated underneath the ribbon. Some lines partially extended into the free space above the substrate. The magnetic field lines encircle the microstrip ribbon and extended above the substrate. The presence of electric field lines in the air reduces the effective dielectric constant seen by the propagating wave. If all the fields existed between the ribbon and ground plane, the dielectric constant would be that of the substrate. Instead, it is somewhat less than that constant, and it depends on ribbon width, substrate dielectric constant, and height.

A large number of studies have been done on, and many analysis techniques have been applied to, microstrip lines [5 - 7]. Because of the engineering importance of the microstrip lines, much effort has been placed on developing closed-form expressions for the transmission line parameters of microstrips. Unfortunately, no single exact formulation for the characteristic impedance and effective dielectric constant has been obtained yet. Nevertheless, widely accepted approximate, moderately accurate, empirical expressions are available, such as

$$Z_0 = \frac{\eta_0}{2\pi\sqrt{\varepsilon_{reff}}}\ln\left\{\frac{8h}{W_e} + 0.25\frac{W_e}{h}\right\} \qquad \text{for } W/h \le 1 \qquad (4.41)$$

$$Z_0 = \frac{\eta_0}{\sqrt{\varepsilon_{reff}}}\left\{\frac{W_e}{h} + 1.393 + 0.667\ln\left(\frac{W_e}{h} + 1.444\right)\right\}^{-1} \qquad \text{for } W/h \ge 1 \qquad (4.42)$$

where $\eta_0 = 120\pi$ and

$$\frac{W_e}{h} = \frac{W}{h} + \frac{1.25}{\pi}\frac{t}{h}\left[1 + \ln\left(\frac{4\pi W}{t}\right)\right] \qquad \text{for } W/h \le 1/2\pi \qquad (4.43)$$

$$\frac{W_e}{h} = \frac{W}{h} + \frac{1.25}{\pi}\frac{t}{h}\left[1 + \ln\left(\frac{2h}{t}\right)\right] \qquad \text{for } W/h \ge 1/2\pi \qquad (4.44)$$

and

$$\varepsilon_{r_{eff}} = \frac{\varepsilon_r + 1}{2} + \frac{\varepsilon_r - 1}{2} F\left(\frac{W}{h}\right) - C \tag{4.45}$$

where

$$F\left(\frac{W}{h}\right) = \left(1 + 12\frac{h}{W}\right)^{-0.5} + 0.04\left(1 - \frac{W}{h}\right)^2 \quad \text{for } W/h \le 1 \tag{4.46}$$

$$F\left(\frac{W}{h}\right) = \left(1 + 12\frac{h}{W}\right)^{-0.5} \quad \text{for } W/h \ge 1 \tag{4.47}$$

$$C = \frac{\varepsilon_r - 1}{4.6} \frac{t/h}{\sqrt{W/h}} \quad (t: \text{ribbon thickness}) \tag{4.48}$$

Graphical illustration of (4.41)-(4.48) is given in Fig. 4.13, where the Z_0 vs. *W/h* ratio for variety of substrate relative permittivity is shown. When the thickness of the strip is negligible (compared to its width), characteristic impedances between 10/20 Ω up to150/250 Ω can be obtained by changing the width, the substrate height, or the substrate relative permittivity.

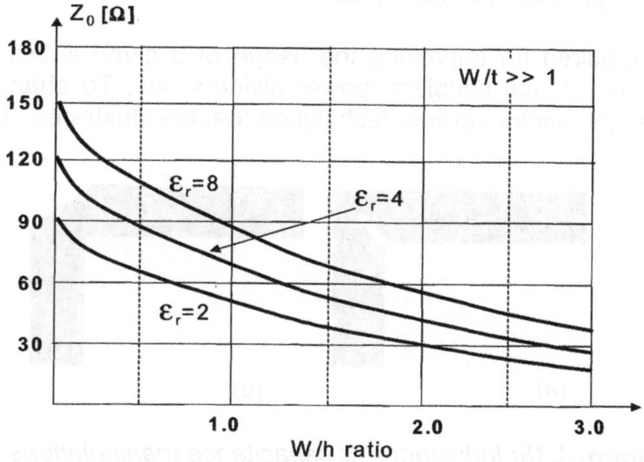

Figure 4.13: *Characteristic impedance vs. W/h ratio of a microstrip line*

As illustrated in the figure

- The higher the substrate relative permittivity, the higher the characteristic impedance.
- The higher the *W/h* ratio, the lower the characteristic impedance.
- The wider the strip width (when substrate height is same), the lower the characteristic impedance.
- The narrower the strip width (when substrate height is same), the higher the characteristic impedance.

Microstrip networks are more than a single straight transmission line. Lines are often bent because of, for example, circuit topology and space limitations. Changes in ribbon width are often used when different characteristic impedances are required. Junctions, intersections of more than two lines, are used in various ways. OC stubs are used for impedance matching. A gap or a notch can be introduced for different purposes. These are all abrupt discontinuities in microstrip lines, which alter the electric and magnetic field distributions. Some mictrostrip discontinuities and their equivalent circuits are presented in Fig. 4.14.

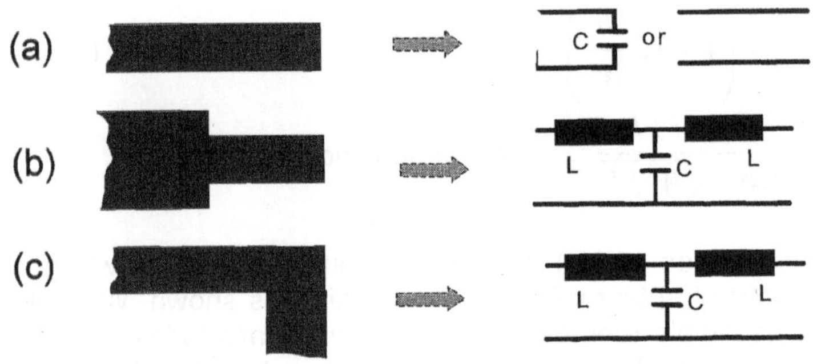

Figure 4.14: *Microstrip discontinuities and their equivalent circuits: (a) OC, (b) step change, (c) bent*

Corners are required for improving the usage of a given substrate area and for networks like directional couplers, power dividers, etc. To change inductance or capacitance of the microstrip line, techniques that are illustrated in Fig. 4.15 can be used.

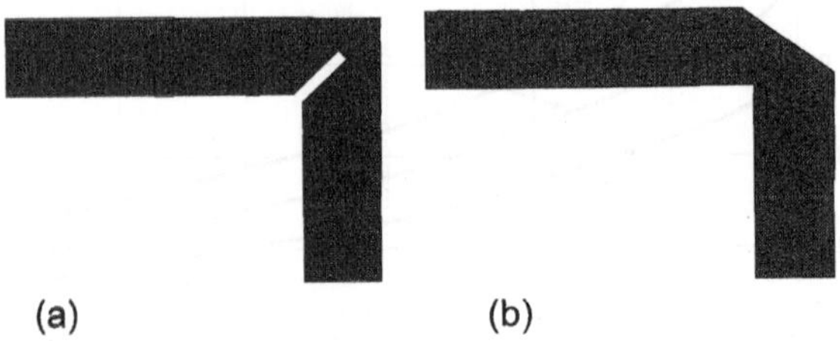

Figure 4.15: *Inductance or capacitance manipulations (a) increase inductance, (b) decrease capacitance*

4.3 SAMPLE MICROWAVE SIMULATORS

Two sample simulators F-MSTRIP and TLMMS are developed for the analysis of microstrip networks. Their structures and capabilities are summarized in Sections 4.3.1 and 4.3.2, respectively. Numerical examples are presented in Section 4.3.3.

4.3.1 The Structure of F-MSTRIP

F-MSTRIP is a sample simulator based on the FDTD method. It can be used in the microstrip network analysis. The structure of the F-MSTRIP is given in Fig. 4.16.

The simulator F-MSTRIP first reads the pre-selected parameters from the parameter file F-MSTRIP.PAR and the user-defined parameters from the input file MSTRIP.INP and runs for short pulse excitation. The output, the TD results, are written to four different files. S parameters S_{11} and S_{21} are calculated off-line, via SPAR routine.

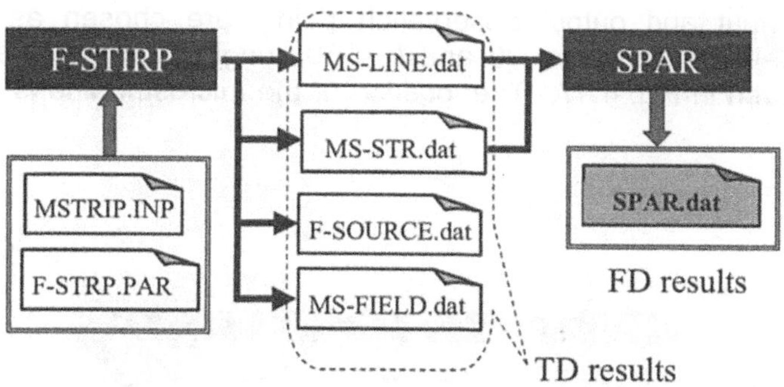

Figure 4.16: *The structure of F-MSTRIP*

F-MSTRIP is prepared to investigate different microstrip structures as illustrated in Fig. 4.17. Here, six different scenarios can be selected for the simulations.

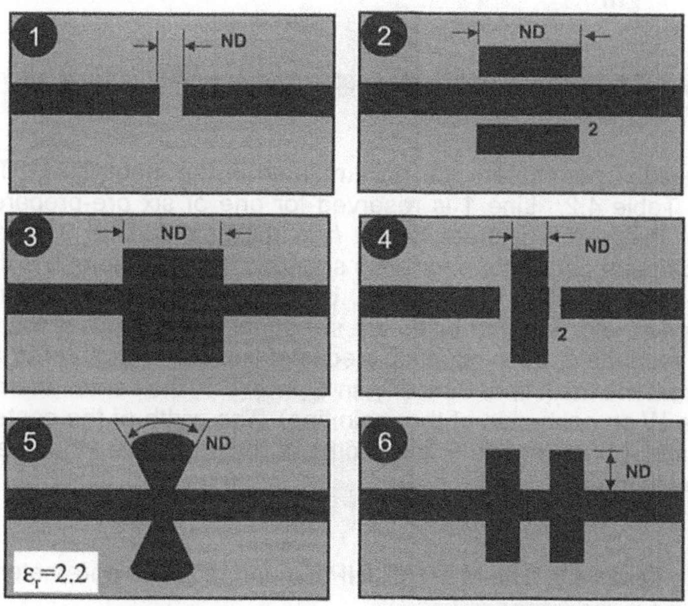

Figure 4.17: *Pre-prepared simple microstrip structures in F-MSTRIP*

Some parameters are fixed and included in the F-MSTRIP.PAR file. A 60 × 24 × 100 FDTD space is reserved for the TD simulations. An 8-cell PML terminates the FDTD computation space. For all the scenarios the user supplies the height h of the dielectric substrate. Relative permittivity is fixed to $\varepsilon_r = 2.2$. The width of the strip, W is set equal to height of the substrate (i.e., $W = h$). Both are discretized with 6 cells. Therefore, cell sizes are automatically chosen as $\Delta x = \Delta y = \Delta z = h/6$. Time step Δt is specified as $c/(2\Delta x)$. A y-directed line source is injected at k = 18 (along z). The source has a Gaussian character in time in order to investigate low-frequency (including DC) effects for quasi-TEM modes. Its bandwidth is automatically determined from Δx to sample minimum wavelength with 40 spatial samples. Input and output observation points are chosen as (30,5,28) and (30,5,70). Cells between k = 30 and k = 60, longitudinally are reserved for the structures given in Fig. 4.17. The location of the microstrip line is pictured in Fig. 4.18.

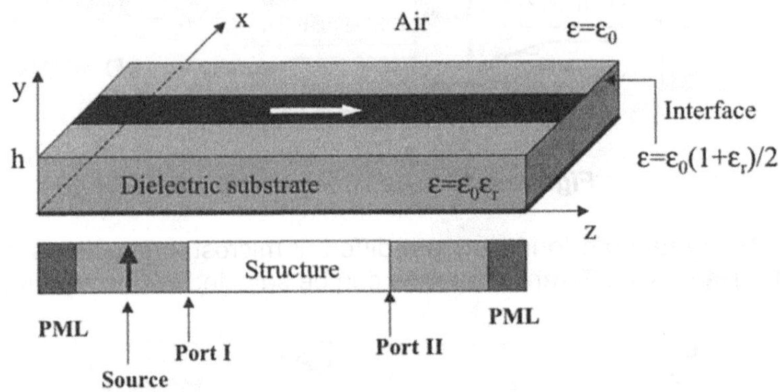

Figure 4.18: *Location of the microstrip line in FDTD computation space*

The user-supplied parameters are accumulated in the input file MSTRIP.INP and are listed in Table 4.2. Line 1 is reserved for one of six pre-prepared scenarios. Line 2 is for dielectric substrate height h in milimeters. Line 3 is for ND, which represents different parameters for each scenario. For example, it is the gap length given by number of cells in Scenario 1, but angle of butterfly stub in Scenario 5, given in degrees, etc. All other sizes are set equal to the width of the strip line. For example, in scenario 2, each coupling pieces of line has a width of W and located 2 cells apart from the main line. Similarly, in Scenario 3, the center line has a width of 3W (extends W on each side of the main line). The width of the center part is also 3W in Scenario 4. In Scenario 6, the widths of the stubs are W and are located W apart from each other.

Table 4.2: The MSTRIP.INP file and required parameters

Row No:	Parameter
1	Scenario type [1, 6]
2	Dielectric substrate height, h [mm]
3	The parameter ND [in number of cells]
4	Maximum iteration time NSTOP [integer]

F-MSTRIP is developed for TD analysis of EM fields under the strip for a chosen scenario. It is prepared to investigate broadband S parameter variations. In F-MSTRIP, the user first supplies required parameters in the input file, MSTRIP.INP. Then, the simulation starts and the procedure continues as follows:

- TD simulations are performed twice inside the algorithm. First, TD variations are obtained for a homogeneous infinite microstrip line. Time variation of excitation pulse is accumulated in SOURCE.dat as two columns. Vertical component of electric field (E_y) are written in an output file FIELD.dat as five columns, time [nsec], E_{y1}, E_{y2}, E_{y3} and E_{y4}, at four different z points k = 30, 45, 60 and 75, respectively.

- Input and output voltages and currents are recorded in MS-LINE.dat, as five columns (time [nsec], V_1, I_1, V_2, I_2). Voltages are calculated from the integration of y-component of electric fields along a vertical line under the strip. Currents are calculated according to Ampere's Law (tangential components along a closed path surrounding the strip).

- In the second run, the user-selected structure is taken into account and input/output voltages and currents are recorded in MS-STR.dat file.

- S parameters are calculated according to (4.22) and (4.23). In order to do that, frequency-domain values of normalized incident and reflected voltage waves at port 1 and reflected voltage wave at port 2 are required. These are obtained from the twice simulated TD data. The normalization is done with respect to the characteristic impedances of microstrip lines of port 1 and 2. Since the same microstrip lines are used in all the scenarios presented, characteristic impedances of the input and output ports are same and normalization is not required.

- The output file MS-STR.dat contains total voltage and current waves at the input port, but reflected voltage wave at the output (which is equal to the transmitted voltage wave of the strip), since the microstrip line is terminated by PML blocks at both ends. This simulates an infinite-length microstrip line.

- Subtraction of the incident voltage wave obtained in the first run, from the total voltage wave obtained in the second run, yields time variation of the reflected voltage wave at the input. From these voltage variations S parameters are obtained via off-line DFT, which is done in SPAR routine.

- In other words, if $V_1^+(t)$, $V_1^t(t)$, and $V_2^-(t)$ are the incidence voltage at the input port, total voltage at the input port, and reflected voltage at the output port, respectively, then after applying DFT, S_{11} and S_{21} are obtained as

$$S_{11}(f) = \frac{V_1^-(f)}{V_1^+(f)} = \frac{V_1^t(f) - V_1^+(f)}{V_1^+(f)}, \qquad S_{21}(f) = \frac{V_2^-(f)}{V_1^+(f)} \tag{4.49}$$

4.3.2 The structure of TLMMS

TLMMS is a sample simulator based on the TLM method. It can also be used in analyzing sample scenarios given in Fig. 4.17. The structure of the TLMMS is presented in Fig. 4.19.

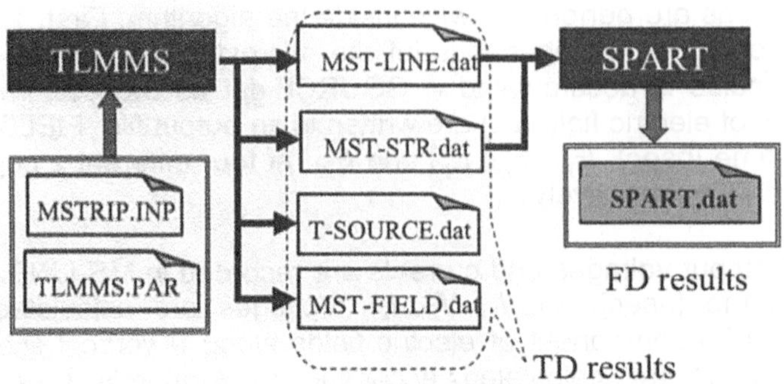

Figure 4.19: *The structure of TLMMS*

All the files used in TLMMS are exactly the same with the F-MSTRIP, except the names contain the letter *T*. For example, the source file "F-Source.dat" in F-MSTRIP package becomes "T-Source.dat" in TLMMS. The same input file, MSTRIP.INP, is used without any change, therefore Table 4.2 is also valid for the TLMMS simulator.

There is a slight difference in FD domain files. F-MSTRIP uses SPAR module for *S* parameter calculation in FD, while TLMMS uses SPART module for the same purpose. Therefore, the reader will find two modules, SPAR and SPART, which do exactly the same job: off-line DFT and *S*-parameter computations.

4.3.3 Sample Scenarios and Typical Results

F-MSTRIP and TLMMS simulators are run for various substrate height and for a variety of scenarios mentioned in Fig. 4.17. Typical results are presented in this section. First, scenario #1 is taken into account with h = 1mm, ND = 4 (0.66 mm gap) and for 3000 time steps. Its INP file is as follows:

MSTRIP.INP (Fig. 4.19) Line 1: 1
1
4
3000

E_y vs. time at four different ranges just under the strip, along the line (recorded in FIELD.dat file) that are obtained via F_MSTRIP, are plotted in Fig. 4.20. This belongs to the infinite strip-line simulation that was obtained in the first run. Pure Gaussian pulse propagation is clearly observed in the figure. Slight deformation in the pulse is because of the numerical dispersion errors.

In the second run, a 4-cell gap is simulated and input/output voltages and currents are recorded. Figure 4.21 corresponds to this case, where voltage vs. time at both input and output ports are plotted.

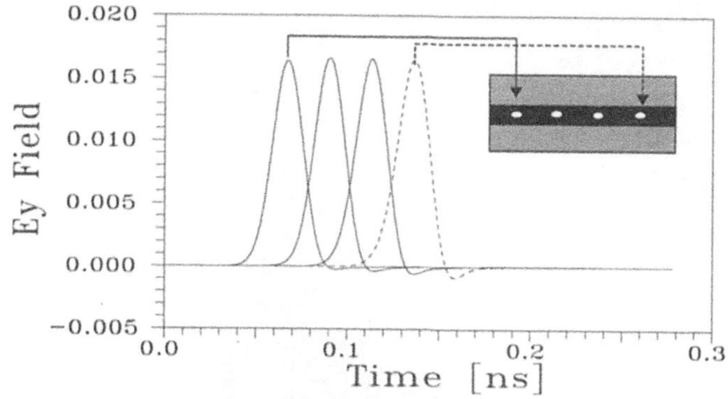

Figure 4.20: *E_y field vs. time under the strip at four different ranges*

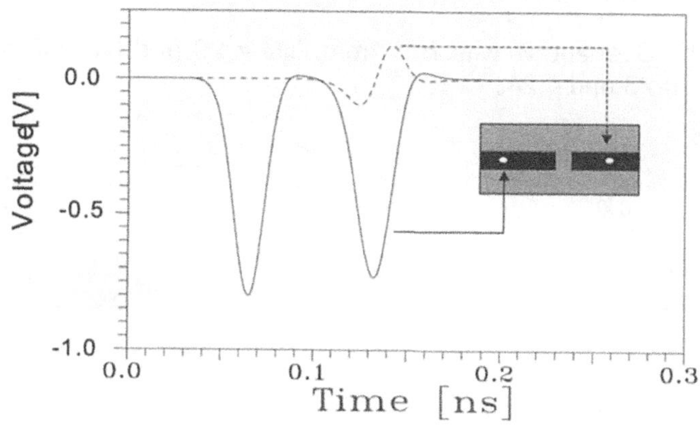

Figure 4.21: *Voltage vs. time at port 1 (solid) and 2 (dashed)*

The solid line corresponds to voltage at port 1. At this port, the propagating incident Gaussian pulse is recorded first. This pulse reaches the gap and most of the energy reflects back at this point and reaches back to the observation point. Since the first part of the strip (up to the gap) is open-ended, equiphase reflection occurs as observed. The dashed line represents time record at port 2. Here, some part of the incident energy passes through the gap and continues propagating along the second part of the strip. Since, DC and lower frequencies cannot propagate, the pulse propagating after the gap is no longer Gaussian; instead it is a once-differentiated Gaussian pulse, which does not contain a DC component.

A series gap in a microstrip line has the obvious property of a DC open circuit, which may be used in a bias-feeding network, but has a high impedance at microwave frequencies because of its small capacitance values. The gap may also be used as a coupling element. *S* parameters of the first scenario, which includes a series gap, are plotted in Fig. 4.22. A full reflection information is hidden in S_{11} for DC and lower frequencies (recall that vertical axis is in dB). As the frequency increases, energy is transferred from the gap, and this is clearly observed from S_{21}.

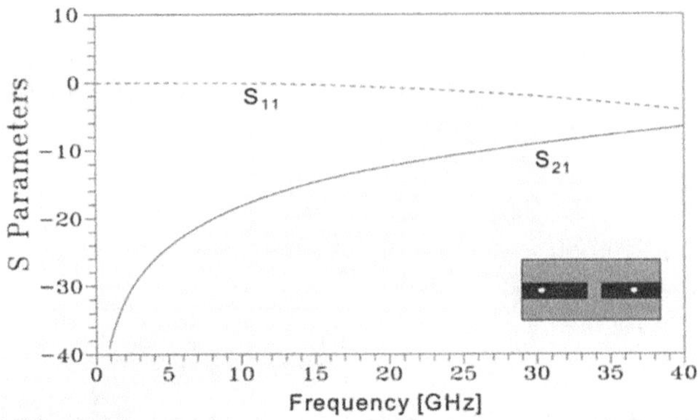

Figure 4.22: *S parameters of the structure given in the first scenario*

Results for the Scenario 3, with h = 1mm, ND = 20 and NSTOP = 5000 are also given in Figs. 4.23 and 4.24.

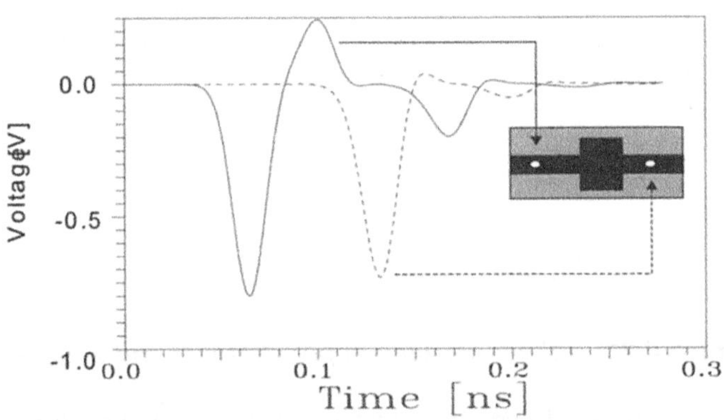

Figure 4.23: *Voltage vs. time at port 1 (solid) and 2 (dashed) for Scenario 3*

In Fig. 4.23, time variations of the input and output voltages are plotted. Since the structure supports DC transmission, the propagating pulses are Gaussian as injected. The incident and two reflected pulses (from the leading and back edges of the center strip line) are clearly observed. At the output, only slightly deformed propagating pulse is seen. If late responses were recorded, reflected pulses would also have appeared at port 2.

In Fig. 4.24, frequency variations of S parameters are plotted. S_{21} is close to unity (i.e., 0 dB) at all frequencies, since the structure supports direct energy transfer. S_{11} is very small and periodically reaches to zero at certain frequencies. Depending on ND, this structure behaves as a kind of quarter-wavelength impedance transformer. At quarter-wavelength frequencies (determined by the length ND), no reflection occurs and full energy transfer is supplied.

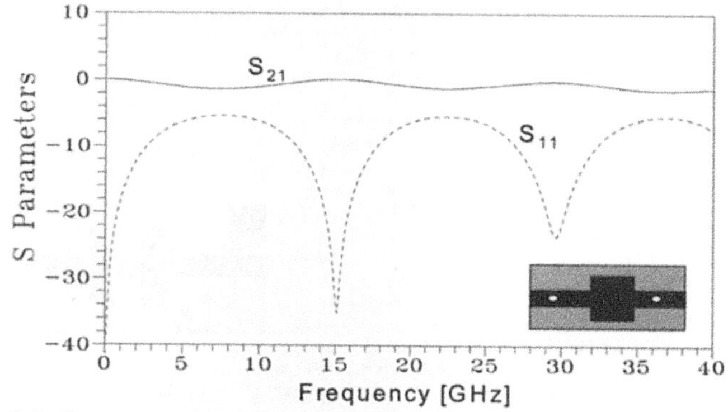

Figure 4.24: *S parameters of the structure given in the third scenario*

In the following three figures the results of both simulators are compared over different scenarios given in Fig. 4.17. The parameters are mentioned separately at each figure captions.

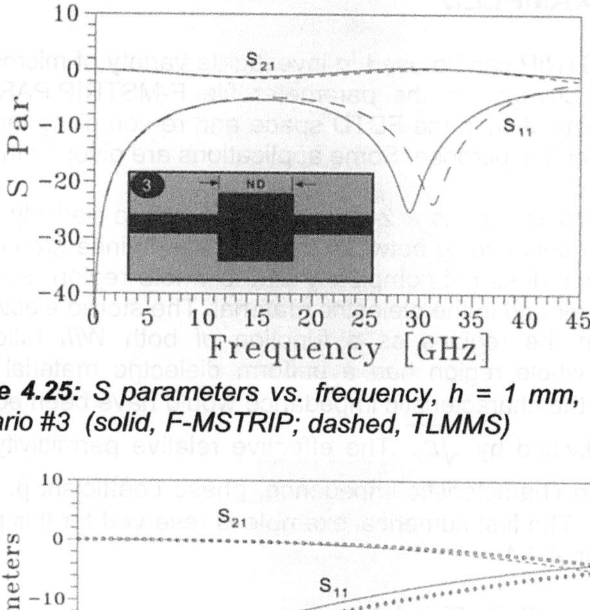

Figure 4.25: *S parameters vs. frequency, h = 1 mm, ND = 20, Scenario #3 (solid, F-MSTRIP; dashed, TLMMS)*

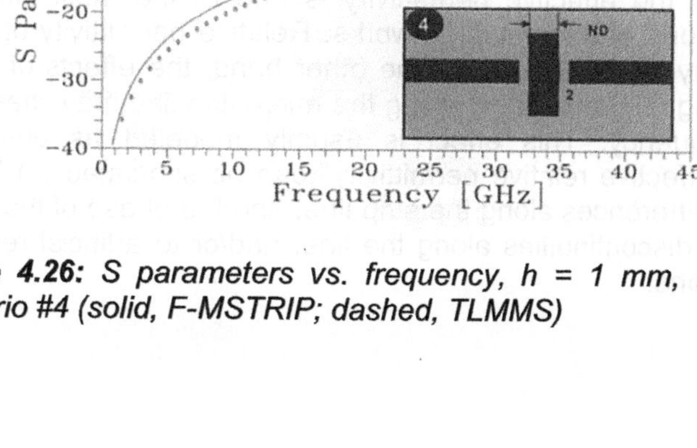

Figure 4.26: *S parameters vs. frequency, h = 1 mm, ND = 8, Scenario #4 (solid, F-MSTRIP; dashed, TLMMS)*

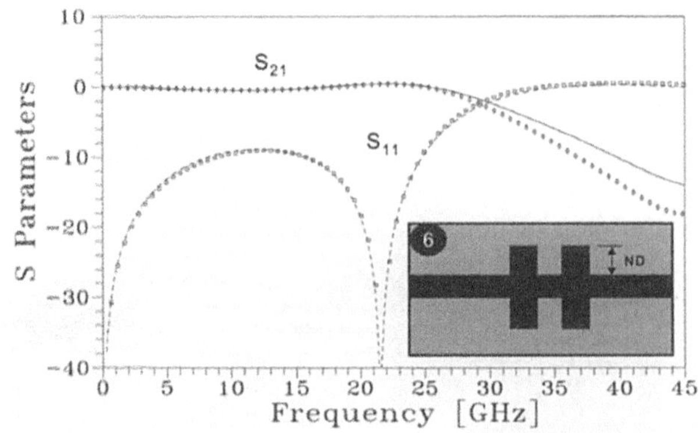

Figure 4.27: *S parameters vs. frequency, h =1 mm, ND = 8, Scenario #6 (solid, F-MSTRIP; dashed, TLMMS)*

4.4 NUMERICAL EXAMPLES

The simulator F-MSTRIP can be used to investigate variety of microstrip networks. Making necessary changes in the parameter file F-MSTRIP.PAR, locating the network under investigation in the FDTD space and re-compiling the code are the only requirements for this purpose. Some applications are given in this section.

A microstrip transmission line is a 2-conductor line with a partially filled dielectric material, with the material being between the strip and infinite ground plane. Since the dielectric material does not completely fill the whole region, electric fields are present both in the air and in the dielectric material. The stored electrostatic energy is divided between the regions as a function of both *W/h* ratio and relative permittivity. If the whole region had a uniform dielectric material with relative permittivity ε_r, then the characteristic impedance would have been equal to its free-space value (Z_0) divided by $\sqrt{\varepsilon_r}$ .The effective relative permittivity is, therefore, required to evaluate characteristic impedance, phase coefficient β, and microstrip line wavelength λ_s. The first numerical example is reserved for this purpose and is presented in Section 4.4.1.

4.4.1 Effective Permittivity Simulations

The calculation of the effective permittivity is one of the most difficult tasks in numerical simulations of microstrip networks. Relative permittivity of the dielectric substrate is usually its DC value. On the other hand, the effects of this dielectric substrate on the signal propagating along the microstrip line may change a lot from frequency to frequency. This effect is usually modeled as effective relative permittivity. The effective relative permittivity can be simulated via F-MSTRIP by observing phase differences along the strip line, and the phase of the signal is very sensitive to slight discontinuities along the line, and/or to artificial reflections from the PML terminations.

The propagating voltage pulses (or E_y fields) as illustrated in Fig. 4.19 at two points along the strip line, $V_1(t)$ and $V_2(t)$, can be Fourier transformed and $V_1(f)$ ve $V_2(f)$ are obtained. If the distance between these two observation points is d, then

$$V_2(f) = V_1(f) \times e^{-j\beta d} \tag{4.50}$$

can be written. Since the propagation constant of the strip line is

$$\beta = \frac{\omega\sqrt{\varepsilon_{reff}}}{c} \tag{4.51}$$

the effective relative permittivity can be calculated as

$$\varepsilon_{reff}(f) = \left[j\frac{c}{2\pi f d} \ln\frac{V_2(f)}{V_1(f)} \right]^2 \tag{4.52}$$

for broad range of frequencies that the injected pulse contains. A typical example is given in Fig. 4.28.

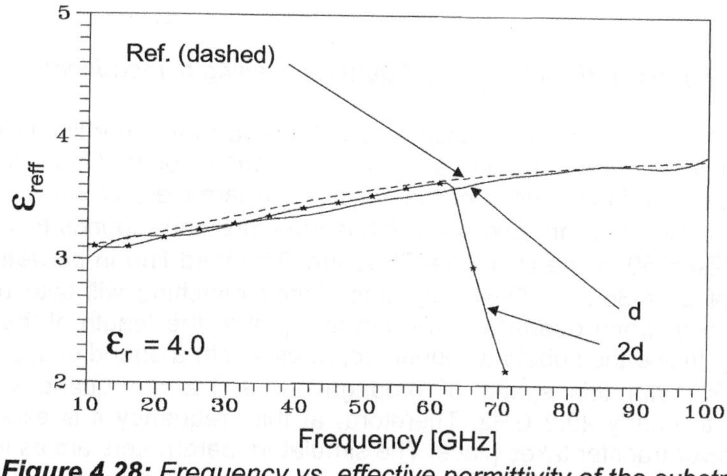

Figure 4.28: *Frequency vs. effective permittivity of the substrate*

Here, a strip line with W =1 mm, h = 0.5 mm and ε_r = 4.0 is located inside a 44 × 22 × 120 FDTD volume with $\Delta x = \Delta y = \Delta z$ = 0.125 mm. The excitation Gaussian pulse has a bandwidth of 250 GHz. In the figure, the curve named reference (dashed) is the result of analytical approximate computation [5]. The other two belongs to TD simulation results for $d = 10\ \Delta z$ and $2d = 20\ \Delta z$. These two results illustrate that the distance between the two observation points along the line determines the frequency range of validity. The distance between the two observation points must be less than half-wavelength in order to follow phase variations. Everything along a transmission line is periodic with half-wavelength periods (that corresponds to a complete turn on Smith chart). For $2d$ = 2.5 mm ($d = 20\ \Delta z$), the results are no longer valid beyond 60 GHz. At this frequency wavelength is 5 mm. Beyond this frequency the propagating signal undergoes more than 2π phase change inside the distance 2d, and phase information is lost. Obviously, this does not happen when the distance is $d = 10\,\Delta z$, which means the simulations are valid up to 120 GHz.

4.4.2 Microstrip Impedance Matching

Characteristic impedance of a line is mostly real. When two lines with different characteristic impedances are aimed to match a quarter wavelength-long third line, a characteristic impedance value, equal to the square root of the multiplication of the other two, may be used. Steps of this quarter-wavelength impedance matching was given in Fig. 4.6. Here, a typical microstrip line structure is given in Fig. 4.29 for quarter-wavelength impedance matching.

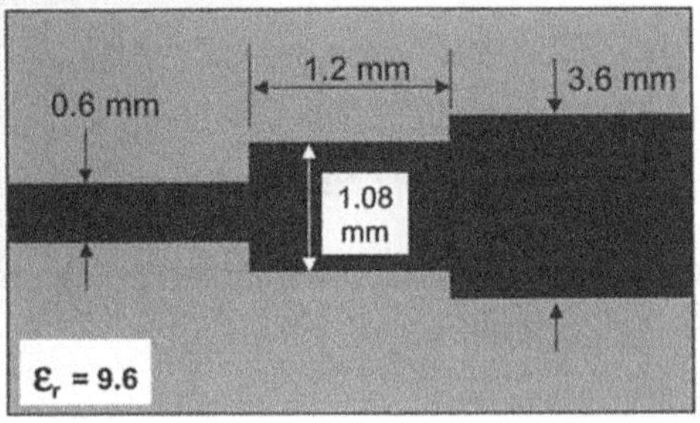

Figure 4.29: *A microstrip quarter wavelength impedance matcher*

A 0.6-mm-width microstrip line is going to be matched to another line with 3.6-mm width. A third line of 1.08-mm width and 1.2-mm length is located in between these lines. Characteristic impedances of the lines are determined by the widths of the lines. With $h = W = 0.6$ mm and $\varepsilon_r = 9.6$ this structure corresponds to impedance matching of $Z_{01} = 50\ \Omega$ line to a $Z_{02} = 75\ \Omega$ line. The third line in between has the impedance of $Z_0 = 36.2\ \Omega$. Obviously, impedance matching will take place at a single frequency, when quarter-wavelength is equal to the length of the third line (i.e. 1.2 mm). Inside the substrate signal propagates with a speed of approximately 0.96×10^8 m/s. With this speed of propagation, a 1.2-mm quarter-wavelength corresponds to nearly 20.2 GHz. Therefore, at this frequency it is expected that maximum power transfer takes place. The simulators parameters are as follows:

FDTD space:	$80 \times 16 \times 100$
Time step:	$\Delta t = 11$ps
Cell sizes:	$\Delta x = 0.406$ mm, $\Delta y = 0.265$ mm, $\Delta z = 0.423$ mm
Simulation time:	NSTOP = 700
Substrate height:	$h = 0.6$ mm
Strip width:	$W = 0.6$ mm
Substrate rel. perm.:	$\varepsilon_r = 9.6$

Source is injected at $k = 20$, and observation points are chosen at $k = 50$ and $k = 80$. The result is given in Fig. 4.30 as S_{11} vs. frequency.

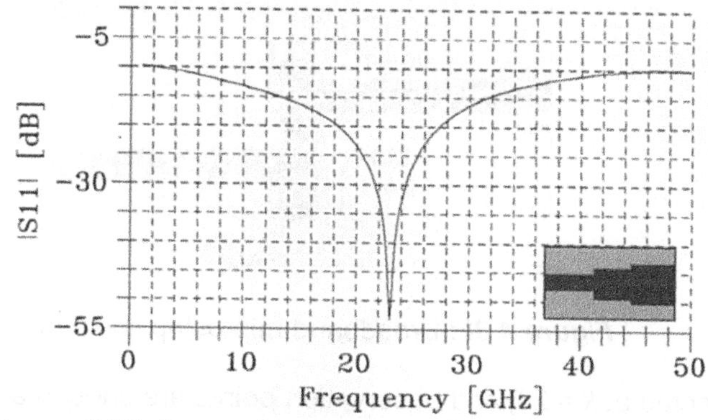

Figure 4.30: *S_{11} vs. frequency of the structure in Fig. 4.23*

As expected S_{11} goes almost to zero (-55 dB) at the frequency that corresponds to a quarter-wavelength. This frequency is read to be approximately 22.8 GHz, which is quite different than 20.2 GHz calculated above. This discrepancy is because of the effective dielectric constant of the substrate. From this figure, effective permittivity of the substrate is calculated to be nearly 7.51.

4.4.3 Microstrip Broadband Filter

Microstrip filters are widely used in electronic networks at microwaves and above. There are variety of ways to design microstrip line filters. One simple method is to use narrow-wide strip lines in stepped form as illustrated in Fig. 4.31. Narrow lines implement series inductors, and wide lines correspond to shunt capacitors. Lengths are adjusted to produce the desired component values. Stepped impedance filters are applicable in cases where relatively large errors are tolerable.

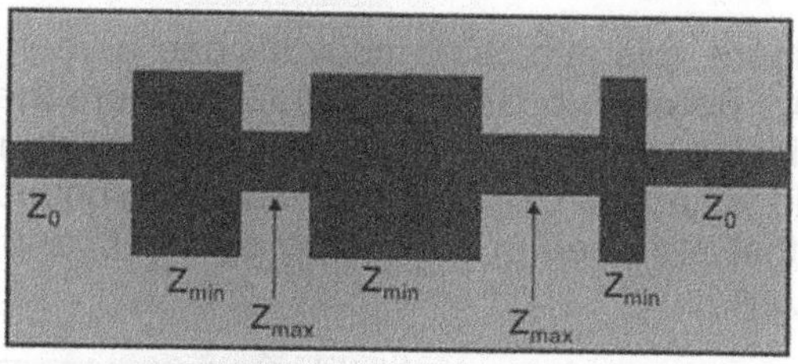

Figure 4.31: *A microstrip line stepped impedance filter*

Here, a broadband low-pass filter is chosen as a second example. The structure and dimensions of the microstrip line filter are given in Fig. 4.32. The simulation parameters are as follows:

FDTD space:	$80 \times 16 \times 100$
Time step:	Δt = 11ps
Cell sizes:	Δx = 0.406 mm, Δy = 0.265 mm, Δz = 0.423 mm
Simulation time:	NSTOP = 4000
Substrate height:	h = 0.8 mm
Strip width:	W = 2.4 mm
Substrate rel. perm.:	ε_r = 2.2

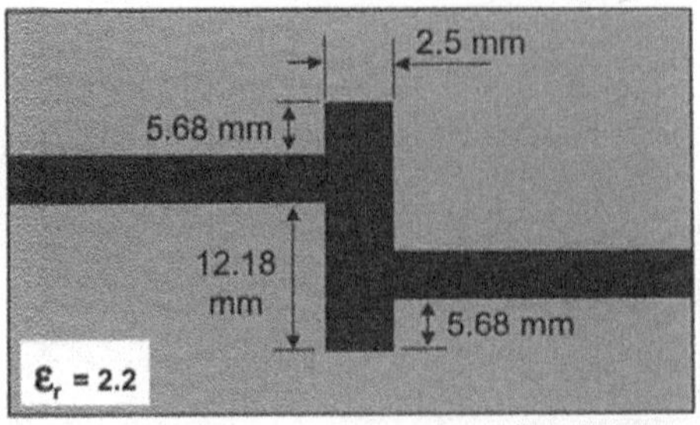

Figure 4.32: Broadband microstrip line filter

Source is injected at $k = 20$, and observation points are chosen at $k = 50$ and $k = 80$. The result is given in Fig. 4.33 as S_{21} vs. frequency.

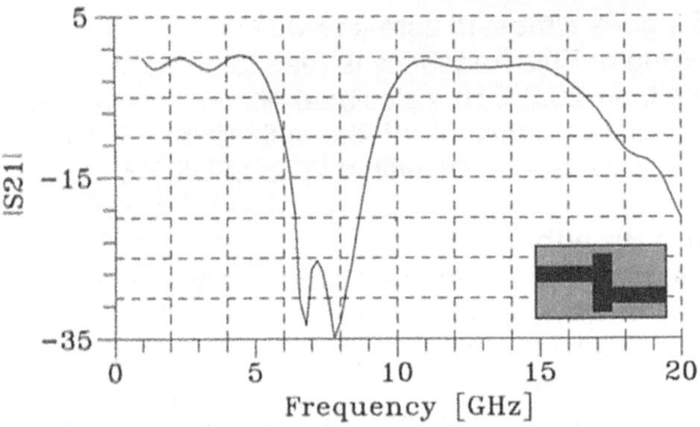

Figure 4.33: *S_{21} vs. frequency of the structure in Fig. 4.25*

As seen in the figure, this microstrip network has a very good transfer character from DC to 5 GHz. Between 5 GHz and 10 GHz, more than 30 dB suppression is obtained. The ripples around 0 dB is because of numerical dispersion and unwanted reflections caused by imperfect absorbing boundary terminations. Some other structures that are used in realization of broadband microstrip line filters is pictured in Fig. 4. 34.

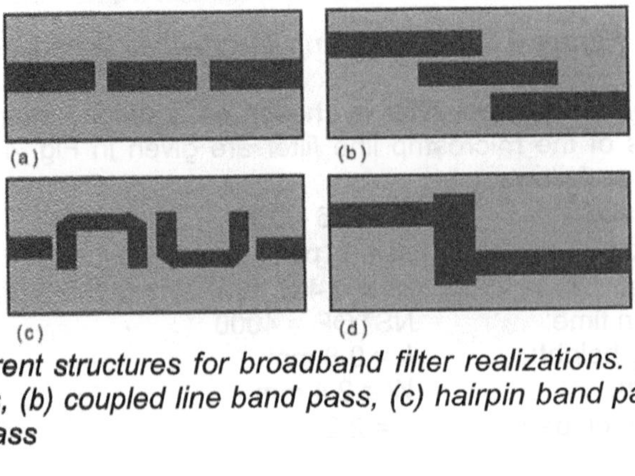

Figure 4.34: *Different structures for broadband filter realizations. (a) Capacitively coupled band pass, (b) coupled line band pass, (c) hairpin band pass, (d) stepped discontinuity low pass*

4.4.4 Microstrip Coupler

Another example simulated with one of the simulation packages (F-MSTRIP) given in Fig. 4.35 is a microstrip line coupler.

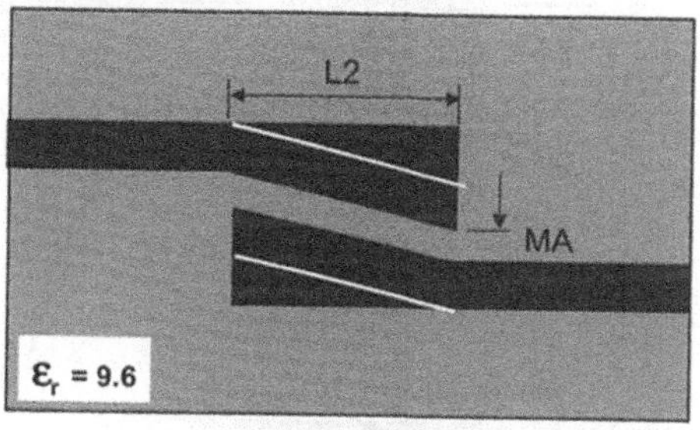

Figure 4.35: *Structure of the microstrip line coupler*

Simulation parameters are as follows:

FDTD space:	$80 \times 16 \times 100$
Time step:	$\Delta t = 11$ps
Cell sizes:	$\Delta x = 0.406$ mm, $\Delta y = 0.265$ mm, $\Delta z = 0.423$ mm
Simulation time:	NSTOP = 4000
Substrate height:	$h = 0.6$ mm
Strip width:	$W = 0.6$ mm
Substrate rel perm.:	$\varepsilon_r = 9.6$

Transfer characteristics of this structure are plotted in Figs. 4.36-4.38 for different parameter sets. First, L2 is taken $30\Delta z$ (12.6 mm) and S_{21} is simulated for two different coupling gaps (i.e., 0.84 mm and 1.68 mm). The results are plotted in Fig. 4.36.

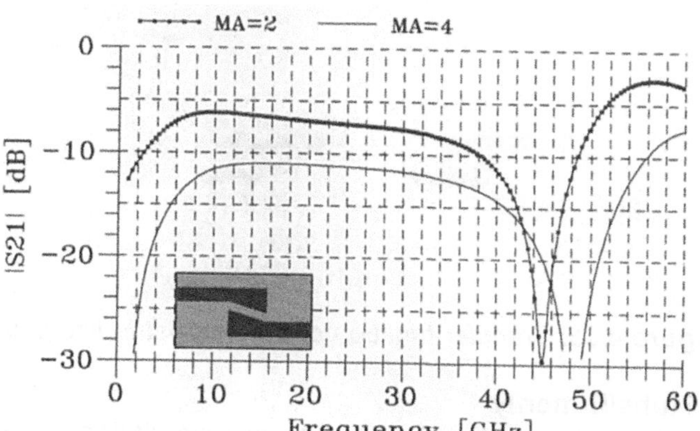

Figure 4.36: *Transfer function of the microstrip line coupler (MA is a parameter)*

As shown in the figure, increasing coupling gap decreases the amount of transferred energy, but increases cut-off frequency. Then, the coupling gap is fixed to 2 cells (0.84 mm) and simulations are repeated for two different coupling distances (i.e., L2 = 30Δz and L2 = 20Δz). The results are presented in Fig. 4.37.

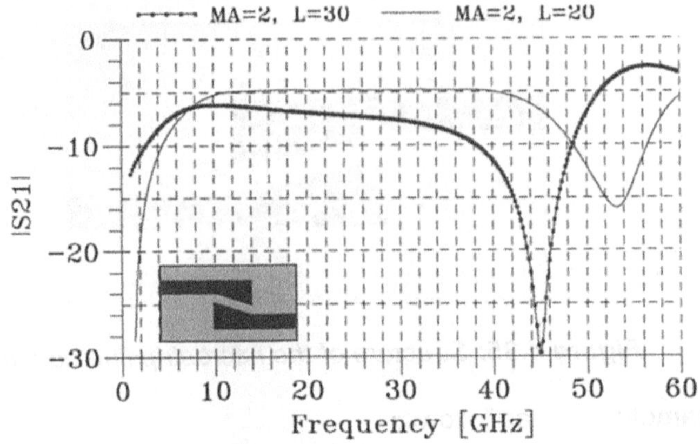

Figure 4.37: *Transfer function of the microstrip line coupler (L2 is a parameter)*

Finally, coupling gap and length are fixed to 2Δz (0.84 mm) and 30Δz (12.6 mm), respectively, and the structure is slightly changed. The simulation results of S_{21} vs. frequency are pictured in Fig. 4.38. As observed from the figure, avoiding sharp edges and smoothing the structure reduces transmission loss and increases power transfer.

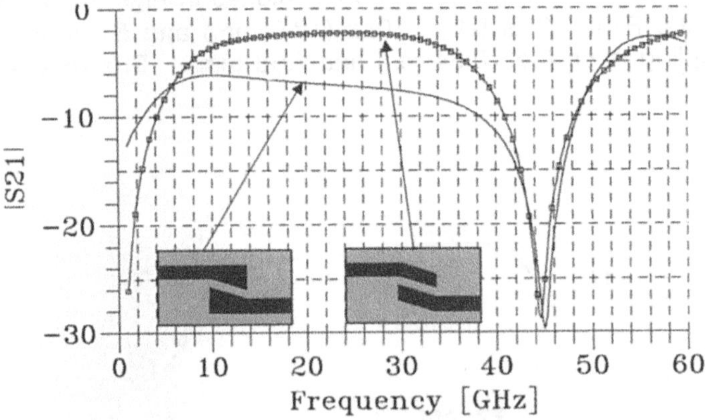

Figure 4.38: *Transfer function of the microstrip line coupler*

4.4.5 Radial Stubs/Butterfly

The last example in this section belongs to the simulation of radial stubs. The purpose of a radial stub is to provide a broadly resonant short circuit in a microstrip network. Radial stubs are widely used as bypass capacitors in the bias circuit of microstrip amplifiers, couplers, dividers, etc. They are also used multi-layer microstrips in energy coupling between layers. An example is given in this section, for the simulation parameters given as follows:

FDTD space:	$70 \times 16 \times 100$
Time step:	$\Delta t = 9.5$ ps
Cell sizes:	$\Delta x = 0.3$ mm, $\Delta y = 0.079$ mm, $\Delta z = 0.3$ mm
Simulation time:	NSTOP = 4000
Substrate height:	$h = 8\Delta y = 0.53$ mm
Strip width:	$W = 2\Delta x = 0.6$ mm
Substrate rel. perm.:	$\varepsilon_r = 10.0$

In Fig. 4.39 S_{11} vs. frequency of the butterfly network is plotted.

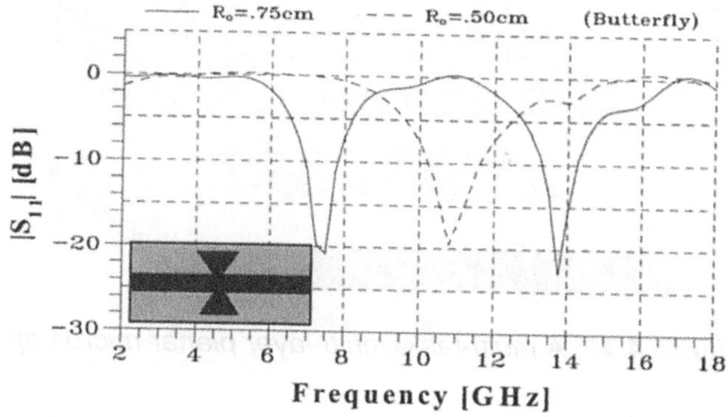

Figure 4.39: *S_{11} vs. frequency of the microstrip butterfly network*

In Fig. 4.40, imaginary part of the input impedance vs. frequency are plotted for three different radial stub angles.

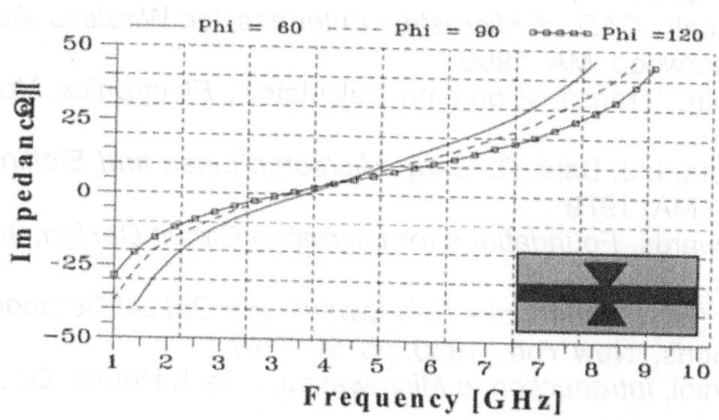

Figure 4.40: *Imaginary part of input impedance vs. frequency of the butterfly*

4.5 CONCLUSIONS AND DISCUSSION

Microwave network design and transmission lines are discussed in this chapter. Emphasis is placed on planar microstrip line networks. Scattered parameter representation of networks are outlined. A sample TD simulator, the FDTD-based compact simulator F-STRIP, is developed and presented, with which a variety of single-layer structures can be analyzed. Typical examples are also included.

The reader may enjoy practicing TD simulations of pulse propagation, reflections from discontinuities, designing impedance matchers, filters, couplers, etc., which are building blocks of complex planar networks. Once experienced, he/she may modify the source codes included and, may design and analyze multi-layer complex networks, such as presented in Fig. 4.41.

Figure 4.41: *A multi-task, multi-layer planar microstrip structure*

References

[1] D. D. Grieg, H. F. Englemann, "Microstrip-A New Transmission Technique for the Kilomegacycle Range", *Proc. IRE*, Vol.40, pp.1644 - 1650, Dec. 1952

[2] G. A. Deschamps, "Microstrip Microwave Antennas", 3rd *USAF Symposium Antennas*, University of Illinois, Urbana, IL, 1953

[3] R. A. Sainati, *CAD of Microstrip Antennas for Wireless Applications*, Artech House, Norwood, MA, 1996

[4] P. H. Smith, "Transmission-Line Calculator", *Electronics*, Vol. 12, pp. 29 - 31, Jan. 1939

[5] K. C. Gupta, I. J. Bahl, R. Garg, *Microstrip Lines and Slotlines*, Artech House, Norwood, MA, 1979

[6] T. C. Edwards, *Foundations for Microstrip Circuit Design*, John Wiley & Sons, New York, 1981

[7] K. Chang (Ed.), *Handbook of Microwave and Optical Components*, Vol. 1, John Wiley & Sons, New York, 1989, pp. 96 - 107

[8] F. E. Gardiol, *Introduction to Microwaves*, Artech House, Dedham, MA, 1984

5

EMC/BEM Modeling

with **M. ORHAN ÖZYALÇIN**

5.1 INTRODUCTION

Electromagnetic compatibility (EMC) and electromagnetic interference (EMI) are terms which have been used in electronic industry and academia for the last two decades [1, 2]. The worldwide interest in EMC has grown significantly particularly in the commercial and domestic equipment sectors. Another term that has increasingly gained attention is bioelectromagnetics (BEM), which deals with the interaction of fields with human beings [3]. Briefly speaking, device-to-device interaction is the issue of the EMC, while device to human interaction is the BEM issue.

EMC is a dynamic field where evolution has accelerated in the 1990s with the introduction of European Community EMC Directive [4]. The directive attempted to harmonize regulations, which applies to almost all electrotechnical equipment and results in EMC standards. The standards are continuously updated by international expert bodies, such as CISPR (International Special Committee on Radio Interference) or IEC (International Electrotechnical Committee), whose role is to issue international standards.

EMC engineering is a multi-disciplinary activity involving electrical, mechanical, and system engineers, as well as physicists and chemists. EMC problems are mostly hidden in the details, which are not thought at universities and which mostly do not obey the well-known field and wave theories. It is totally application-oriented discipline and should be handled case by case, including "trial and error" approach. It is usually experience, not the knowledge, which results in the solution.

On the other hand, BEM problems are more complex than EMC problems, since human tissues are non-homogeneous, time-varying, non-linear media. Very rapid developments in mobile communication have drawn attention to possible health risks of EM energy emitted from the transmitters of mobile phones as well as from the base stations in the past few years. The debate on EM pollution has greatly focused on the question of "whether these two have adverse health effects in either short- or long-term periods" [5 - 9].

The non-governmental organization accepted as an independent expert body by the United Nations (UN), the World Health Organization (WHO), the International Labor Organization (ILO), the European Union (EU), and most of the developed countries is the *International Commission on Non-Ionizing Radiation Protection* (ICNIRP) [10]. The Commission was established for the purpose of advancing non-ionizing radiation (NIR) protection for the benefit of people and the environment and in particular to provide guidance and recommendation on protection from NIR exposure. In 1998, the ICNIRP published its latest guidelines for exposure limits, which were based on the evaluation of all available and known analytical/numerical/experimental study results [10]. In these guidelines as well as in others [11,12], for example, electrical field strength at 900 MHz and 1800 MHz (where major contribution arises from cellular systems) should not exceed 42 V/m and 59 V/m, respectively, anywhere in public residence.

Typical field strengths measured inside buildings or on streets a few meters below and off-side of base station directional antennas are of the orders of a few V/m. On the other hand, at points 2.2 cm away from an antenna, maximum values of the electric field are measured to be about 400 V/m for a 2-W, 900-MHz phone and about 200 V/m for a 1-W, 1800-MHz phone. Given the much lower exposures to radiation from base stations than from handsets, the greater public concern from the people all around the world is paradoxical. It presumably arises because *individuals can choose whether or not to use a mobile phone*, whereas *they have little control over their exposures from base stations*. Furthermore, people derive a personal benefit from the use of a phone, but gain nothing directly from the presence of a base station close to their home or place of work. If anything, they may suffer a loss of amenity and perhaps a reduction in the value of their property.

Major areas of current BEM studies may be listed as [10]

- physical characteristics and dosimetry of EM fields
- laboratory (in vitro and in vivo) studies
- epidemiological studies
- health risk assessment and protection principles
- development of guidance on exposure limits

The only area where numerical simulation techniques can be used as a tool to conduct research is the dosimetry, where thermal effects are of interest. EM energy absorbed by human tissues (lossy dielectrics) causes temperature increase and this absorbed energy can be calculated by simulating or measuring electric field inside the tissues. Simulations are basically performed via powerful TD techniques, such as FDTD and TLM, while measurements are done by phantoms that are prepared via chemical gels, which have tissue-equivalent electromagnetic properties (at those frequencies of interests).

There are two important parameters in dealing with both EMC and BEM problems via numerical simulations as well as laboratory measurements. These are the Shielding effect (SE) and the dosimetry [1].

SE is a parameter which determines the degree of isolation of EM energy in two environments separated by any obstacle (e.g., a PEC box isolating inside from outside). An example is depicted in Fig. 5.1. Here, the SE of an enclosure simulation is shown. For SE calculations, two simulations are performed:

- The field (E_0) at the observation point which is created by a source (here a cellular phone) is first obtained.
- Then, the source is enclosed by an enclosure and the same procedure is repeated and the field (E) at the same observation point is monitored once again. The ratio of the Fourier transforms of the former to the latter in dB

$$SE = 20\log_{10}\left|\frac{E_0}{E}\right| \text{ [dB]} \tag{5.1}$$

is SE of the equipment under test (EUT). Here EUT is the enclosure. It should be noted that the simulations/measurements are assumed to be in TD. Broadband FDTD and TLM simulations are used to obtain $E_0(t)$ and $E(t)$. By applying DFT and then using (5.1) one obtains broad band SE vs. frequency variation for the EUT.

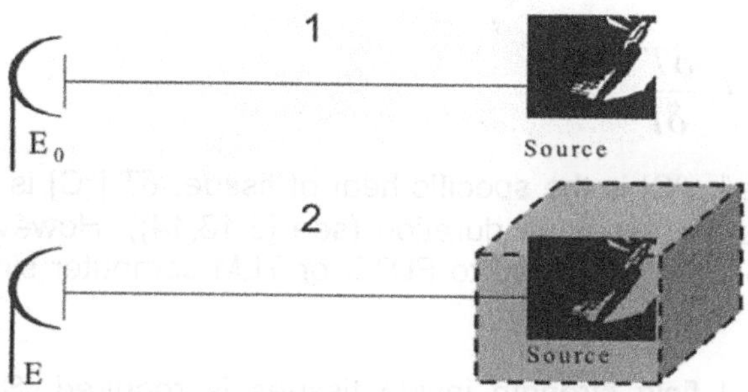

Figure 5.1: *Definition of shielding effectiveness*

On the other hand, the dosimetric quantities used in these guidelines, taking into account different frequency ranges and waveforms, are as follows [10]:

- Current density, J, in the frequency range up to 10 MHz,
- Current, I, in the frequency range up to 110 MHz,
- Specific energy absorption rate (SAR), in the frequency range 100 kHz to 10 GHz,
- Specific energy absorption (SA), for pulsed fields in the frequency range 300 MHz to 10 GHz,
- Power density, S, in the frequency range 10 – 300 GHz.

In the frequency range of 10 MHz to a few GHz (including 900-MHz and 1800-MHz cellular frequencies), SAR is the main physical parameter. Available experimental evidence indicates that the human body cannot regulate itself under a permanent temperature increase of more than 1°C, which corresponds to a whole-body SAR value between 1 and 4 W/kg under 30-min EM exposure for resting humans. From this 4-W/kg value, a general public, continuous exposure limit of 0.08 W/kg is derived with a safety factor of 50.

Where appropriate, the derived limits called as reference levels are obtained from the basic restrictions by mathematical modeling and by extrapolation from the results of laboratory investigations at specific frequencies [10]. Reference levels are easy to measure and are applicable for base station controls. Unfortunately, basic restrictions should be taken into account for cellular phone EM exposure.

In tissues, SAR is proportional to the square of the internal electrical field caused by external device (e.g., cellular phone). It is given as

$$\text{SAR} = \frac{\sigma}{\rho}|E|^2 \quad \text{[W/kg]} \tag{5.2a}$$

where E [V/m], σ [S/m] and ρ [kg/m^3] are electric field strength, conductivity and density of the tissue, respectively. Often, SAR values for a group of tissues are of interest, in which case SAR is calculated as the ensemble average. Using its definition as

$$\text{SAR} = C\frac{\delta T}{\delta t} \tag{5.2b}$$

where C [joule/kg°C] is the specific heat of tissue, δT [°C] is the temperature rise, and δt [sec] is the exposure duration (see [3,13,14]). However, in this method of determining SAR is not suited to FDTD or TLM computer simulations and will not be used here.

Since electrical field strength inside tissues is required, SAR can be obtained either via laboratory studies [10] (where EM phantoms equivalent to human tissues are used) or with computer simulations [13,15 - 17]. In either case, standardization process has just been completed yet [14].

The most recognized EM exposure standards, including ICNIRP, adopt the specific absorption rate (SAR), averaged over the whole body (SAR_{WB}), as the basic parameter to establish the safety of an exposure [10]. The value of 4 W/kg is accepted worldwide as the threshold for the induction of biological thermal effects, and introducing a safety factor of 50 for general public exposure gives rise to a basic limit on SAR_{WB} equal to 0.08 W/kg.

Sometimes, although SAR_{WB} goes below the basic limit, the local SAR values can be rather high in a confined body region, where the power absorption takes place, such as a human head near an antenna of a mobile phone. For such conditions, limits on local SAR averaged over tissue masses of 1 or 10 g have been introduced in the standards, as 1.6 W/kg over 1 g or 2 W/kg over 10 g.

Parallel to an increase in computer speed and memory, EMC and BEM problems have also begun to be simulated directly in a four-dimensional (i.e., in space and time) real world. The TLM [19] and the FDTD [20] methods have become almost the most important TD techniques used in EMC and BEM simulations.

5.2 EMC AND SHIELDING EFFECTIVENESS (SE)

Electromagnetic interference and/or leakage solutions are dominated by shielding, grounding, and filtering. Grounding and filtering treat conducted interference/leakage, while shielding treats the control of radiated EM fields. Here shielding is taken into account in numerical simulations. Shielding is a process of preventing EM radiation (leakage) from coupling into or out of defined areas or regions. Basically, this is achieved by using shielding materials such as metals or metalized plastics. In some cases, special solutions (e.g., cut-off waveguides) may be used.

SE depends upon a number of factors, such as the EM wave impedance, frequency, polarization, thickness of the barrier, composition of the barrier and the type of the radiating element, among which the factor related to the radiating element is usually dominant. There are two types of radiating elements: sources that behave as electric dipoles (wires) and sources that behave as magnetic dipoles (loops). Electric dipoles have dominant electric-type near fields and high impedance, but magnetic dipoles have dominant magnetic-type near fields and low impedance (see Fig. 5.2).

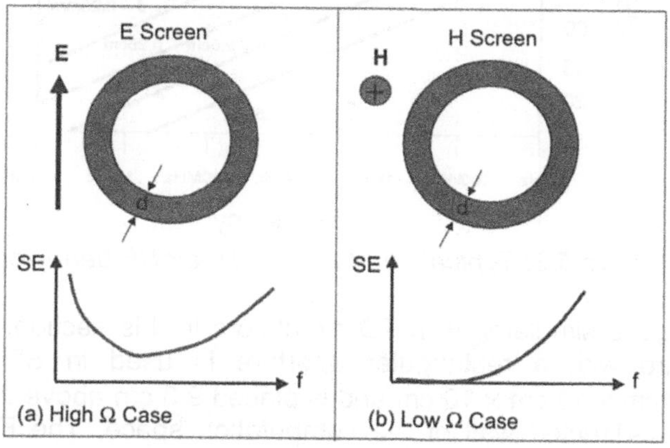

Figure 5.2: *EM sources and SE vs. frequency against them*

The latter presents a more difficult shielding problem. Of course, radiation fields of both types of sources become similar and can be approximated by plane waves at distances which are a few wavelengths away from these elements.

In shielding practice, depending on the criticality of the problem under consideration, SE values of 30 - 50 dB are considered to be acceptable, whereas SE values in the range of 60 - 90 dB represent very high quality shields. The shielding performance of a metal box with no holes or seams exceeds 100 dB, which is usually too high to measure. If the barrier has holes, slots, joints, vents, windows, or other discontinuities, the shielding effectiveness can only be as good as allowed by such shielding imperfections. One rule of thumb is to find the longest dimension of these discontinuities and determine the frequency at which that length represents a half-wavelength. As a worst-case approximation the barrier can be assumed to yield 0 dB shielding effectiveness beyond that frequency.

Shielding boxes are very common in practice since they furnish effective shielding while at the same time providing a structure for supporting circuit boards, easing cooling requirements, and isolating different subsystems.

5.2.1 Aperture Leakage and SE

Aperture leakage is directly specified by the aperture length/signal frequency relation. In Fig. 5.3, typical frequency variations of typical apertures are depicted. As illustrated, a 25-cm hole may have an SE of up to 80 - 90 dB at 10 kHz (when wavelength is on the order of tens of kilometers), but reduces to almost 0 dB after a few hundred MHz (when wavelength is on the order of centimeters).

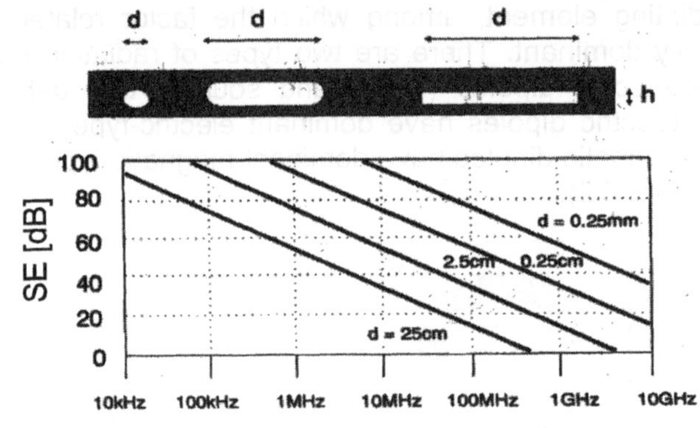

Figure 5.3: *Typical apertures and their SE behaviors*

Aperture leakage is simulated with TD simulators in this section. A cubical metal (PEC) enclosure with a rectangular aperture is used in SE modeling. The enclosure is 10 cm × 10 cm × 10 cm and is placed 9.3 cm above the floor, 11.3 cm away from the *x*-directed wall of the computation space. The EUT has a (two) rectangular aperture(s) on its sides with 2.7 cm × 4 cm (4 × 6 cells). A vertically polarized, 7.3-cm-long dipole antenna inside the EUT is excited and the emission is monitored at a distance of 6 cm away from the front face of the EUT. The feed point of the dipole and the observation point are placed on a line parallel to the floor (14.7 cm above the floor).

This configuration is modeled inside the computation space of 60 × 45 × 45 cells. The cell size is 0.67 cm in each direction. All the parameters are same in both TLM and FDTD calculations. Only absorbing boundary simulators are different (PML in TLM and second order MUR in FDTD). The dipole inside the EUT is excited by a pulse having the (5%) bandwidth of 3 GHz. The antenna is vertically polarized (i.e., the *y*-component of the electric field is excited at point (30,22,22)). An SE of this structure is obtained at the following three observation points: OP1 (41,22,22), OP2 (30,38,22), and OP3 (10,22,22). The three observation points used in the SE calculations are chosen as follows:

- OP1 is in front of the EUT and directly faces the aperture.
- OP2 is above the EUT.
- OP3 is behind the EUT.

Locations of different observation points are indicated in the insets of respective figures for convenience.

An example of near-field variations (obtained via FDTD) at different time instants is given in Fig. 5.4. Here, the excitation inside the enclosure at the beginning and energy leakage at later times is clearly observed from the plots.

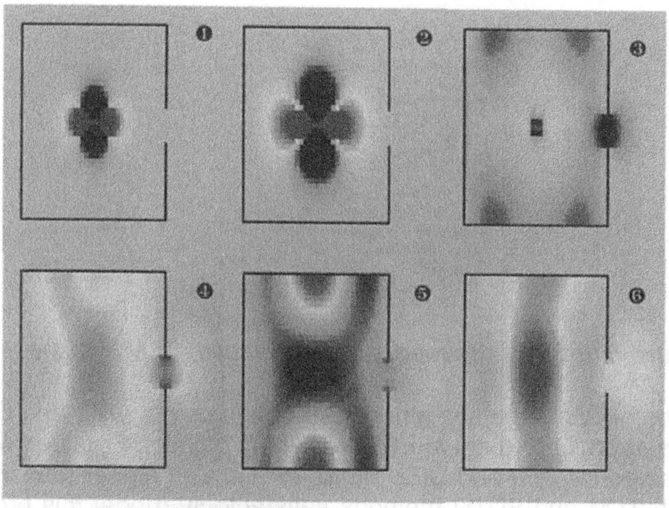

Figure 5.4: *Near fields inside the PEC box, leaking outside*

The simulations are performed with both FDTD and TLM simulators. The first comparison is given in Fig. 5.5, where vertical fields E_y vs. time at OP1 are plotted.

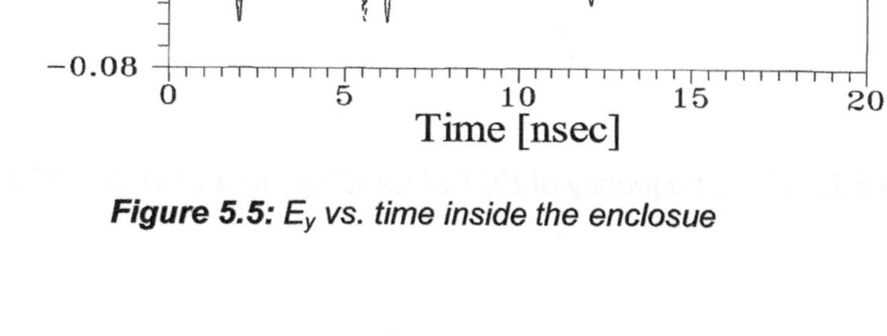

Figure 5.5: *E_y vs. time inside the enclosue*

A once differentiated Gaussian pulse is used as a delta-gap source. Poor agreement in late time responses occur because of the numerical dispersion effects. Note that the ratio of minimum wavelength to the mesh size ($\lambda_{min} / \Delta\ell$) is 15 in this problem, that is not good enough. As much as $\lambda_{min} / \Delta\ell$ =30 may be required to obtain good agreement here (this may be more than 100 in some RCS problems).

In Fig. 5.6, SE vs. frequency obtained via both FDTD and TLM simulators are plotted. As shown in the figure, although time variations do not agree very well, frequency results are good.

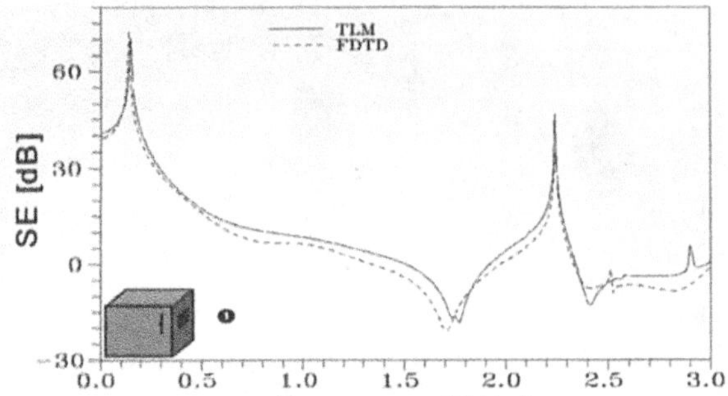

Figure 5.6: *SE vs. frequency at OP1 (solid) TLM, (dashed) FDTD*

The variations of SE at the two other observation points OP2 and OP3 are shown in Fig.5.7. As seen, OP3 (behind EUT) results in higher SE values when compared to those obtained for OP1 and OP2. SE at OP2 is also higher than that at OP1. We see that both TLM and FDTD methods satisfactorily model this problem over the entire frequency range. We note that SE drops sharply at all observation points around 1.8 GHz, which approximately corresponds to a resonance of the aperture (aperture size is $\lambda/4$ at 1.875 GHz).

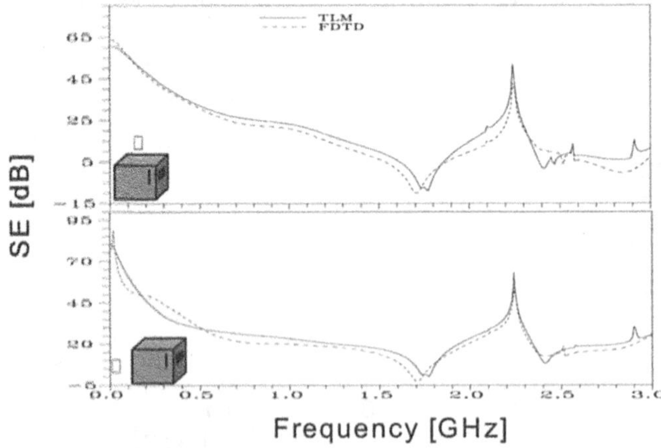

Figure 5.7: *SE vs. frequency of EUT at (a) OP2 (30,38,22), (b) OP3 (10,22,22).*

5.2.2 Multiaperture Leakage and SE

As a final example, simulated SE results for an EUT which has two apertures is given. These calculations are performed using exactly the same parameters as those given in the previous single-aperture example. The only difference is that EUT now has a second aperture on the top or in the back face in addition to the one in the front face. Apertures are of the same size and are located at the centers of the respective faces. The following three cases are considered:

(i) One aperture in the front face and one in the back face placed parallel to each other, i.e., with the longer dimension along the *z*-axis
(ii) One aperture in the front face and one on the top face, placed parallel to each other, i.e., with the longer dimension along the *z*-axis
(iii) The same as (ii) above, but the aperture on the top face is rotated so that its longer dimension is now along the *x*-axis

The SE calculations at the observation point OP2 (30,38,22) are given in Fig. 5.8. The positions of the apertures are also indicated in the insets of these figures for convenience. In the same figures the SE variations at OP2 in the case of a single aperture in the front face is given in Fig. 5.7a, as the TLM results are replicated in dashed lines, for comparison purposes. These results clearly indicate that the interaction of EM waves emanating from the two apertures may result in a decrease or increase of SE at an observation point depending on characteristic parameters, especially the frequency.

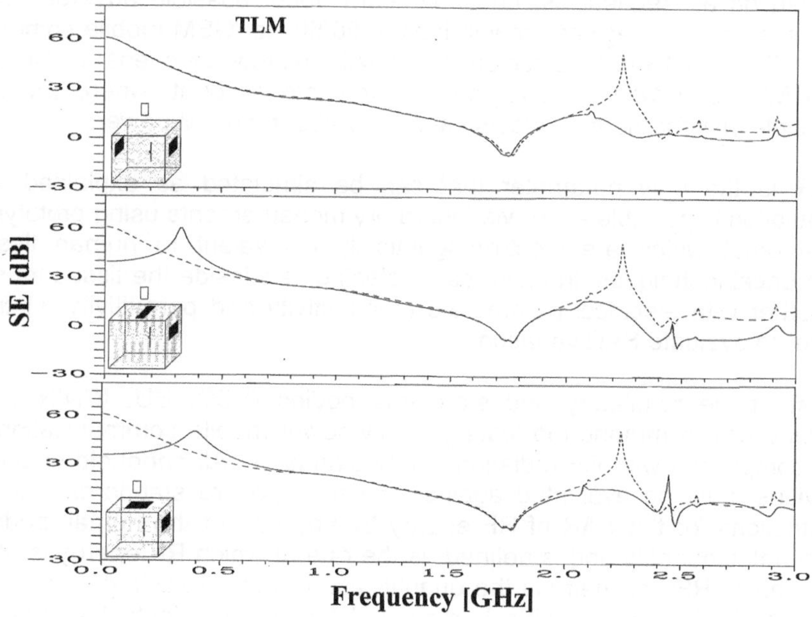

Figure 5.8: *SE of EUT with multi-apertures located at different faces in free space. Observation point is taken at top (OP2). Dashed lines show the plot of structure with one aperture given in Fig. 5.7a.*

With the aid of the results obtained in the various test problems investigated with these TD simulators, it is possible to state some characteristic features of SE:

- The TD simulators are very flexible and can be used to obtain accurate estimates of the SE at all observation points for almost all structures (geometry, source characteristics, medium parameters) of practical significance,
- SE of an enclosure, aperture, or a barrier is critically dependent on the problem parameters. In other words, the estimation of the frequency variation of SE for a given realistic structure is a task, which cannot be addressed using even the best EMC engineering expertise. The test cases presented here illustrate that SE estimations can only be performed either by measurements or via simulations covering all frequencies of interest.

These examples explain the difficulties in dealing with EMC problems, where every option must be checked and tested individually. There is no a general rule in EMC problems [21].

5.3 BEM AND SPECIFIC ABSORPTION RATE (SAR)

The interaction of EM fields with human tissues, as documented by the ICNIRP, covers the range of DC to 300 GHz, which is called the *non-ionizing radiation* frequency region. Although the frequency range is very broad and emphasis is placed on all frequencies, public concern about possible adverse health effects focuses on three regions: power lines at 50/60 Hz, GSM mobile communication at 900 MHz and 1800-MHz region, and finally microwave ovens at 2.45 GHz. Here, 900 MHz and 1800-GHz region is taken into account, where EM modeling is possible only for thermal effects caused by absorbed EM power.

SAR is the only parameter that can be simulated as explained above. The simulation is possible either via laboratory measurements using prototypes (human phantoms) which are electromagnetically equivalent to human tissues or via numerical techniques. In either case, electric field inside the tissue must be known together with electrical parameters (conductivity and permittivity) of the tissue in order to evaluate SAR variation.

Most of the regulatory and standards bodies (FCC, EU, ICNIRP, CENELEC) require or recommend the testing of low-power mobile communication equipment for compliance with RF radiation safety standards. All handheld or body-mounted devices must be evaluated against human exposure standards which set basic restrictions for the SAR of RF energy by any part of the human body. SAR is a dosimetric quantity and is defined as the rate at which RF energy is absorbed per unit mass. RF dosimetry is the quantification of the magnitude and distribution of absorbed EM energy within human and biological objects that are exposed to electromagnetic radiation (EMR).

A precision RF Dosimetric Assessment System DASY3 developed by the Swiss Federal Institute of Technology is used to perform precise SAR measurements (see, for example, Fig. 5.9 [22]).

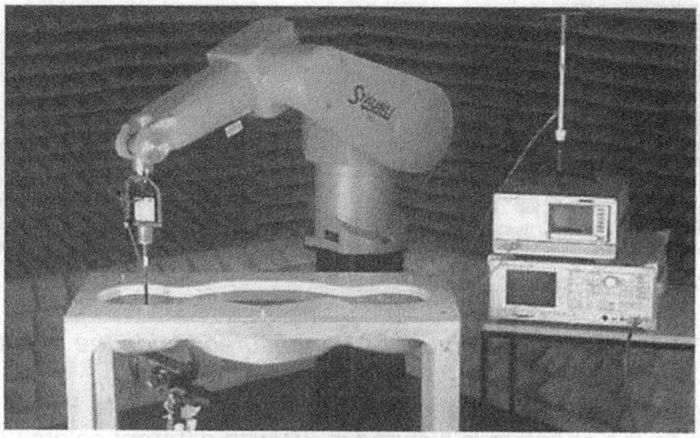

Figure 5.9: *SAR measurement system*

It consists of a computer controlled, high precision robotics system, robot controller, extreme near-field probes, probe alignment sensor, and generic phantoms containing the brain equivalent material, as shown in Fig. 5.10. The six axis robot precisely positions the probe at the points of maximum EM field. The system also provides precise measurements up to 10 GHz in free space as well as inside tissue simulating liquids and cadavers for BEM research purposes. The flexible data evaluation and visualization capabilities allow EM design and performance evaluations in other applications including SAR optimization, EMI, EMC, printed circuit board (PCB), wireless communications, biomedical applications and industrial RF machines [22 - 24].

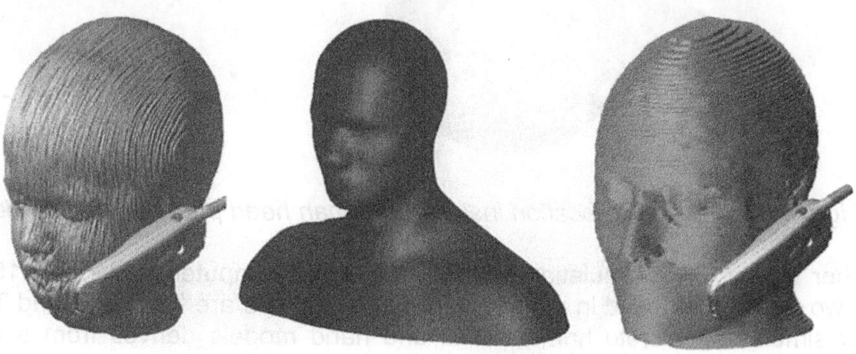

Figure 5.10: *Human head phantoms (left to right: child, generic, adult)*

A typical SAR laboratory result is given in Fig. 5.11, where maximum SAR detected inside a human head phantom at 900 MHz for different hand-held device positions are measured [22, 23].

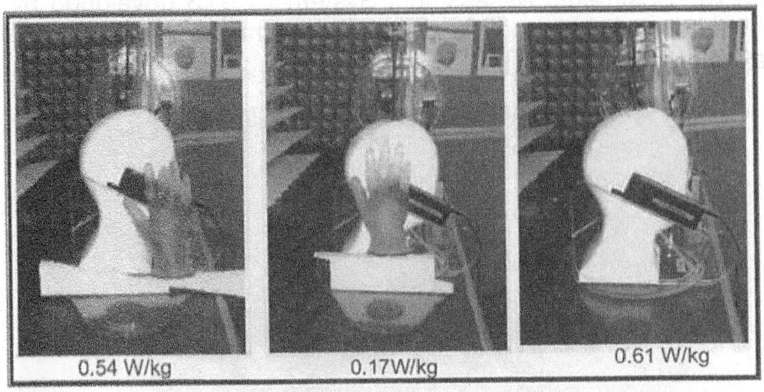

Figure 5.11: *Maximum SAR measured inside the phantom*

The visualization capability of the laboratory set is given in Fig. 5.12, where SAR intensity inside the phantom is pictured in 3D with different colors [22]. It should be noted that laboratory SAR simulations require precise measurements and sensitive electric field probes. Electrical parameters of the tissues should also be determined precisely.

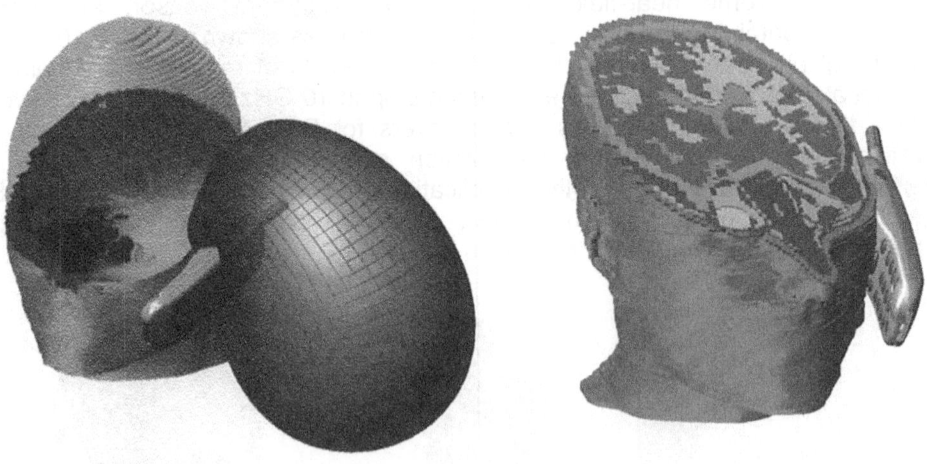

Figure 5.12: *SAR vs. location inside the human head phantom and far fields*

Another way of SAR simulation is to use powerful computer models [13,15 - 17]. The two techniques used in numerical SAR simulations are the FDTD and TLM. In either simulator, discrete human head and hand models derived from a nuclear magnetic resonance (NMR) image [22], as shown in Fig. 5.13, are used.

At this point it would be appropriate to discuss some features of power calculation in TD simulations. The most important steps in numerical SAR modeling are the calculation of radiated and absorbed power.

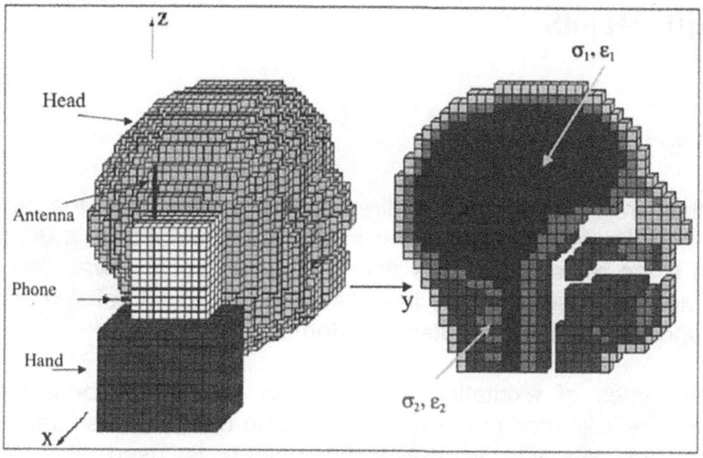

Figure 5.13: *Discrete human head and hand model*

5.3.1 Calculation of Absorbed and Radiated Powers

In TLM, matched stubs are used to model medium losses. One extra stub per electric field component is required at each cell in order to represent dielectric losses. This results in a cell structure with a total number of 21 ports. Any pulses reflected down the new ports, 19 to 21, are considered lost. Hence no pulses are incident on these lines. This gives a 21 × 18 scattering matrix *S*, in the following form

$$\begin{bmatrix} V^r \\ V_{19} \\ V_{20} \\ V_{21} \end{bmatrix} = SV^i . \quad (5.3)$$

Total loss (absorbed) power dissipated in the node of TLM is then calculated via

$$P_{\text{absorbed}} = G_x V_{19}^2 + G_y V_{20}^2 + G_z V_{21}^2 \quad (5.4a)$$

where G_x, G_y and G_z are the conductances of the loss stubs along *x*, *y* and *z*, respectively, and are defined as

$$G_i = \sigma \frac{\Delta j \Delta k}{\Delta i} \qquad (i,j,k) \in (x,y,z) \quad (5.4b)$$

It should be remembered that the same loss power can also be calculated via

$$P_{\text{absorbed}} = \sigma \times \left(E_x^2 + E_y^2 + E_z^2\right) \quad (5.5)$$

(as performed in FDTD) where σ is the conductivity parameter.

Radiated power is calculated (in a similar manner in both FDTD and TLM) by integrating the Poynting vector over a closed virtual surface *S* as

$$P = \oiint_S (\mathbf{E} \times \mathbf{H}) \cdot \mathbf{dS} \tag{5.6}$$

where

$$\mathbf{dS} = dS \cdot \mathbf{n} \tag{5.7}$$

and **n** is the unit normal vector directed outwards from the surface. Since rectangular coordinate system is used in the TD simulations, a virtual surface is also chosen to be rectangular. It is also chosen in such a way that includes the EUT and is not too close (typically at a distance of 8 to 10 cells) either to the EUT or to the boundaries of the computational domain.

Two different types of excitation are used in the simulations. First one is a Gaussian-type band-limited pulse and the second one is sinusoidal excitation. But in the latter, the Hanning windowing technique is used to avoid excitation instabilities and sinusoidal steady-state error in the system as explained in [25]. Windowing technique is applied to the wave for the duration of two periods in order to effectively suppress the transients. The time evolution of the resultant sinusoidal wave is shown in Fig. 5.14. The mathematical expression of this sine wave is as follows;

$$\text{Source} = AMP \times \sin(2\pi ft) \times 0.5 \times (1 - \cos(\pi \times IT / NP)) \tag{5.8}$$

where AMP and *IT* are the amplitudes and the iteration number, respectively and *NP*, suppression factor [25] is given by

$$NP = \frac{2}{f \times \Delta t} \tag{5.9}$$

where f and Δt are frequency and time step, respectively.

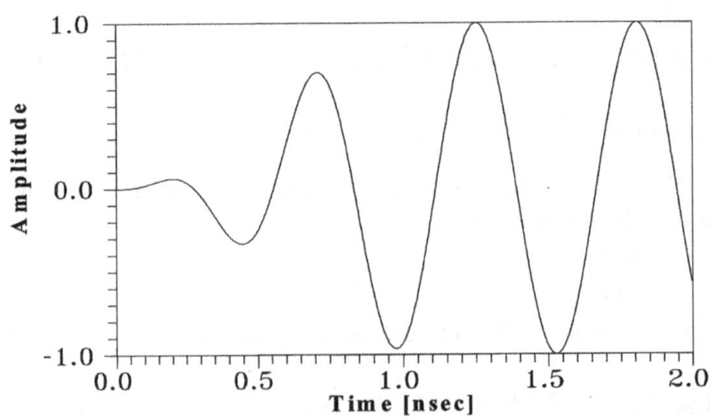

Figure 5.14: *Sine wave of which two periods are suppressed by a half Hanning window*

To test the radiated and absorbed power subroutines used in SAR calculations, the structure depicted in the inset of Fig. 5.15b is considered. This structure is composed of two lossy dielectric cubes, one of which is located inside the other. The computation volume of 14 cm × 14 cm × 14 cm, is modeled with 70 × 70 × 70 cells using the unit cell size, $\Delta\ell$ = 0.2 cm. Dimensions of the outer dielectric cube used in calculations is 6 cm × 6 cm × 6 cm (30 × 30 × 30 cells), while the dimensions of inner cube are 2.6 cm × 2.6 cm × 2.6 cm (13 × 13 × 13 cells). The electrical parameters of the outer and inner cubes are ε_{r2} = 5 and σ_2 = 0.2 S/m and, ε_{r3} = 2.5 and σ_3 = 0.5 S/m, respectively. The outer cube is placed between (*x*:23,52; *y*:23,52; *z*:23,52), while the inner one is between (*x*:31,43; *y*:31,43; *z*:31,43). The *y*-component of the electric field is excited at point (18,18,18). Time variation of the radiated power is shown in Fig. 5.15 for the case when the computation space is terminated by PEC boundaries which form a resonator. Figure 5.15a corresponds to the response of the resonator. The source is a once-differentiated Gaussian pulse having 4 GHz bandwidth. The variation for the integration of Poynting vector in Fig. 5.15a corresponds to a radiating source inside a PEC resonator. The positive power values correspond to power radiated outward from the surface (from source towards the PEC walls of the resonator), while the negative values show power flow in the opposite direction, i.e., reflected power traveling inward the virtual surface. The same situation occurs in Fig. 5.15b where the pointing integral results for the same resonator including lossy structure are plotted. However, the radiated power decays with time, indicating dissipation of the energy of source pulse by the lossy dielectric cubes.

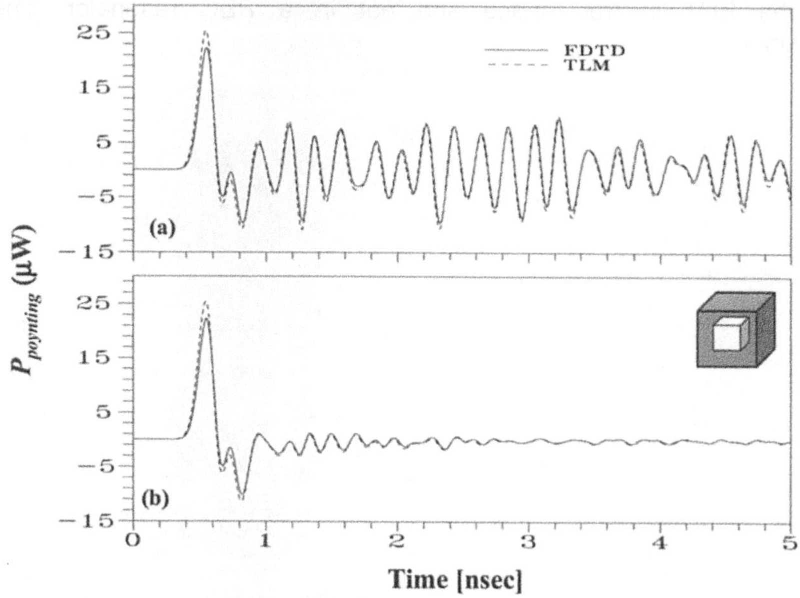

Figure 5.15: *(a) Poynting power for an empty resonator. (b) EUT in a PEC resonator.*

The radiated and absorbed power calculations are repeated for the sinusoidal excitation with a frequency of 1.8 GHz. The results obtained when EUT is enclosed into a PEC resonator (i.e., PEC terminations at the boundaries of the computational domain) are shown in Fig. 5.16. Since the sinusoidal radiating element continuously pumps power into the resonator, the amplitude of the Poynting and absorbed powers initially increase with time as observed in the figure until a steady-state regime is established at later times. Again, positive and negative radiating powers correspond to outward and inward propagating powers, respectively. Note that for the radiated power the average value is zero since whatever EM energy radiates away from the virtual surface used in calculations is reflected back by the PEC walls. On the other hand, absorbed power is always positive as expected. It is clearly seen from Figs. 5.15 and 5.16 that there is an excellent agreement between the results obtained via TLM and FDTD simulators.

A number of tests are performed with similar configurations, i.e., using different electrical parameters, excitation and observation locations, geometries, etc. In all cases the results obtained via TLM and FDTD exhibited an almost perfect agreement. It is therefore safe to assume that

- in the absence of geometrical discretization (cubical structures) errors and
- in the presence of PEC boundaries,

TLM and FDTD can provide very accurate results for absorbed power values in reasonable computer times. However, SAR simulations are usually performed by placing the EUT in free space and not in a PEC resonator. Therefore, computations

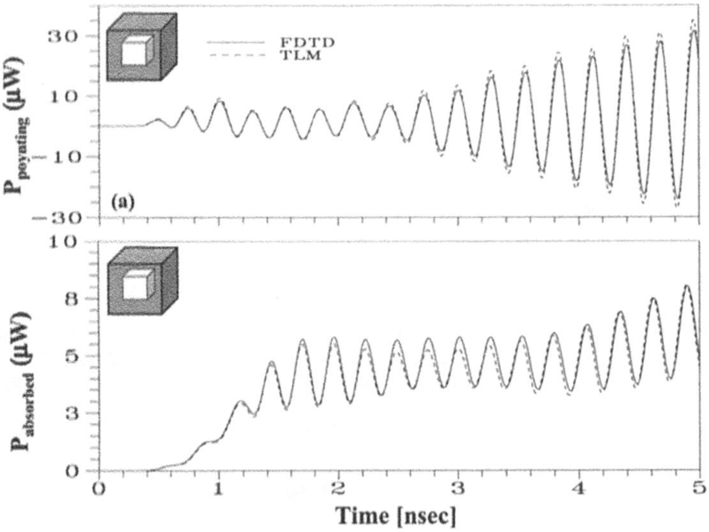

Figure 5.16: *(a) Poynting power for EUT. (b) Absorbed power in EUT for sinusoidal excitation.*

are performed when TD simulators are terminated by PML blocks. An output is shown in Fig. 5.17, where results obtained for a pulse excitation are given. In this test the same geometrical and electrical parameters are used as for Figs. 5.15 and 5.16, except that PML-type ABCs are implemented at the boundaries of the computational space and a λ/3 (at 2 GHz) thin wire antenna is used.

The solid curve shown in Fig. 5.17a shows the variation of the radiated power in the absence of the EUT. As described above, the radiated power is calculated by applying Gauss' theorem over a virtual surface which includes the source and is chosen between the EUT and the boundaries of the computational domain. The solid drawn curve in Fig. 5.17a, therefore, corresponds to the time variation of the total radiated energy from the source distribution used in the absence of any interaction from a nearby scatterer. Clearly, energy emitted from the antenna may be modified with respect to its distribution under "free space" conditions when some scatterers (such as the EUT in this example) are placed in its vicinity. The dashed curve in Fig. 5.17a is obtained by adding up the power radiated away from the domain in the presence of the EUT with the power absorbed by the EUT. These two constituents are shown separately in Figs. 5.17b and 5.17c.

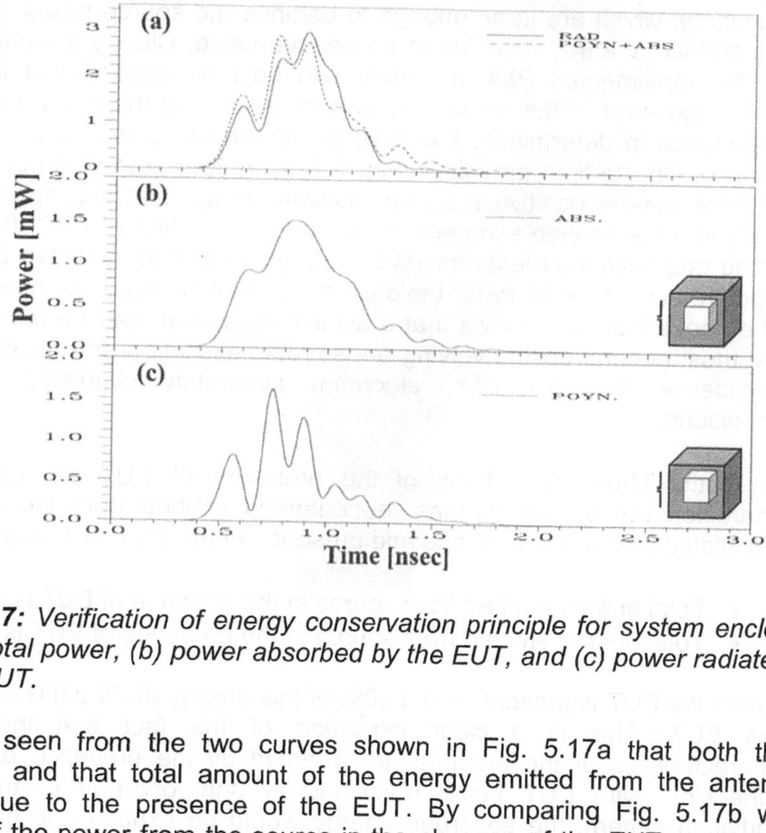

Figure 5.17: *Verification of energy conservation principle for system enclosed by ABC. (a) total power, (b) power absorbed by the EUT, and (c) power radiated away from the EUT.*

It can be seen from the two curves shown in Fig. 5.17a that both the time distribution and that total amount of the energy emitted from the antenna are modified due to the presence of the EUT. By comparing Fig. 5.17b with the variation of the power from the source in the absence of the EUT as given by the solid curve in Fig. 5.17a, it becomes evident that the two curves differ in both amplitude and time extent. Clearly, EM waves intercepted by the EUT slow down; on the other hand, a substantial part of the energy is absorbed by the EUT which is closer to the source than a major part of the virtual surface used in the calculations. Thus when the two cases are compared, one with and the other without the EUT, then it may be concluded that when EUT is present the energy corresponding to the absorbed part travels a shorter average distance but at a lower speed. Due to the combined effect of these two factors (e.g., lower speed, shorter distance), it is in general not possible to predict the difference between the time dependences of the pulses with and without the EUT. The remaining discrepancies between the two curves can be attributed to contributions from the following two possible sources:

- Imperfect modeling of ABC
- The variation of the power radiated by the source as a function of presence/absence of the EUT

These are investigated in more detail below:

- As noted before, a negative value in the radiated power curve given in Fig. 5.17a would mean a net power input through the virtual surface used in the calculations. This would be an indication of reflections from the PML boundary, which are large enough to balance the source power incident on the surface, thereby resulting in a negative value. Clearly the effectiveness of the implemented PML algorithm can best be observed at later times where the level of the source pulse diminishes and the reflections gain on importance in determining the direction of the net power flow through the surface. On the time arrivals in Fig. 5.17b, note that the sign of the power remains always positive (i.e., net outward flow). This finding proves that there is no appreciable imperfectness in the modeling of ABC. It should be noted that validation tests for PML ABC given in [26] indicated that a PML algorithm works perfectly for the case of paraxial propagation in waveguides. The above test now shows that a similar conclusion can be drawn also for the most general case involving an isotropic excitation and gives us further confidence that our PML algorithm accurately simulates free-space conditions.

- To better show the effects of the presence of EUT on the emission characteristics of the antenna the energies emitted from the source are calculated both in the absence and presence of the EUT as follows:

    - Total energy emitted from source in the absence of EUT is 1.21 pJ
    - Total energy emitted from source in the presence of EUT is 1.32 pJ.

When the EUT is present, about 60% of this energy (0.78 pJ) is absorbed by the EUT. This is a clear indication of the fact that the emission characteristics of the antenna are affected by the presence of the EUT, otherwise, with EUT being contained in only one-half of the antenna radiation pattern, the absorbed energy would be expected to be less than 50% of the source energy. On the other hand, it is to be expected that the imaginary part rather than the real part of the dielectric constant of the EUT would be more effective in modifying the antenna characteristics in the above example (note that for the two media constituting the EUT $\varepsilon = \varepsilon_o(\varepsilon_r - j\sigma / \omega\varepsilon_o) = \varepsilon_o(\varepsilon' - j\varepsilon'')$, $\varepsilon' - j\varepsilon''$ assume values 2.5 - *j*10 and 5 - *j*4, respectively). Indeed, the total emitted energy for another test is calculated wherein the shape of the EUT and the relative dielectric constants of the cubes were kept in tact, but the conductivities of the media were set to zero. In this case, it is observed that the total energy was within 1% of the total energy emitted in the absence of the EUT (i.e., 1.22 pJ).

5.3.2 Peak SAR Calculations

After the above-described preliminary tests, SAR calculations are taken into account. First, the results of SAR calculations for some simpler structures are presented, similar to but more complicated than those investigated above, and subsequently address much more complicated structure, which represents a discrete, approximate model of the human head. Figure 5.18 shows the first structure used in SAR simulations. It is formed by four concentric dielectric lossy cubes with electrical parameters given in Table 5.1.

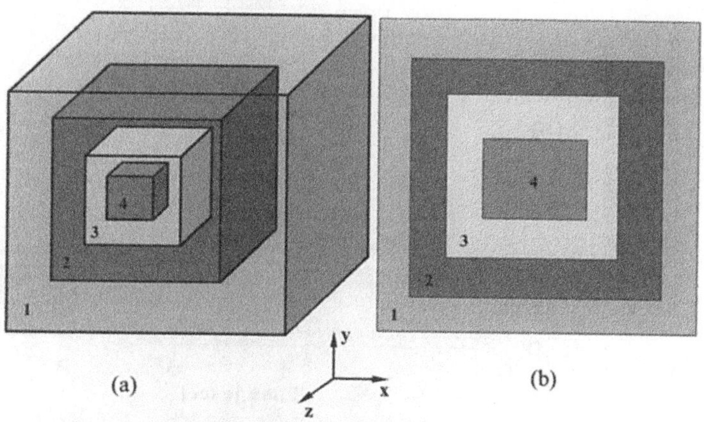

Figure 5.18: *The structure used in SAR calculations. (a) Perspective, (b) front view*

Computation volume is taken as 17.2 cm × 17.2 cm × 17.2 cm. Mesh size, $\Delta\ell$ is equal to 0.2 cm (i.e., computation volume contains 86 × 86 × 86 = 636056 cells). The dimensions of the dielectric cubes are 9.2 cm × 9.2 cm × 9.2 cm (46 × 46 × 46 cells) for cube-1 (see Fig. 5.18), 6.8 cm × 6.8 cm × 6.8 cm (34 × 34 × 34 cells) for cube-2, 5.2 cm × 5.2 cm × 5.2 cm (26 × 26 × 26 cells) for cube-3, and 3.6 cm × 3.6 cm×3.6 cm (18×18×18 cells) for cube-4. Number of cells in each dielectric region is given in the last column of Table 5.1.

Note that total number of cells in the computation volume, whose electrical parameters differ from free space, is about 15% of the number of cells on the whole computation volume (i.e., 97336). A λ/3 dipole driven by a 900-MHz sinusoidal wave with averaged power normalized to 600 mW is placed near the structure. The *y*-component of electric field is excited at point (15,43,43). A 5-layer PML ABC is used to simulate free space in these calculations. The time variation of the absorbed power in this structure as obtained via TLM and FDTD is plotted in Fig. 5.19.

Table 5.1 Electrical parameters of the EUT pictured in Fig. 5.18a at 900 MHz

Dielectric	ε_r	σ [S/m]	ρ [kg/m^3]	*# of Cells*
1	58.5	1.21	1040	58032
2	34.5	0.6	1100	21728
3	8	0.11	1850	11744
4	55	1.23	1030	5832

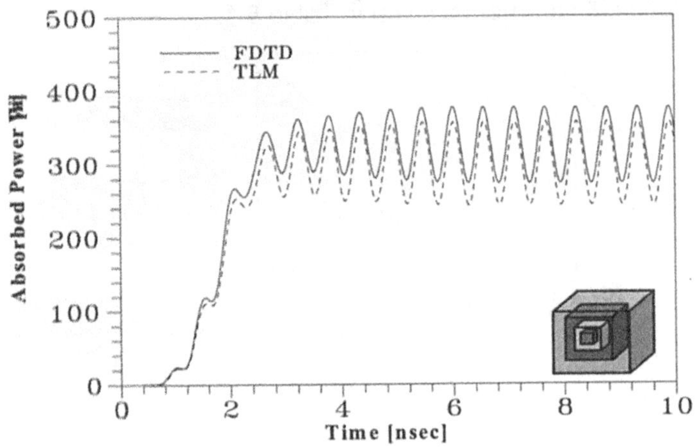

Figure 5.19: *Absorbed power in a cubical test structure*

It should be noted that the average of the absorbed powers is very close for both methods, while minimal points of power levels differ slightly indicating numerical dispersion effects in two algorithms are somewhat different.

At this point it would be appropriate to recall the close relationship between SAR and absorbed power calculations. Indeed, when viewed on a cell-by-cell basis, both results differ only by multiplicative factors corresponding to the density of tissue ρ, represented by the cell (see (5.2a)) and the inverse of the number of cells used in absorbed power calculations (e.g., average value per cell). Hence Fig. 5.19 can also be interpreted as the "whole structure SAR" for a homogeneous density distribution (i.e., $\rho = 1$ for all cells). Figure 5.19 shows the evolution of the absorbed power in time. Transient response following the onset of the sinusoidal signal will of course die away and steady-state conditions will settle in as time progresses. In SAR calculations this steady-state value is of interest. Moreover, international standards are stated in such a way that SAR is related to the peak absorbed power rather than its average value (note that this relation involves a spatially varying renormalization using appropriate ρ values). As noted in Section 5.1, often local SAR values rather than global SAR values are of interest. The reason for this is that although the SAR value obtained for the whole structure (global SAR) may well be within the limits dictated by the standards, SAR calculated in certain sub-domains (local SAR) may exceed these limits.

An example of local SAR calculations is given in Fig. 5.20, where SAR values averaged over a tissue mass of 1 g (SAR_{1g}) are calculated both with TLM and FDTD. Clearly there is a perfect agreement between these results. The slight shift between the curves in Fig. 5.20 is due to numerical dispersion and can be reduced by decreasing the cell size. The calculations are performed in the 1-g-cell group (5 × 5 × 5 = 125 cells in this case) located centrally within the innermost dielectric cube and SAR_{1g} is obtained as the average of SAR_{1cell} values over the 1-g-cell group. In the figure the variation of SAR is given for the steady-state case.

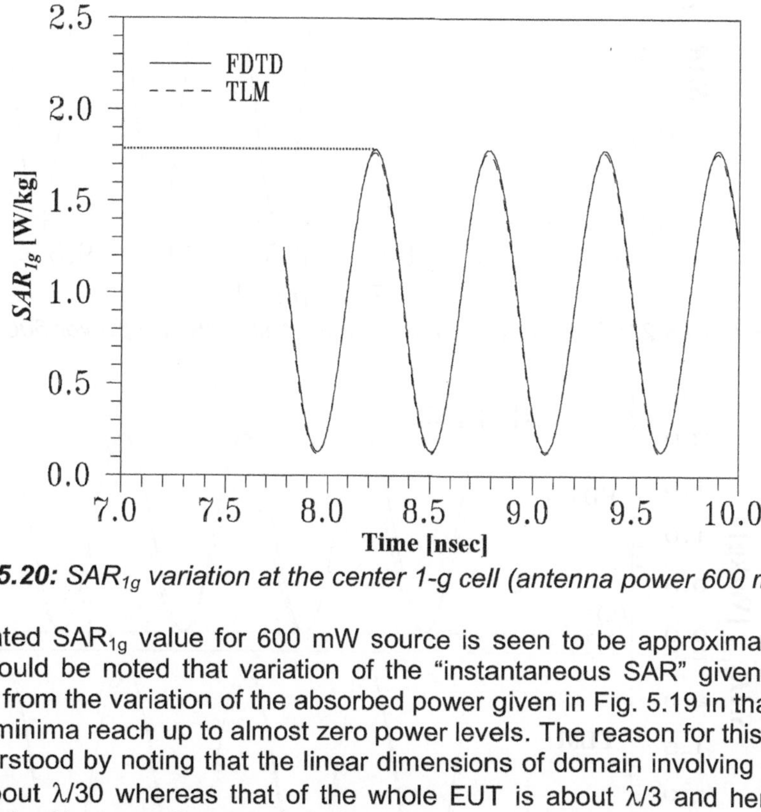

Figure 5.20: *SAR_{1g} variation at the center 1-g cell (antenna power 600 mW)*

The calculated SAR_{1g} value for 600 mW source is seen to be approximately 1.8 W/kg. It should be noted that variation of the "instantaneous SAR" given in Fig. 5.20 differs from the variation of the absorbed power given in Fig. 5.19 in that in the former the minima reach up to almost zero power levels. The reason for this can be easily understood by noting that the linear dimensions of domain involving 1-g-cell group is about λ/30 whereas that of the whole EUT is about λ/3 and hence the power absorbed in 1-g cells can mimic the input signal closely, whereas the power absorbed on the whole structure effectively corresponds to a time-averaged value.

Another example which displays the variation of local SAR at the finest granularity (e.g. on the cell basis) is depicted in Fig. 5.21. Here the calculations performed in the central cell of the cell group used in Fig. 5.20 are given. One can clearly see that the SAR_{1cell} value is rather close to the SAR_{1g} value as expected, but it somewhat differs from it, obviously an indication of the non-uniform illumination in the region containing the 1-g-cell group. The slight differences between TLM and FDTD solutions in Fig. 5.21 obtained with a mesh size of $\Delta\ell$ = 0.2 cm, can be attributed to the effects of numerical dispersion discussed in [26]. This point is further illustrated in Fig. 5.22 where TLM and FDTD solutions obtained at half-resolution (i.e., using a

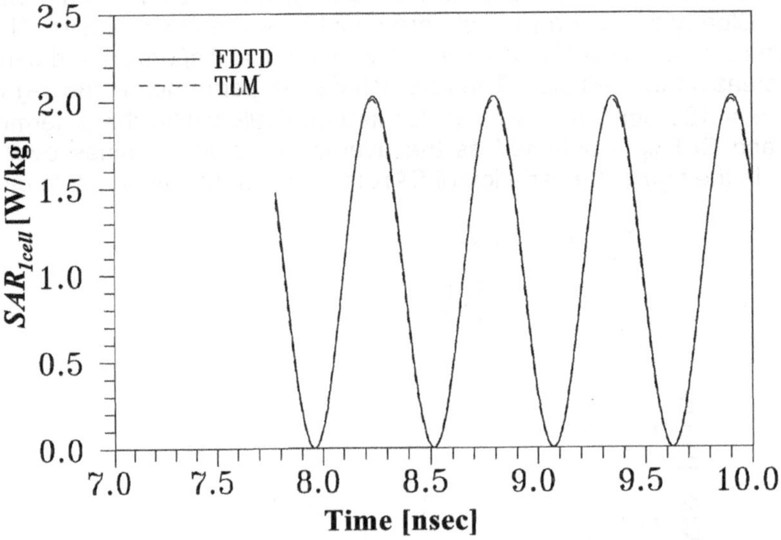

Figure 5.21: *SAR vs. time at the center cell (antenna power 600 mW)*

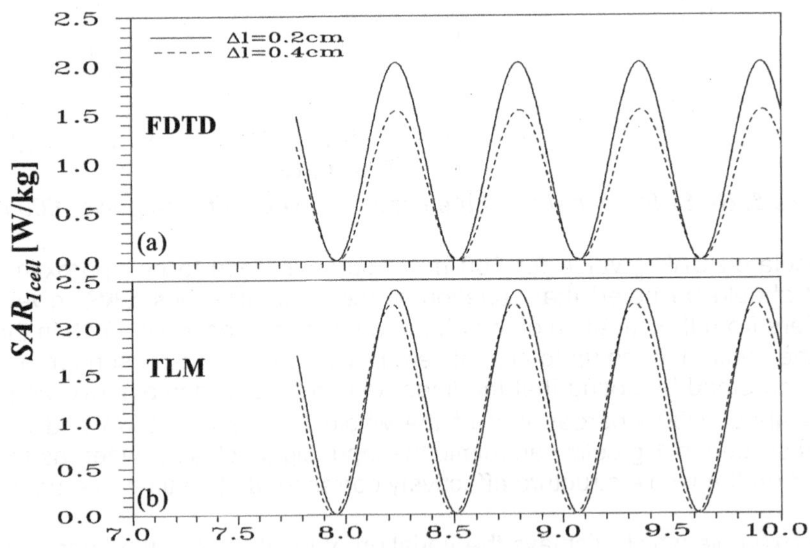

Figure 5.22: *SAR vs. time at the center cell. (a) FDTD, (b) TLM*

mesh size of $\Delta\ell$ = 0.4 cm) are displayed. It can be seen from these figures, that for this test problem TLM has more favorable convergence characteristics (note that this conclusion does not apply generally, see discussions in [26]).

In a large structure containing many ten thousands of cells, the determination of critical sub-domains which may be subject to over-exposures, requires the investigation of the variation of local SAR values over the entire structure. On the other hand, since absorbed powers at each cell are readily available at all time steps in TLM or FDTD TD calculations, they can be conveniently displayed in the form of color SAR density maps. Two such examples are given in Figs. 5.23 and 5.24, which display the variation of SAR in each cell in slices taken along horizontal and vertical directions, respectively. These figures show snapshots taken at the same arbitrary time step after the steady-state conditions are established. Looking at these SAR density maps, one can easily identify "critical areas" which, as expected, are located in the high loss regions (see Table 5.1) in the vicinity of the antenna.

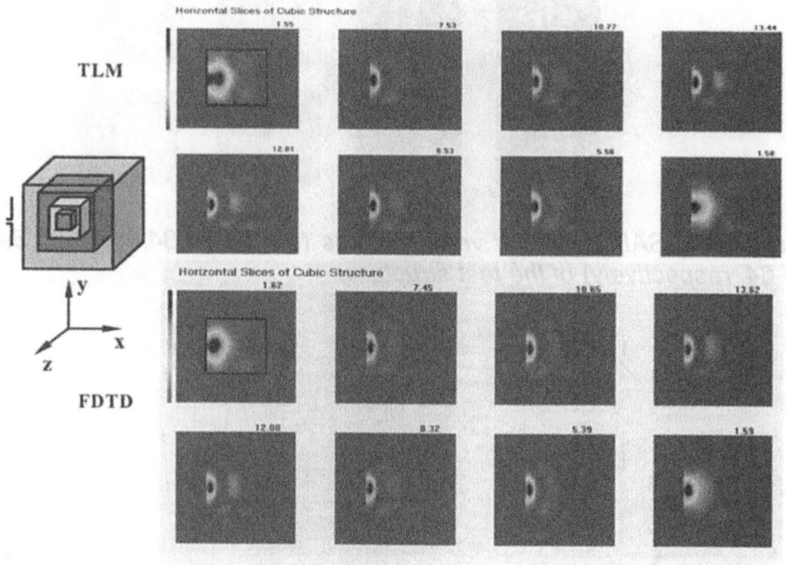

Figure 5.23: *SAR values at horizontal slices (defined in the order from left to right in both rows by cell numbers y = 24,30,34,40,49,55,59, and 64, respectively) of the test structure. The rectangular lines on the first windows depict the boundaries of the EUT. The maximum values of* SAR_{1cell} *are indicated on top of each slice*

Although SAR values obtained by taking as the reference domain the "whole EUT", "1-g sub-domain", and "an individual cell", as depicted in Figs. 5.19, 5.20, and 5.21, respectively, are all approximately equal to 2 W/kg, it is seen from Figs. 5.23 and 5.24 that values as much as 6 times larger than the "whole structure" SAR can exist in some cells which are located in the outermost dielectric cube and close to the antenna. This is further illustrated in Fig. 5.25.

The results obtained via TLM for global and local SAR values and the close agreement between these and the FDTD results in the fairly simple test problems considered above, demonstrate that our TLM algorithm is properly implemented and "fine-tuned" for performing accurate SAR calculations.

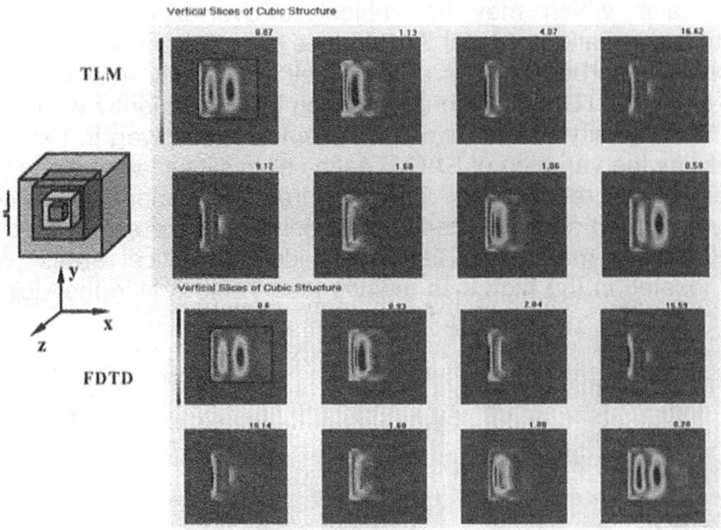

Figure 5.24: *SAR values at vertical slices (z = 24,30,34,40,49,55,59, and 64, respectively) of the test structure.*

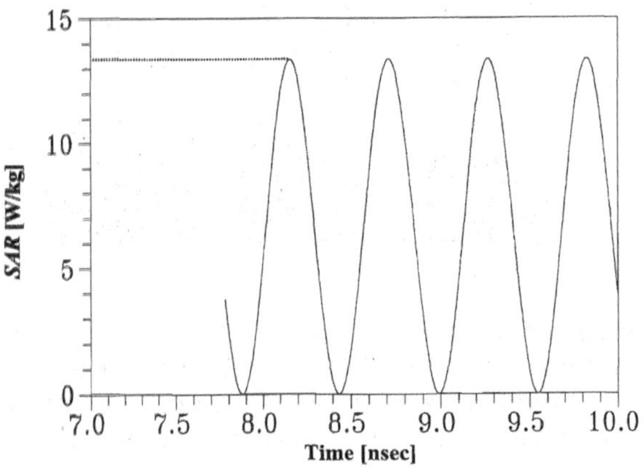

Figure 5.25: *SAR variation in extreme case (cell # is (25,43,43))*

5.3.3 SAR Simulations with a Discrete Human Head Model

The complex problem of modeling and simulation of variation of SAR values in the human head is now going to be addressed. The human head model used in the computations is taken from the literature [22]. This model is derived from actual NMR images. In this model, 7 different tissue types are used and their electrical parameters (ε_r,σ) and densities (ρ) are specified at the frequencies of interest (900 MHz) for cellular phone applications.

These parameters are given in Table 5.2. It should be noted that a scaling of the dimension of the head by a factor of 0.5 is used (i.e., the head model used in computations is half the size of a mature human head and corresponds approximately to the dimensions of the head of a newborn baby), in order to meet the memory requirements with the available computational resources. The last column of Table 5.2 shows the number of cells in the head used in calculations.

Table 5.2 Electrical parameters of the human head tissues at 900 MHz

Tissue	ε_r	σ [S/m]	ρ [kg/m^3]	# of Cells
Skin/fat	41.4	0.87	1010	18544
Muscle	55.0	0.94	1040	27920
Bone	21.0	0.32	1810	17504
Brain	35.7	0.61	1040	32528
Humor	74.1	1.97	1010	128
Lens	51.3	0.89	1100	64
Cornea	52.1	1.22	1170	848

Apart from the 7 tissues listed in Table 5.2, the model also includes locations and extends of the air-filled regions (nasal cavity etc.) within the human head. The total number of cells in the computation domain is 688896 ($78 \times 96 \times 92$ cells). But, the number of cells that is tabulated below is only 97536, whereas the number of cells used for representing the human head is about 20% of the total number of cells. The remaining 80% of the cells are used to define the computational domain, which contains the antenna and is terminated by PML ABC.

The results obtained using both TLM and FDTD codes for $SAR_{1MAXbrain}$ calculations are given in Fig. 5.26.

Before discussing the salient features of the curves in the figure, it would be appropriate to give the definition of the quantity calculated, i.e., $SAR_{1MAXbrain}$, and explain how the calculations are performed. SAR values averaged over a tissue mass of 1-g (SAR_{1g}) were defined in the previous section. The calculated value of SAR_{1g} is a point function (i.e., defined at each cell point) which involves an average effect of absorbed power and density variations in a certain (in this case corresponding to 1g) neighborhood of the point in question. It is therefore clear that an SAR_{1g} value obtained for one cell will be a function of the position of the cell and the density distribution (tissue type) in a certain neighborhood of this cell, as well as of the time step used in calculations. In case of sinusoidal excitation the time dependence will exhibit a periodic behavior once the steady-state conditions are reached.

With the aid of the above explanations, the definition of $SAR_{1MAXbrain}$ can now be stated as "the maximum of SAR_{1g} values calculated for all brain cells whose entire

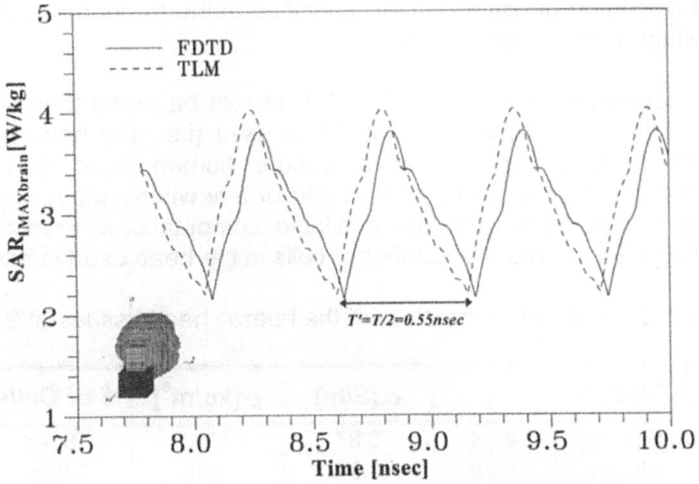

Figure 5.26: *The time variation of maximum SAR in the brain, as averaged over 1 g. Period of the variation is half of the period of source that is 1.11 nsec*

1-g neighborhoods are contained in brain region." As noted above, the $SAR_{1MAXbrain}$ value is also a function of time. $SAR_{1MAXbrain}$ calculations are performed as follows:

- The number of cells in 1-g neighborhood is 125
- The number of brain cells is 32528
- The number of internal cells whose 1-g neighborhood is contained in the brain region (i.e., excluding boundary cells which are closer than 2 cells to the boundaries of other tissues) is approximately 25,000,

At a given time step,

i) Choose an internal cell, and calculate SAR_{1g}.
ii) Repeat (i) for all internal cells and utilize their maximum value to define $SAR_{1MAXbrain}$ at this time step.
iii) Repeat the above calculations for the number of time steps required for obtaining steady-state time variation (e.g., 3000 times here) of $SAR_{1MAXbrain}$ depicted in Fig. 5.26.

It is evident from the above that the TLM calculations leading to Fig. 5.26 are rather time-consuming and impose some severe memory requirements. For the above example approximately 15 hours is required on a Pentium-3, 850-MHz PC with 256-MB RAM.

Since the curves depicted in Fig. 5.26 correspond to times after the establishment of the steady-state regime, the time variation of the absorbed power in the brain tissue does, of course, exhibit a periodic behavior with a period of 0.55 ns corresponding to signal frequency. We further note that the $SAR_{1MAXbrain}$ as obtained with TLM and FDTD varies in a quasi-sinusoidal fashion, respectively, between 2.3 - 4 W/kg and 2.2 - 3.8 W/kg, with average values of 3.15 W/kg for TLM and 3.0 W/kg for FDTD. Thus, the $SAR_{1MAXbrain}$ calculations reveal the following:

- Time variation of TLM and FDTD results exhibit essentially identical patterns,
- For the discretization level used in these calculations the actual values obtained with TLM and FDTD differ from each other by about 5%. It should be noted that in this problem the $\lambda/\Delta\ell$ value, which yields a measure for the fineness of discretization; ranges from about 160 in free space to about 18 in the humor region. Although the value of 160 can be considered to represent an excellent modeling, the value of 18 yields a resolution which may, and evidently is, too coarse for modeling the problem with sufficient accuracy.

Similar conclusions can be drawn from the curves shown in Fig. 5.27. Here TLM and FDTD calculations corresponding to normalized SAR_{1cell} values performed utilizing the same head model and same computational parameters are given for two tissue types: brain in Fig. 5.27a and bone in Fig. 5.27b. It should be noted that also in this case both algorithms result in almost identical time variations, however with about 8% difference between the calculated amplitudes and averages.

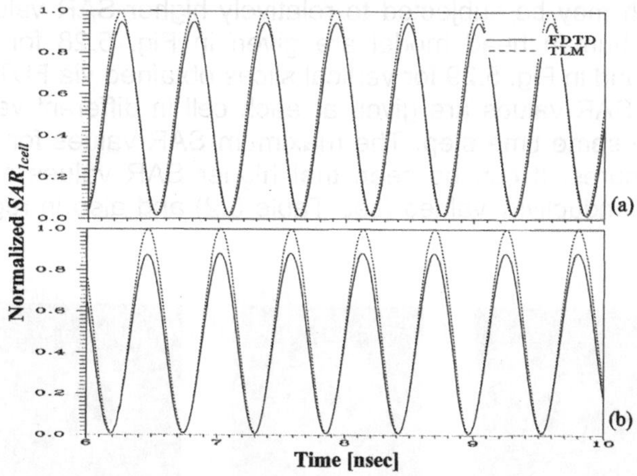

Figure 5.27: *Normalized SAR vs. time in (a) brain, and (b) bone. Period of the wave is 0.55 nsec*

Guided by the above observations it is safe to conclude that TLM and FDTD are capable of representing the physics of the extremely complex problem of calculating the SAR values in real-life situations. Numerical computations show that utilizing the readily available computational resources of today (Pentium-4, 2 GHz, 1-GB RAM), these calculations can be performed for solving even more realistic problems (full-size human head models involving more than 7 tissue types used here [13]) and will yield solutions with accuracies which should be considered at least as good as "adequate" if not "excellent" for most practical applications (note that all computer head models are necessarily of a generic nature and the experimental approach which is the only alternative way of obtaining SAR values is prone to quite a number of error sources). On the other hand the numerical errors in calculations can, in principle, be reduced to much lower levels. Note that apart from the problems in realizing PML ABC discussed above, the main difficulty stems from the problems posed by the "sufficiently fine" sampling requirements in the spatial domain. The accumulation of expertise in numerical EM over the last few decades has revealed that in order to be able to adequately represent a scattering surface, it must be sampled, as a rule of thumb, along each dimension with at least 10 points or at intervals of $\lambda/10$, whichever is smaller. For the problem of SAR calculations in the human head, using the above considerations it can be concluded that in order the calculations be acceptable the granularity of the model (i.e., minimum dimensions of regions containing different tissue types) should be not less than 2 cm (i.e., ten times the cell size, 0.2 cm). This condition was violated in various parts of the numerical head model used in these calculations. About one order of magnitude smaller cell sizes are required to resolve a realistic human head model in keeping with the above criteria. However, some simple calculations show that in this case the problem will be numerically intractable and remain this way also in the foreseeable future since this refinement of the grid size would require the computational work to increase by 2/3 orders of magnitude.

Color density maps can be generated for better visualization of critical areas in the human head which may be subjected to relatively higher SAR values. Two such examples for the human head model are given in Fig. 5.28 for vertical slices obtained via TLM and in Fig. 5.29 for vertical slices obtained via FDTD methods. In these figures, the SAR values are given at each cell in different vertical slices of head model at the same time step. The maximum SAR values for each slice are indicated in the figures. It can be seen that higher SAR values are obtained in tissues with high conductivity values (see Table 5.2) and also in regions closer to the antenna.

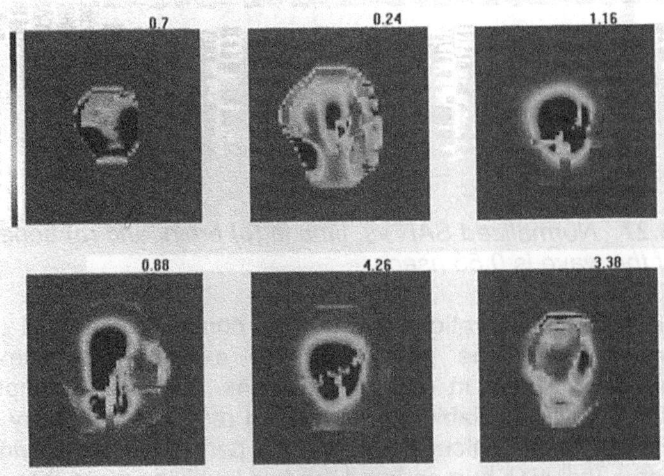

Figure 5.28: *Peak SAR distribution at different vertical slices (TLM)*

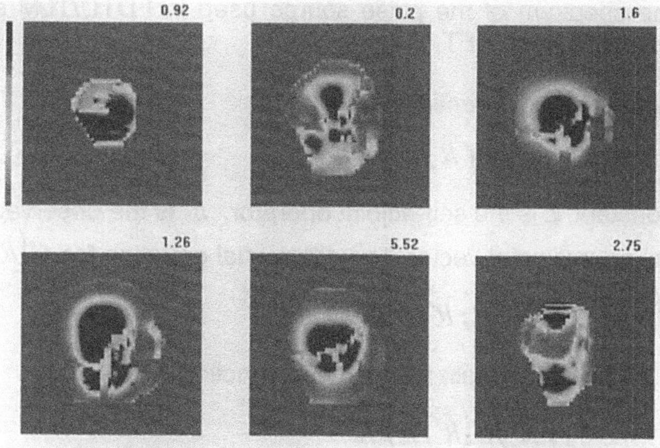

Figure 5.29: *Peak SAR distribution at different vertical slices (FDTD)*

5.4 CALIBRATION AGAINST ANALYTICAL EXACT SOLUTIONS

In this section, validity of the TD simulators are discovered. Cases where quite interesting discrepancies occur between the results of FDTD and TLM techniques are met, such as in SAR calculations, in EMC simulations, in radar cross section (RCS) modeling [28] and in wireless propagation in urban region [29]. The common property in these applications is that they posses first- and second-order scattering effects. It is frequently difficult to interpret the discrepancies in these applications, since they occur only at certain observation points and/or during a small fraction of observation period. It may be speculated that they are mostly related to secondary (diffraction) effects, but, canonical tests and calibration are still a requirement.

In this section, this requirement is targeted. The FDTD and TLM algorithms are tested against an exact solution on a canonical EM problem. A PEC rectangular cavity is taken into account, inside where the effects of a pulsed source is observed at different places. This problem has an exact analytical solution in terms of Green's functions, where EM fields are represented via a 3D infinite series. The rectangular cavity Green's function has very poor convergence, especially when the source and observation points are located closely to each other, and different techniques, such as Poisson summation [30] have been applied for rapid and accurate calculations. It may therefore be speculated that calibration is two-way; FDTD and TLM against the exact Green's function approach (basically in FD), and vice versa in TD.

5.4.1 A PEC Resonator and Green's Function Representation

The rectangular cavity is built in FDTD and TLM algorithms, and TD solutions are obtained directly for a broadband pulse excitation. FD solutions are obtained via DFT. On the other hand, the Green's function formulation is used to obtain solutions in FD for a given frequency. Analytical exact TD solutions are obtained by calculating multi-frequency results from the Green's function formulation, weighing them with the spectrum of the pulse source used in FDTD/TLM simulations and finally by applying inverse DFT.

Consider the following differential equation first:

$$L\Phi(\vec{R})+\lambda\omega(\vec{R})\Phi(\vec{R})=f(\vec{R}) \tag{5.10}$$

where λ is constant, L is the self-adjoint operator, $\vec{R}$ is the observation point vector and $\vec{R}'$ is the source point vector. The differential equation for $G(\vec{R};\vec{R}')$

$$\nabla^2 G(\vec{R};\vec{R}')+k^2 G(\vec{R};\vec{R}')=-\delta(\vec{R}-\vec{R}') \tag{5.11}$$

is proposed, so one can obtain the unknown function $\Phi(\vec{R})$ via

$$\Phi(\vec{R})=-\int_{\vec{R}'} f(\vec{R}')G(\vec{R}';\vec{R})dR' \tag{5.12}$$

and can solve (5.11) [31]

$$G(x,y,z;x',y',z')=\sum_m\sum_n \phi_m(x)\phi_m^*(x')\varphi_n(y)\varphi_n^*(y') \times G_z(z,z';\lambda_z=k^2-\lambda_m-\lambda_n) \tag{5.13}$$

Definition of a Green's function is given in terms of the inhomogeneous differential equation (5.11) which must satisfy the same boundary conditions with the unknown function in (5.10). Associated with the operator L of the Green's function there is a set of eigenfunctions $\{\phi_m(x),\varphi_n(y)\}$ and eigenvalues λ_m and λ_n determined by the homogeneous differential equations

$$\begin{aligned}
&[L+\lambda_m\omega(x)]\phi_m(x)=0 \qquad 0\le x\le a\\
&[L+\lambda_n\varpi(y)]\varphi_n(y)=0 \qquad 0\le y\le b
\end{aligned} \tag{5.14}$$

Equation (5.14) and boundary conditions are commonly referred to as the Sturm-Liouville system.

$$\begin{aligned}
&G(0,y,z;x',y',z')=G(a,y,z;x',y',z')=0 \qquad 0\le x,x'\le a\\
&G(x,0,z;x',y',z')=G(x,b,z;x',y',z')=0 \qquad 0\le y,y'\le b\\
&G(x,y,0;x',y',z')=G(x,y,d;x',y',z')=0 \qquad 0\le z,z'\le d
\end{aligned}$$

$$\left[\frac{\partial^2}{\partial x^2}+\lambda_m\right]\phi_m(x)=0 \qquad 0\le x\le a \quad \text{with} \qquad \phi_m(0)=\phi_m(a)=0$$

$$\left[\frac{\partial^2}{\partial y^2}+\lambda_n\right]\varphi_n(y)=0 \qquad 0\le y\le b \quad \text{with} \qquad \varphi_n(0)=\varphi_n(b)=0 \tag{5.15}$$

$$\left[\frac{\partial^2}{\partial z^2}+\lambda_z\right]G_z(z,z';\lambda_z)=-\delta(z-z') \qquad 0\le z,z'\le d$$

$$\text{with} \qquad G_z(0,z')=G_z(d,z')=0$$

One can easily solve (5.15) with given boundary conditions as follows:

$$\phi_m(x)=\sqrt{\frac{2}{a}}\sin\sqrt{\lambda_m}x \quad ; \quad \lambda_m=\left(\frac{m\pi}{a}\right)^2 \qquad m=1,2,3,\ldots \tag{5.16a}$$

$$\varphi_n(y)=\sqrt{\frac{2}{b}}\sin\sqrt{\lambda_n}y \quad ; \quad \lambda_n=\left(\frac{n\pi}{b}\right)^2 \qquad n=1,2,3,\ldots \tag{5.16b}$$

and G_z can be solved in the same manner via the equation

$$G_z(z,z';\lambda_z)=-\sum_p\frac{\Theta_p(z)\Theta_p^*(z')}{\lambda_z-\lambda_p} \tag{5.17}$$

where λ_p and Θ_p are the eigenvalues and eigenfunctions of the homogeneous differential equation

$$\left[\frac{\partial^2}{\partial z^2}+\lambda_p\right]\Theta_p(z)=0;\ 0\le z\le d;\ \Theta_p(0)=\Theta_p(d)=0 \tag{5.18}$$

One solves (5.18) for λ_p and Θ_p as follows

$$\Theta_p(z)=\sqrt{\frac{2}{d}}\sin\sqrt{\lambda_p}z \quad \text{where } \lambda_p=\left(\frac{p\pi}{d}\right)^2 \qquad p=1,2,3,\ldots \tag{5.19}$$

Substituting (5.16a), (5.16b), (5.17), and (5.19) into (5.13), one obtains the Green's function as

$$G(x,y,z;x',y',z') = -\frac{8}{abd} \times \sum_{m=1}^{\infty}\sum_{n=1}^{\infty}\sum_{p=1}^{\infty} \frac{\sin\left(\frac{m\pi x}{a}\right)\sin\left(\frac{m\pi x'}{a}\right)\sin\left(\frac{n\pi y}{b}\right)\sin\left(\frac{n\pi y'}{b}\right)\sin\left(\frac{p\pi z}{d}\right)\sin\left(\frac{p\pi z'}{d}\right)}{k^2 - \left(\frac{m\pi}{a}\right)^2 - \left(\frac{n\pi}{b}\right)^2 - \left(\frac{p\pi}{d}\right)^2} \tag{5.20}$$

5.4.2 PEC Resonators and TE_{10p} Solutions

In keeping with the above considerations it is assumed as the driving source distribution a *y*-directed current sheet given by

$$\mathbf{J_y}(x',y',z') = \mathbf{a_y} J_y(x',y',z') = \mathbf{a_y} \sin\left(\frac{\pi}{a}x'\right)\delta(z-z') \tag{5.21}$$

where $x' \in (0,a)$, $z' \in (0,d)$, and the primed coordinates refer to source location. It is straightforward to show that this source distribution will only excite *y*-independent fields polarized along *y*.

In this case the problem of determining the electric field $\mathbf{E} = E_y \mathbf{a_y}$ reduces to the solution of the scalar non-homogeneous wave equation

$$\nabla^2 E_y + k^2 E_y = -j\omega\mu J_y \tag{5.22}$$

with $\partial / \partial y = 0$, under the boundary conditions,

$$\begin{aligned} E_y(0, 0 \le y \le b, 0 \le z \le d) &= E_y(a, 0 \le y \le b, 0 \le z \le d) = 0 \\ E_y(0 \le x \le a, 0 \le y \le b, 0) &= E_y(0 \le x \le a, 0 \le y \le b, d) = 0 \end{aligned} \tag{5.23}$$

$E_y(x)$ at *z* due to the current sheet (5.21) at z' can then be expressed as [31]

$$E_y(x,z;z')_{10p} = -\frac{4}{d} j\omega\mu_o \sin\left(\frac{\pi}{a}x\right) \sum_{p=1}^{\infty} \frac{\sin\left(\frac{p\pi}{d}z\right)\sin\left(\frac{p\pi}{d}z'\right)}{k^2 - \left(\frac{\pi}{a}\right)^2 - \left(\frac{p\pi}{d}\right)^2}, \tag{5.24}$$

where the subscripts (1,0,*p*) are used as a reminder to show that the terms of the series can be identified as TE_{10p} modes of the cavity with associated excitation coefficients.

When wave number *k* is real, the cavity is at resonance at the frequencies determined by

$$f_{1,0,p} = \frac{c}{2\sqrt{\varepsilon_r}}\sqrt{\left(\frac{1}{a}\right)^2 + \left(\frac{p}{d}\right)^2} \quad (5.25)$$

Equation (5.24) is not valid at the resonance frequencies which render the denominator zero. These singularities are avoided when we assume small losses and define k as

$$k = \frac{\omega}{c}\left[\varepsilon_r - j\frac{\sigma}{\omega\varepsilon_o}\right]^{1/2} \quad (5.26)$$

where ω, c, σ, ε_o, and ε_r are the angular frequency, speed of the light, conductivity, permittivity, and relative dielectric constant, respectively.

It should be noted that calculations at observation points very close to the source plane and for media with very small losses should be avoided. It is clear that the series representation given in (5.24) will converge rather slowly. Therefore, several tests are run in order to check the dependence of numerical accuracy on the number of terms kept in the calculations. Such a test result is given in Table 5.3 for a = 100 m, b = 100 m, d = 100 m, assuming that the resonator is filled with a lossy dielectric (ε_r = 1.1 and $\sigma = 10^{-5}$ S/m). The values given in Table 5.3 are calculated for f = 2.6 MHz which is not close to a resonance. As can be seen from the table, with the chosen set of parameters, while only 100 terms are needed for 3-digit accuracy, more than 5000 terms are required for 6-digit accuracy. It should be noted that as losses are reduced in (5.24) or observer approaches to the source plane, the required number of terms increases drastically.

Table 5.3 The convergence analysis of (5.24) at f = 2.6 MHz

Mode Number (*p*)	**Field Amplitude**
10	35.7125473
100	35.536068
500	35.5360374
1000	35.5360374
5000	35.53619
6000	35.5366554
10000	35.5366554

5.4.3 FDTD and TLM Codes and Validation

The TD algorithms are tested against the Green's function representation for a 100-m × 100-m× 100-m rectangular cavity, filled with different lossy dielectrics. The cavity is represented by 100 × 100 × 100 cubical cells in TLM and 101 × 101 × 101 cells in FDTD (when terminated with a PEC wall, the termination cell is included in TLM, but excluded in FDTD; therefore an extra cell is required in FDTD). The cell sizes are taken as $\Delta x = \Delta y = \Delta z = \Delta\ell$ = 1 m in both algorithms. The time step is chosen to be 1.8 nsec. The source has the spatial distribution as given in Fig. 5.30 (represented by (5.21)). The temporal distribution of the source is represented by a once-differentiated Gaussian pulse. Its time extends between the points where pulse amplitude reduces to 10^{-6} of its maximum value, which is about 400 nsec. This corresponds to a bandwidth of 6 MHz in FD (here bandwidth is defined as the range in FD between the points where pulse amplitude reduces to 10% of its maximum value). The temporal distribution of the pulse and its frequency spectrum is given in Figure 5.31.

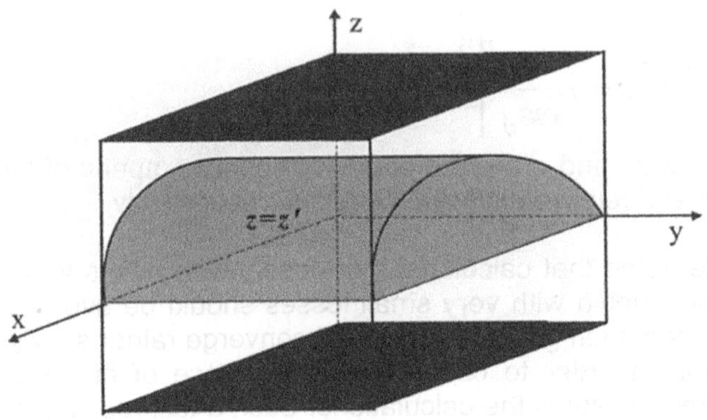

Figure 5.30: *Rectangular cavity and spatial source distribution at z = z' plane.*

The investigation of this canonical test problem provides a convenient setting for calibrating the TD algorithms and to assess their accuracy. For this problem these comparisons reveal that there is an excellent agreement between results obtained via TLM and FDTD. The following favorable properties of the canonical test problem under consideration can be worth to note:

- Since the PEC cavity has a rectangular shape, it can be discretized without geometrical approximation; therefore there will be no discretization errors.
- The TLM and FDTD techniques can be compared directly with a highly accurate reference solution. Therefore, discrepancies between the reference and computed results will be solely due to the limitations of these techniques (e.g., cell structure and resulting numerical dispersion).
- Choosing the source as in Fig. 5.31 reduces number of resonances. In this case there are only three resonances (i.e., at 2.02 MHz, 3.2 MHz, and 4.52 MHz) within the bandwidth of interest (see Fig. 5.32).
- The source distribution given in (5.21) does not excite E_x and E_z field components. This gives another possibility to assess the validity of the calculations; whether these components are not excited or appear as artifacts in TLM and FDTD algorithms can be checked.

The tests and comparisons are carried out in both the FD and the TD. In FD, reference solution is directly obtained from (5.24) for each frequency of interest. The TLM and FDTD results in FD are obtained indirectly. First, TD responses are accumulated at the chosen observation point for given source at a given excitation point, and frequency responses are then obtained via DFT. On the other hand, when TD results are of interest, FDTD and TLM yield these directly, whereas reference solution has to be obtained via inverse DFT.

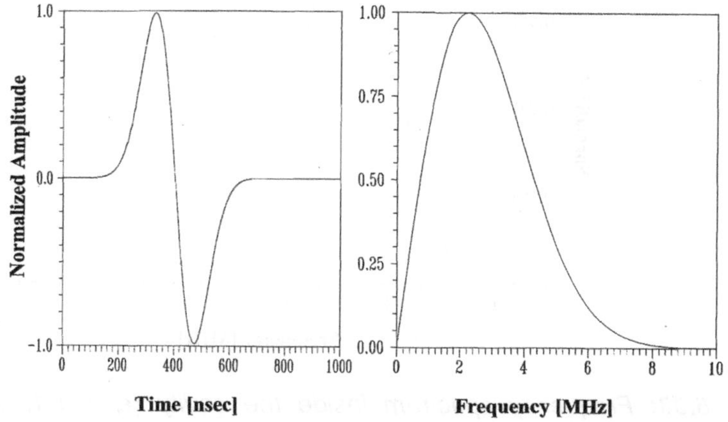

Figure 5.31: *(a) Once-differentiated Gaussian pulse vs. time. (b) Its frequency spectrum*

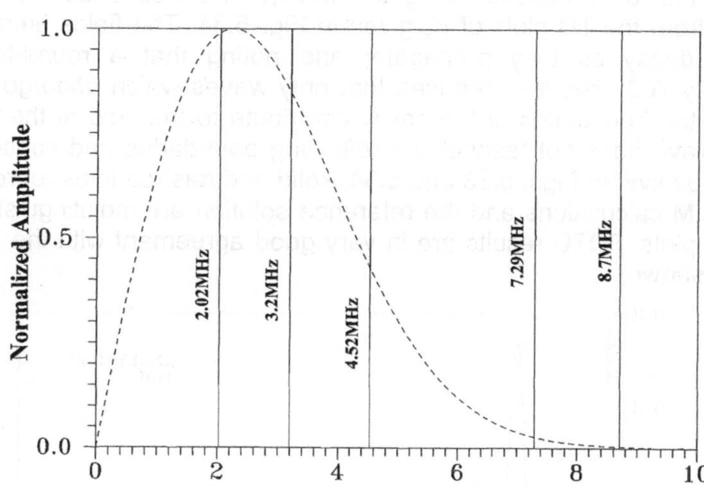

Figure 5.32: *Frequency spectrum of the source (dashed line) and TE_{10p} resonance frequencies computed by (5.25) for the PEC resonator filled by dielectric, $\varepsilon_r = 1.1$ (solid line)*

The first comparisons are given in FD and TD in Figs. 5.33 and 5.34, respectively, for $\varepsilon_r = 1.1$ and $\sigma = 10^{-4}$ S/m.

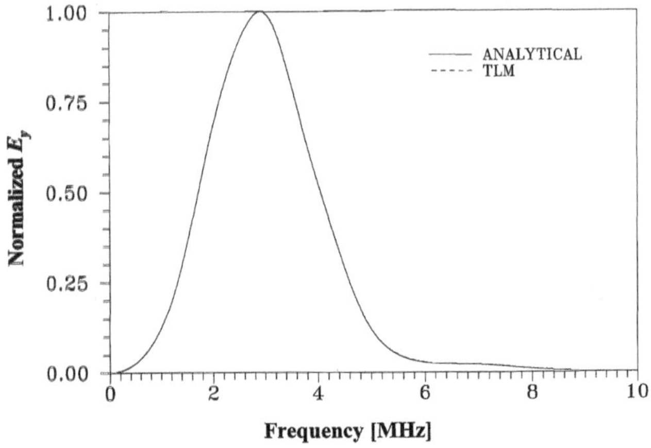

Figure 5.33: *Frequency spectrum inside the cavity (ε_r = 1.1, σ = 10^{-4} S/m). Frequency resolution is 10 kHz.*

Here, the source is located at the cell (50,50,25) and the observation is chosen at (50,50,75). As shown in Fig. 5.33, no resonances are observed in this case because of the lossy material filling the cavity. The loss effects can be better understood from the TD plots of E_y given in Fig. 5.34. The fields emanating from the source decay as they propagate; and noting that a round-trip requires approximately 700 nsec it is deduced that only waves which undergo one or two bounces on the boundaries of the cavity contribute to the field at the observation point. The wave does not "sense" the reflecting boundaries and consequently no resonances evolve. In Figs. 5.33 and 5.34, solid and dashed lines representing the results of TLM calculations and the reference solution are indistinguishable in the scale of the plots. FDTD results are in very good agreement with the TLM results and are not shown.

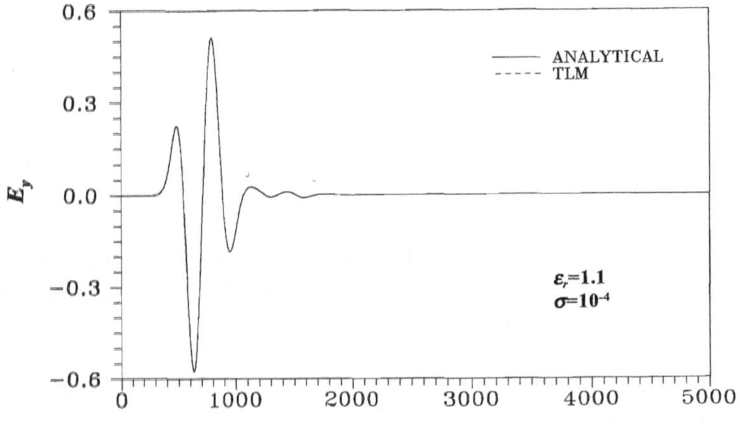

Figure 5.34: *E_y vs. time recorded at (50,50,75) when the source is at (50,50,25)*

The second comparisons are given in FD and TD in Figs. 5.35 and 5.36, respectively, for $\varepsilon_r = 1.1$ and $\sigma = 10^{-5}$ S/m, with all other parameters same as in Figs. 5.33 and 5.34. As observed in Fig. 5.35, the resonance character appears when σ is reduced by an order (from $\sigma = 10^{-4}$ S/m to $\sigma = 10^{-5}$ S/m). This is also observed in TD variations in Fig. 5.36.

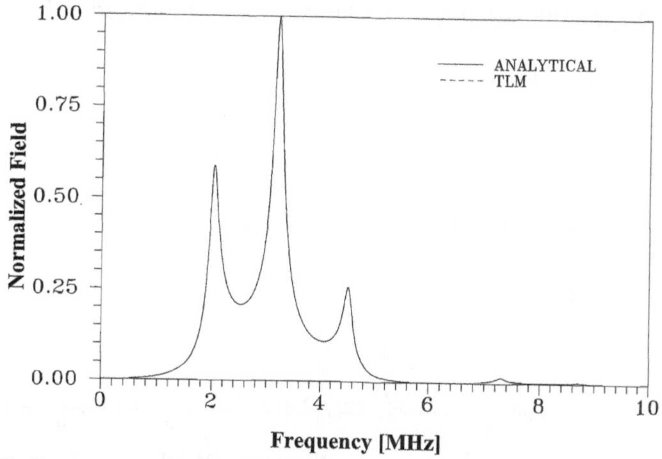

Figure 5.35: *Frequency spectrum inside the cavity ($\varepsilon_r = 1.1$, $\sigma = 10^{-5}$ S/m)*

Here, the field at the observation point diminishes around 8000 -10000 nsec, which means that fields have the opportunity to bounce back and forth between the opposite walls of the cavity more than 10 times. In practice, to obtain the resonance signature from the TD simulations in either TLM or FDTD, it is found that 5 or 6 reflections are enough. Again, solid and dashed lines, representing the results of TLM/FDTD and the Green's function formulations, are indistinguishable in the scale of the plots.

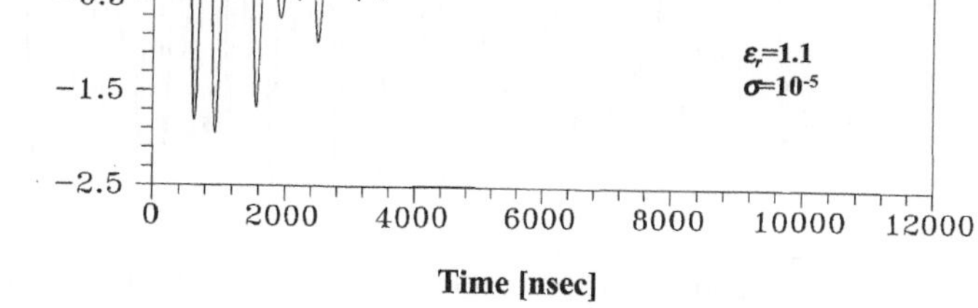

Figure 5.36: *E_y vs. time at (50,50,75) when the source is at (50,50,25)*

The FD variations in Figs. 5.33 (for $\sigma = 10^{-4}$ S/m) and 5.35 (for $\sigma = 10^{-5}$ S/m) demonstrate that the spectrum inside the cavity is continuous and dominant contributions arise near resonances. It is analytically known that when $\sigma = 0$, pure resonances occur inside the cavity, where the only field contributions come from the resonance frequencies. Unfortunately, the Green's function representation in (5.24) has singularities at these frequencies, which means we can only approach these frequencies numerically in the limit. This is illustrated in Fig. 5.37.

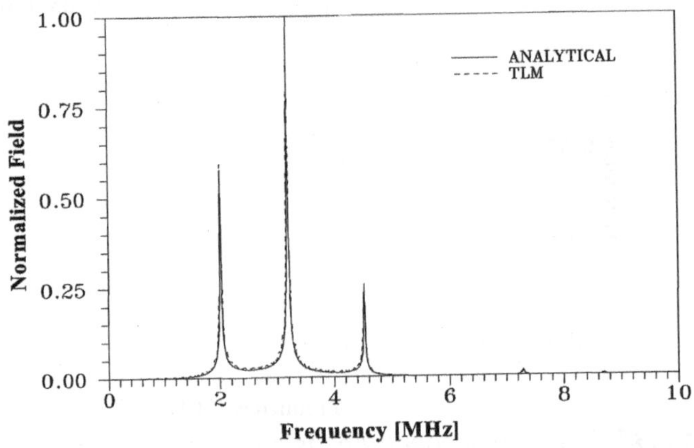

Figure 5.37: *The spectrum of cavity ($\varepsilon_r = 1.1$, $\sigma = 10^{-6}$ S/m, $\Delta f = 10$ kHz)*

Here, field strength versus frequency in FD is shown for $\varepsilon_r = 1.1$ and $\sigma = 10^{-6}$ S/m, where discrete nature of the resonance contributions are clearly observed. Perfect agreement between the results of TLM/FDTD and the Green's function computations are worthwhile to note. The calculations and comparisons are repeated for $\sigma = 10^{-7}$ S/m and $\sigma = 10^{-8}$ S/m (when $\varepsilon_r = 1.1$) and indistinguishable results are obtained. A comparison in TD for $\varepsilon_r = 1.1$ and $\sigma = 10^{-8}$ S/m is given in Fig. 5.38. The agreement continues to be same even for the late time responses.

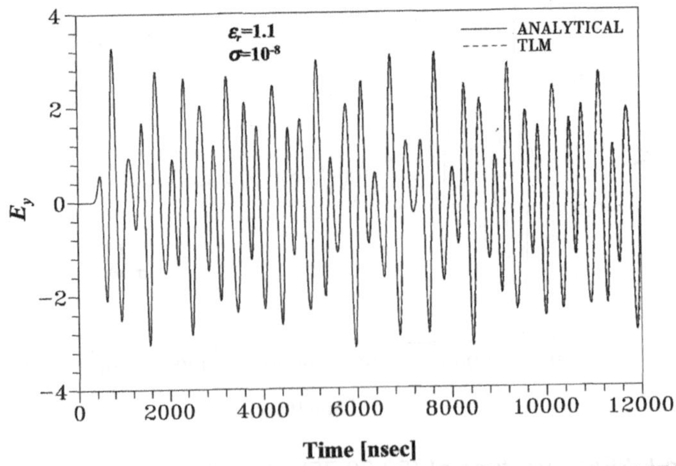

Figure 5.38: *E_y vs. time at (50,50,75) when the source is at (50,50,25)*

It is well known that running simulations with either TLM or FDTD techniques require parameter optimization. Spatial mesh sizes, Δx, Δy, Δz, time step Δt, simulation period ($T_{obs} = n \times \Delta t$), source bandwidth B, and pulse duration τ are characteristic parameters that should be optimally selected a priori the simulation. On the other hand, the Green's function representation is exact, but similar steps should be followed when numerically computed. Table 5.4 lists the requirements of the Green's function representation in TD comparisons.

Table 5.4 Requirements of (5.24) in TD analysis (ε_r= 1.1).

σ [S/m]	Δf	T_{obs} [μs]
10^{-4}	0.5 MHz	2
10^{-5}	0.1 MHz	10
10^{-6}	10 kHz	100
10^{-7}	1 kHz	1,000
10^{-8}	100 Hz	10,000

As seen from the table, the loss factor (σ) of the medium determines the frequency resolution (Δf), which in turn specifies the observation period T_{obs}. Theoretically, zero frequency resolution can be obtained in the Fourier transformation, since the integration range in TD is infinite (i.e., from $-\infty$ to $+\infty$). On the other hand, everything is finite and discrete in DFT, which introduces well-known periodicity in both TD and FD, aliasing effects and spectral leakage.

5.5 TIME-DOMAIN SIMULATORS

Four different packages are prepared for the reader who wants to exercise different SE and SAR problem scenarios. The FDSE and TLMSE are for SE simulations, and FDSAR and TLMSAR are for SAR calculations. SE and SAR are very important, complex and complicated problems in EM engineering. Therefore the user can easily solve the problems explained in the following section. One who deals with more complex problems may modify algorithms and make use of them.

5.5.1 FDSE and TLMSE Packages

The TD SE simulators, FDSE and TLMSE can be used for aperture leakage simulations. Their structures are given in Figs. 5.39 and 5.40.

FDSE and TLMSE use the *.PAR file for fixed parameters and the SE.INP input file for user specified parameters. They run TD simulations twice, one for free space (without the EUT between the source and observation point) and one with the EUT in between them. The TD output are recorded in three files. SOURCE.dat contains source vs. time in two columns. In FREE.dat and EUT.dat, field vs. time are given in two columns for the first and second TD runs, respectively. The prefix "F" and "T" are used to mention FDTD- and TLM-based simulations, respectively. The second common routine SEDFT calculates off-line SE vs. frequency response by using FREE.dat and EUT.dat files. SEDFT also outputs the frequency spectrum of the short pulse used as excitation.

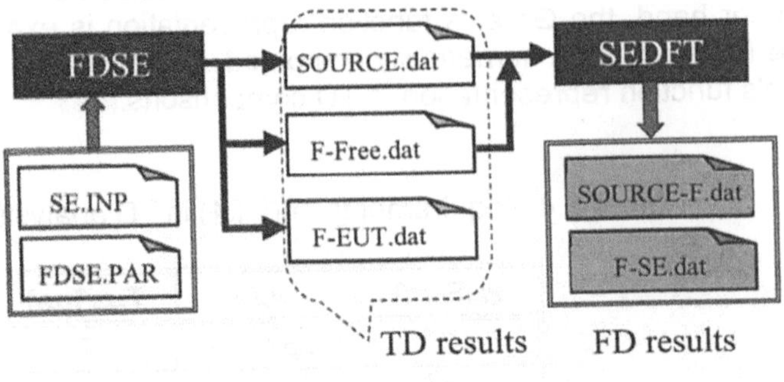

Figure 5.39: *Structure and reserved files of the FDSE simulator*

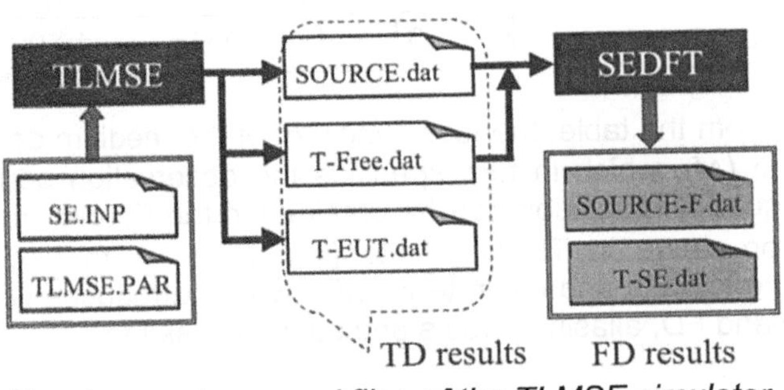

Figure 5.40: *Structure and reserved files of the TLMSE simulator*

The computational domain includes 80 × 80 × 60 cells and mesh size of the unit cells is 1cm. A 40 × 40 × 20 PEC resonator is located inside the computation space. A $\lambda/10$ (at 500 MHz) dipole is used to excite the system at (40,40,30). A once-differentiated Gaussian pulse is used to excite vertical E field. The bandwidth of this pulse is 1 GHz, which corresponds to $\lambda_{min}/\Delta\ell = 30$. An 8-layer PML block is used to simulate free-space. These are all specified in the *.PAR files.

The TD SE simulators simulate aperture leakage for the scenarios pictured in Fig. 5.41. Different apertures may be selected on the vertical walls of the resonator.

They are a single horizontal aperture, a single vertical aperture and triple vertical apertures. These are specified in the SE.INP file by the user.

The formation of the SE.INP file is given in Table 5.5. The first line is for the type of the scenario. Four integers are supplied in this row. The integers [0,3] are reserved for no aperture, single horizontal aperture, single vertical aperture, and triple vertical apertures, respectively. The first integer is for the first wall, the second is the second wall, etc., as mentioned in Fig. 5.41.

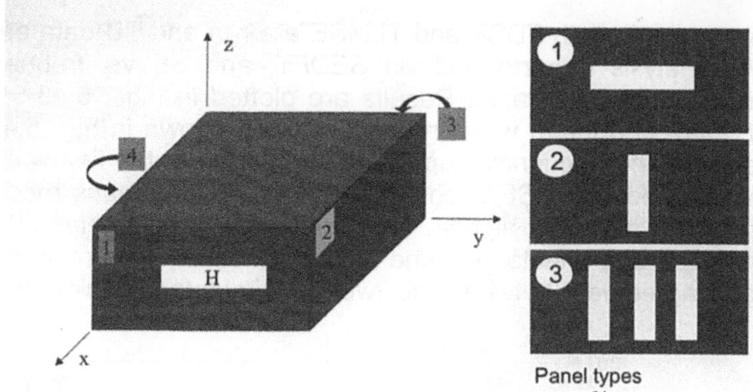

Figure 5.41: *The structure and possible scenarios in the SE simulators*

Line 2 of SE.INP is reserved for the number of observation points. Up to four points may be specified for SE calculations. Lines 3 to 6 are for the (x,y,z) coordinates of the observation points. These are supplied as three integers, such as (65,40,30), etc., that represent number of cells along (*x,y,z*) directions, respectively. The other two lines are for maximum time step and for whether windowing is going to be applied or not. A typical SE.INP file and the corresponding scenario is depicted in Fig. 5.42. Here, four observation points are used. They are located at the center and five cells away from each wall. The TD simulation lasts 9000 time steps and Hanning window is going to be applied when SEDFT is run.

Table 5.5 SE.INP file and ranges of user specified parameters

Row No:	**Parameter**
1	Panel types of four vertical walls [0-3]
2	Number of observation points
3 / 6	Observation point (*x,y,z*)
7	Number of simulation time step, NSTOP
8	Hanning window [1, YES; 0, NO]

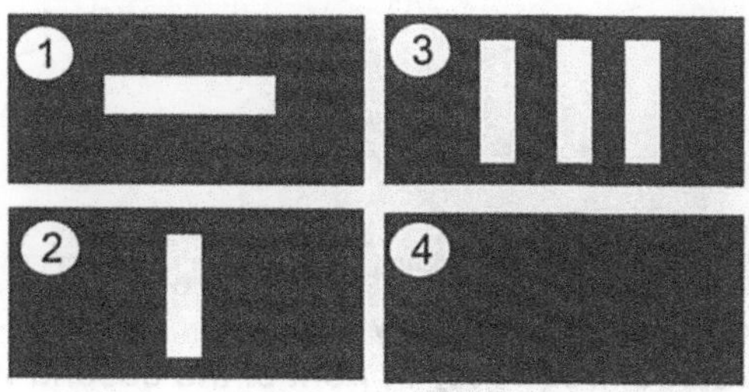


```
SE.INP
1 2 3 0
4
65 40 30
40 65 30
15 40 30
40 15 30
9000
1
```

Figure 5.42: *A typical SE.INP file and corresponding scenario*

With this scenario both FDSE and TLMSE are run and TD data are recorded. Off-line DFT analysis is performed via SEDFT, and SE vs. frequency at the four chosen points are calculated. Results are plotted in Figs. 5.43-5.46, for the first, second, third, and fourth walls, respectively. As shown in Fig. 5.43, SE is almost positive for all the frequencies up to 1 GHz in front of the first wall, except at 0.52 GHz. At this frequency, SE becomes negative, which means resonator with these apertures amplifies the signal. The resonance frequencies of the resonator, aperture sizes and orientations and field polarization determine this frequency. A very good agreement between the two TD simulators is clearly observed in the figure.

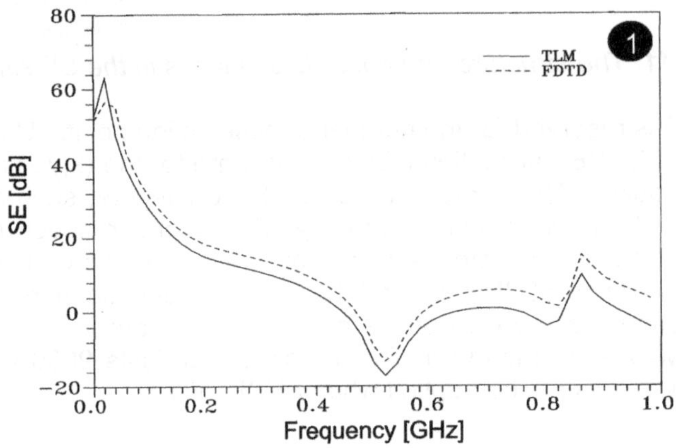

Figure 5.43: *SE vs. frequency in front of the first wall*

Similar SE behaviors and agreement between FDSE and TLMSE are observed in front of the other panels as illustrated in Figs. 5.44-5.46.

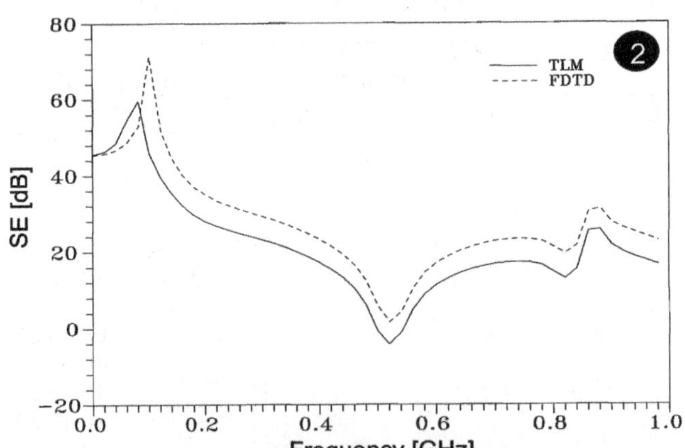

Figure 5.44: *SE vs. frequency in front of the second wall*

It should be noted that the selection of the TD simulation time is crucial in these simulations. Apertures at the walls of a resonator cause multi-reflections inside, and cause long-lasting leakage outside. These ringings must be recorded enough to get realistic SE behaviors after DFT analysis. To illustrate this, the simulators are run again for a shorter TD period (3000 time steps) and the SE vs. frequency at the selected points are re-calculated. The result in front of the first wall is plotted in Fig. 5.47.

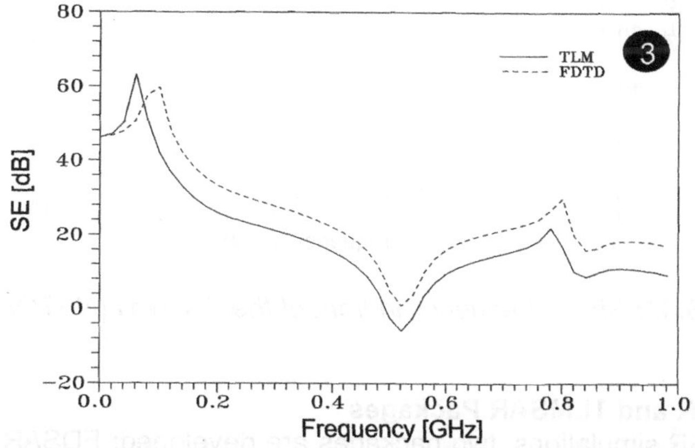

Figure 5.45: *SE vs. frequency in front of the third wall*

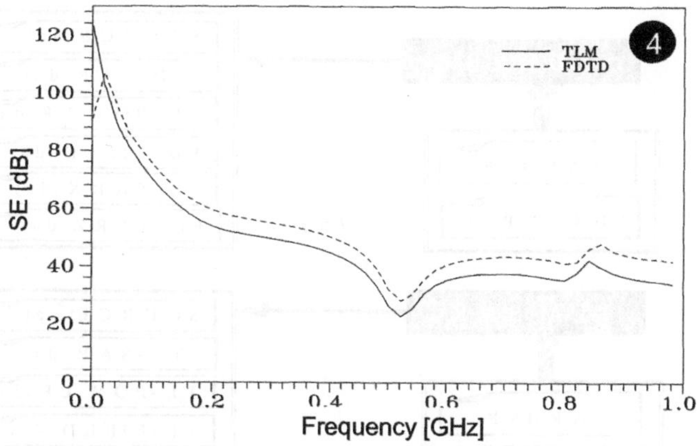

Figure 5.46: *SE vs. frequency in front of the fourth wall*

When Fig. 5.43 (for 9000 time steps) and Fig. 5.47 (for 3000 time steps) are compared, it is observed that

- having shorter simulation times still yields the general SE variation,
- but the details are lost (for example, negative SE at 0.52 GHz).

Since SE analysis is usually done to find out leakage at every single frequency, these details are very important to trace. Therefore, the simulation period must be long enough to capture all the signatures that specify the SE behavior.

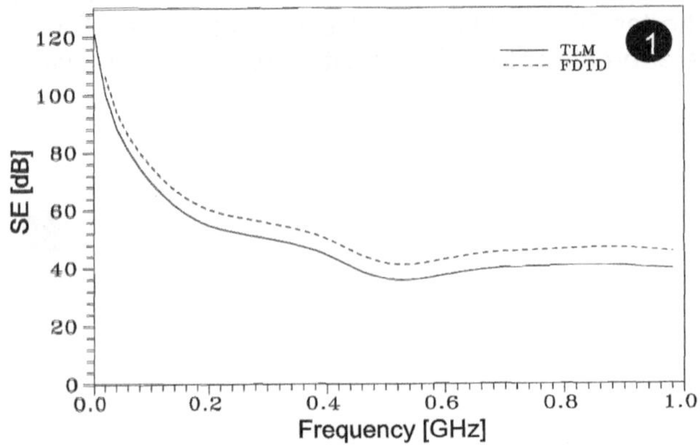

Figure 5.47: *SE vs. frequency in front of the first wall (NSTOP = 3000)*

5.5.2 FDSAR and TLMSAR Packages

To perform SAR simulations, two packages are developed: FDSAR and TLMSAR, which are based on FDTD and TLM techniques, respectively. Their structures are given in Fig. 5.48.

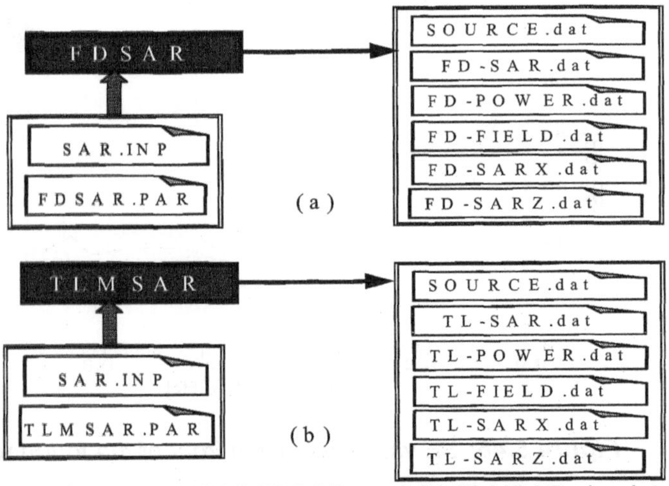

Figure 5.48: *Structures of (a) FDSAR, and (b) TLMSAR simulators*

Most of the parameters, such as the size of the computation volume, cell sizes (Δx, Δy, Δz), time step (Δt), number of time steps, and frequency, are fixed and are collected inside FDSAR.par and TLMSAR.par files. User-supplied input parameters are read from the input file SAR.INP. The structure analyzed with FDSAR and TLMSAR packages is pictured in Fig. 5.49, together with a sample input file. Here, the computation volume is chosen as $70 \times 70 \times 60$ and cubical unit cell size is 0.5 cm. A 10-layer PML block is used when calculations in free space is required. A $10 \times 10 \times 12$ cell lossy material, called equipment under test (EUT), is located inside the computation volume as shown in the figure. A 900-MHz sinusoidal excitation is supplied by a 6.5-cm-long center-fed dipole. The 1-cell-thick and 13-cell-long dipole may be located vertically or horizontally, and its 1-cell gap source is given at (26,36,31). For example, if vertical polarization is chosen the dipole occupies cells from (26,36,25) to (26,36,37). The EUT may be a single material or double material (along *x* direction).

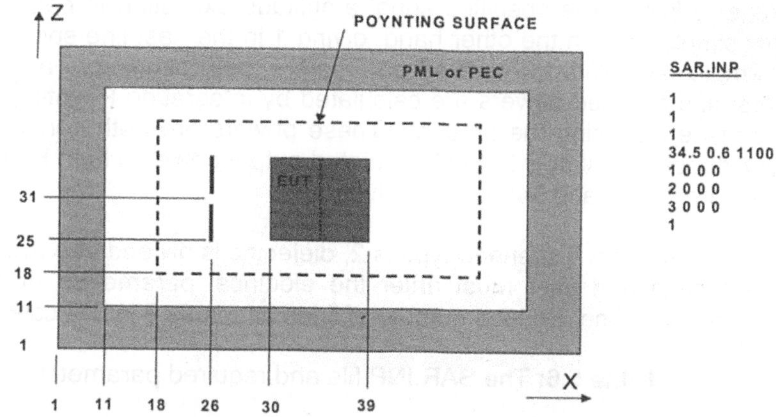

Figure 5.49: *The structure for the scenarios of FDSAR and TLMSAR and a typical SAR.INP file*

The SAR.INP input file contains 8 rows of parameters, which are listed in Table 5.6. The first row is reserved for termination. A variable NTER is reserved inside the code for this line. Giving 0 and 1 in the first line corresponds to SAR calculations inside a PEC resonator and in free space, respectively. The second line is for source polarization. Vertical (*z*) or horizontal (*y*) field components may be chosen by the antenna located in the same manner (i.e., along *z* for V and *y* for H). The third line specifies the EUT. In this line integers 0, 1, and 2 are reserved for no material, single material (scenario 1), and double material (scenario 2) cases, respectively. If 1 or 2 is given in this line corresponding constitutive parameters are specified in the following 4th or 4th and 5th rows, respectively. For the scenario 1 (scenario 2), 5th, 6th and 7th lines (6th, 7th and 8th lines) correspond to the field components and the cell coordinates, which are going to be observed during the simulation. These are specified by four integers in these lines (NPOL, NOX, NOY and NOZ). The first integer specifies which electric field component is going to be recorded (1, E_x, 2, E_y and 3, E_z). The rest are for *x*, *y*, *z* coordinates of observation points (in number of cells), respectively. If NOX = NOY = NOZ = 0 are given, the default observation points are taken into account by the algorithms. The default values for these rows in the program (which may be changed by the user) are as follows:

Line 5:	1 0 0 0	Ex at (35,35,31)
Line 6:	2 0 0 0	Ey at (45,45,31)
Line 7:	3 0 0 0	Ez at (35,35,35)

It should be noted that fixed parameters of the sample simulators are not optimized. They are specified in a way to optimize computer requirements, so that the reader may use them in an any ordinary personal computer. Therefore, a small volume material and coarse antenna discretization are chosen. Because of these, it is observed that the radiated power supplied by the 1-cell-thick and 13-cell-long dipole antenna may be different when excitation is supplied by the same unit amplitude source. Therefore, SAR simulations in these sample simulators may be performed with two different source amplitudes and the last row in the input file is reserved for this purpose. The reader may supply an integer (0 or 1) in this line. The integer 0 in this line specifies equal amplitude excitation in both FDSAR and TLMSAR simulators. On the other hand, giving 1 in the last line specifies different source amplitudes. With these fixed parameters, both simulators are run in free-space first and radiated powers are calculated by integrating Poynting vector on a closed surface enclosing the antenna. These powers are both normalized to 250 mW. Amplitudes that supply 250-mW radiated output power in both FDTD and TLM simulators are 10840 and 9450, respectively.

It should be noted that if scenario type is 2, dielectric is divided into two equal parts along *x*-direction and user must enter the electrical parameters of this second dielectric on the 5th line. So total number of lines becomes 9 in this case.

Table 5.6: The SAR.INP file and required parameters

Row No.	Parameter
1	Termination type NTER, 0: PEC; 1: PML-ABC
2	Polarization 1, E_z; 2, E_y
3	Scenario type 0, No EUT; 1, single dielectric; 2,two dielectrics
4	Dielectric constitutive parameters permittivity, conductivity [S/m] and unit mass density [kg/m^3]
5,6,7	NPOL, NOX, NOY, NOZ
8	IAMP 0, AMP = 1; 1, AMP_{FDTD} = 10840 and AMP_{TLM} = 9450

The simulators run with these two files, and they output the results in six different data files. SOURCE.DAT contains the source variation vs. time [ns] in two columns. FD-SAR.DAT (TL-SAR.DAT) gives SAR variation at two different points inside the loss material vs. time [ns]. The first column is time, and second and third columns are SAR values calculated by the algorithms. FD-POWER.DAT (TL-POWER.DAT) includes time [nsec] vs. absorbing (in the second column) and Poynting powers (in the third column) [mW]. FD-FIELD.DAT (TL-FIELD.DAT) has four columns: time [nsec] and desired *E*-field components at three different observation points. FD-SARX.DAT and FD-SARZ.DAT (TL-SARX.DAT and TL-SARZ.DAT) give SAR variations inside the lossy material along *x* (when $y = 36$, $z = 31$)) and *z* (at $x = 35$, $y = 36$). These files have two columns. The first column is the cell position and the second column is the instant SAR value.

TD SAR simulators are run for various scenarios and are compared with each other. For the SAR.INP scenario given in Fig. 5.49, FDSAR and TLMSAR are run and obtained Poynting powers for two different amplitude cases mentioned above, as shown in Fig. 5.50. As seen in the figure, results agree very well for 250-mW antenna output powers, but not for the equi-amplitude excitation.

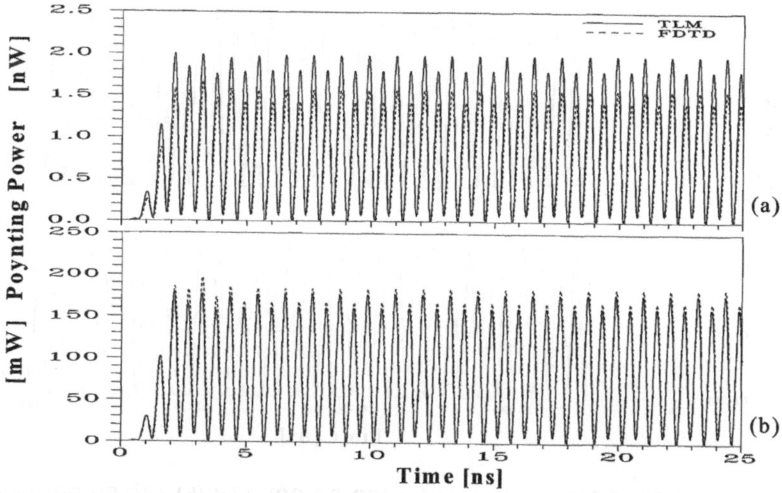

Figure 5.50: *The Poynting powers obtained for (a) unit amplitude and (b) 250-mW output power (different amplitudes)*

Time variations of the absorbed powers for 250-mW excitation are given in Fig. 5.51. Although periodicity is in good agreement in both simulators, absorbed powers are quite different. It oscillates between 110 mW and 220 mW (having 165-mW average value) with a margin of 110 mW in FDSAR. The margin and average are different in TLMSAR (between 50 mW and 220 mW with an average of 135 mW). The reason is that the sizes of EUT and optimization of parameters (numerical dispersion etc.) are not suited completely.

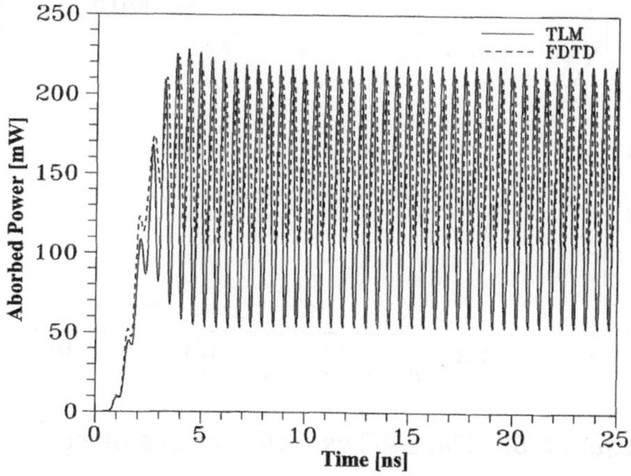

Figure 5.51: *Absorbed power vs time. Solid, TLM; dashed, FDTD*

The SAR values at the specified (default) cell positions are presented in Fig. 5.52. Both algorithms do not result in very good outputs for that given scenario. These plots are obtained with 250-mW antenna output power again. Note that the same algorithms result in perfect outputs for the case given in Section 5.3.2.

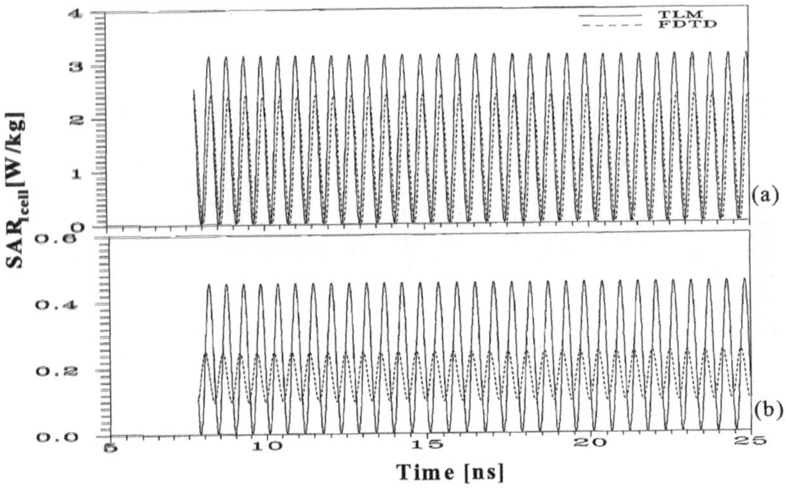

Figure 5.52: *SAR vs. time at (a) (32,32,30) and (b) (36,32,30)*

FDSAR and TLMSAR also compute SAR along a line inside the lossy material. A typical example is given in Fig. 5.53. To record SAR along *x* and SAR along z, the simulators wait until steady state is reached, which is approximately 5-7 periods with these parameters. The period at 900 MHz with these discretization parameters is nearly 133 Δt. The time instant to record SAR along *x* and *z* is fixed to 1097. It corresponds to the first maximum after the 8th period. In Fig. 5.53, SAR vs. *x* (from 30 to 39 at *y* = 36 and *z* = 31) is plotted.

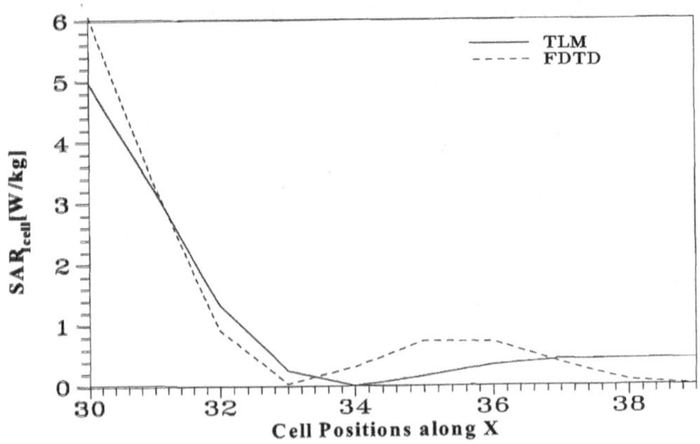

Figure 5.53: *The SAR vs. cell positions along x*

As seen in the figure, SAR decreases exponentially as the recorder moves away from the antenna horizontally.

5.6 CONCLUSIONS AND DISCUSSION

EMC and BEM problems are taken into account, and characteristic examples are presented to discuss numerical simulation techniques. Powerful TD simulators are developed and are presented for the interested reader in this chapter. Some theories together with examples given here will be very helpful for the beginners and intermediate level users. Willing users can modify the algorithms for their own use of purpose.

The algorithms given here can model any types of problems by modifying them. But as already mentioned, one must be aware of some essential numerical requirements during the simulations such as numerical dispersion criteria and source parameters. Once these requirements are taken into consideration, the problems to be solved can be modeled easily and will be ready to discuss with other researchers. But the most important drawbacks of the algorithms are the simulation run times. One must have state-of-the art computers with high memories to overcome this problem.

References

[1] David Morgan, *IEE A Handbook for EMC testing and Measurement*, Peter Peregrinus Ltd., 1996
[2] Don White, *The EMC, Telecom and Computer Encyclopedia Handbook*, EMF-EMI Control, third edition, 1999
[3] IEEE Standards Coordinating Committee 34 on Product Performance Standards Relative to the Safe Use of Electromagnetic Energy, Recommended Practice for Determining the Spatial-Peak Specific Absorption Rate (SAR) in the Human Body Due to Wireless Communications Devices: Experimental Techniques, *IEEE,* P1528, 2001
[4] See http://europa.eu.int/comm/enterprise/electr_equipment/emc/
[5] IEGMP (Independent Expert Group on Mobile Phones), "Mobile Phones and Health", (see http://www.iegmp.org.uk)
[6] GAO, "Research and Regulatory Efforts on Mobile Phone Health Issues", US General Accounting Office Report to Congress, GAO-01-545, May 2001 (see http://www.gao.gov)
[7] J. C. Lin, "The Precautionary Principle: A Rose by Another Name", *IEEE Antennas and Propagation Magazine*, Vol. 42 No.2, pp. 129 - 131, April 2001
[8] See http://www.microwavenews.com
[9] Commission of the European Communities, 2000, Communication on the Precautionary Principle, Brussel, Feb. 2000 (see http://europa.eu.int/comm/off/health_consumer)

[10] ICNIRP "Guidelines for Limiting Exposure to Time-Varying Electric, Magnetic, and Electromagnetic fields (up to 300 GHz)", *Health Physics*, Vol. 74, pp. 4494 - 522, 1998

[11] IEEE Standards for Safety Levels with Respect to Human Exposure to Radio Frequency Electromagnetic Fields, 3 kHz to 300 GHz, *IEEE Standard* C95.1-1991, 1992

[12] "Evaluating Compliance with FCC Guidelines for Human Exposure to Radio Frequency Electromagnetic Fields", FCC, Washington D.C., *OET Bull.* Vol. 65, Aug. 1997

[13] P. Bernardi, M. Cavagnaro, S. Pisa, E. Piuzzi, "Specific Absorption Rate and Temperature Increases in the Head of a Cellular-Phone User", *IEEE Transaction on MTT*, Vol. 48 No. 7, pp. 1118 - 1126, July 2000

[14] See, http://www.fcc.gov

[15] F. Akleman, L. Sevgi, "FDTD Analysis of Human Head – Mobile Phone Interaction in Terms of Specific Absorption Rate (SAR) Calculations and Antenna Design", *Proc. of IEEE-APS, Conference on Antennas & Propagation for Wireless Comm.*, pp. 85 - 88, Nov. 2 - 4, 1998, Waltham, MA

[16] M. O. Özyalçın, F. Akleman, L. Sevgi, "Numerical Modeling, Validation & Verification in Biomedical Engineering: Sar Calculations Caused by Cellular Phones", *ICECOM'01, 16th Int. Conf. on Applied Electromagnetics and Comm.*, Dubrovnik, Croatia, Oct. 1 - 3, 2001

[17] L. Sevgi, S. Paker, "FDTD Based RCS Calculations and Antenna Simulations", *AEU, International J. of Electronics and Commun.*, Vol.52 No.2, pp. 65 - 75, March 1998

[18] Spectrum Science Institute, RF Dosimetry Research Board, "SAR Measurement Requirements", Report No:SSI/DRB-TP-Do1-030, 51 Spectrum way, Nepean Ontario, K2R 1E6, Canada

[19] P. B. Johns, R.L. Beurle, "Numerical Solution of Two-Dimensional Scattering Problems using TLM", *Proc. IEE*, Vol. 118, pp. 1203 - 1208, 1971

[20] K. S. Yee, "Numerical Solution of Initial Boundary Value Problems Involving Maxwell' s Equations", *IEEE Transactions on Antennas and Propagat.*, Vol. 14 No.3, pp. 302 - 307, May 1966

[21] M. O. Özyalçın, L. Sevgi, "Comparisons of FDTD and TLM Methods in EMC-Shielding Effectiveness Analysis", *The Eighth Biennial IEEE Conference on Electromagnetic Field Computation*, Tucson-Arizona, June 1 - 3, 1998.

[22] See, for example, http://www.speag.com, http://www.semcad.com, http://www.dasy4.com, Schmid & Partner Engineering AG, Zurich, Switzerland

[23] See, for example http://www.indexsar.com

[24] See for example http://www.eemctech.com. au

[25] L. Gürel, U. Oguz, "Signal-Processing Techniques to Reduce the Sinusoidal Steady-State Error in the FDTD Method", *IEEE Trans. Antennas and Propagat.*, Vol. 48 No. 4, pp. 585 - 593, 2000

[26] M. O. Özyalçın, *Modeling and Simulation of Electromagnetic Problems via Transmission Line Matrix Method, Ph.D.* Thesis, Istanbul Technical University, Istanbul, Turkey, Nov. 2002

[27] J. L. Dubard, D. Pompei, "Optimization of the PML Efficiency in 3-D TLM Method", *IEEE Transactions on MTT*, Vol. 48, pp. 1081 -1088, 2000

[28] F. Akleman, M.O. Özyalçın, L. Sevgi, "Comparison of TLM and FDTD Techniques in RCS Simulation and Antenna Modelling," *Proc. of AP 2000*

(ICAP + JINA) Millenium Conference on Antennas and Propagation, April 9 - 14, 2000, Davos, Switzerland

[29] F. Akleman, M.O. Özyalçın, L. Sevgi, "Time Domain Ground Wave Propagators based on FDTD and TLM Techniques," *CEM-TD, Proc. of 4th Int. Workshop on CEM in the Time Domain – FDTD, TLM and Related Techniques*, pp.97-104, Sep 16-19, 2001, Nottingham, UK

[30] M.-J. Park and S. Nam, "Efficient Calculation of the Green's Function for the Rectangular Cavity," *IEEE Microwave and Guided Wave Lett.*, Vol. 8 No. 3, pp. 124 - 126, March 1998

[31] C.A. Balanis, *Advanced Engineering Electromagnetics*, John Wiley & Sons, New York, 1989

6

Radar Simulation

6.1 INTRODUCTION

The word *radar* is an acronym (that was used as early as 1940 during World War II) derived from the phrase *radio detection and ranging*. Although radar has many uses, its basic objectives are to detect targets of interest and to derive information such as range, range rate (velocity), angular position, and reflectivity signature [1]. The ability of a radar is not only to detect but also to discriminate, to track, to classify, and even to identify targets. Although these goals can be achieved by either continuous wave (CW) or pulsed radars, this chapter will take only the latter into account. Moreover, unless otherwise is stated, surface targets on sea will be of interest.

This chapter is about computer simulation of a radar and an integrated surveillance system. Computer simulation of a radar is among the toughest numerical EM problems. It includes all aspects discussed separately in the previous five chapters, from propagation to antenna simulation, from RCS modeling to signal processing, etc.

Simulation allows the most complex systems to be studied. A system can be simulated either physically or numerically, or in hybrid form. The most desirable simulation is the one done with the real system, if it is possible and cost effective to alter the system physically and operate under new conditions. However, this is a very rare case. It is usually necessary to build a mathematical and/or computer model as a representation of the system and study it as a surrogate for the actual system.

Why does one need to simulate a radar? Among others, some major answers may be:

- To understand radar signal environment.
- To optimize radar parameters, such as transmitter power, receiver gain, path loss, RCS, noise and clutter, etc.
- To find out under what conditions the radar equation is satisfied for a given set of parameters.
- To evaluate a software, such as detection, a tracking or a sensor fusion algorithm, etc.
- To use the radar simulator in a multi-sensor surveillance system simulation.
- To play different scenarios and try to find out the answers to "What if ..”-type questions.
- And, finally, to use it as a system design tool.

One needs to decide what kind of radar is going to be simulated first. Is it going to be a detection, or a tracking, or a surveillance radar? Is it going to have profiling capability? Or, will it be a SAR (synthetic aperture radar) or an ISAR (inverse synthetic aperture radar)? Will classification and/or identification capabilities be added? What frequency range is going to be used? Answers to all these questions may totally change the simulation requirements and the picture. For example, total signal environment in MW radars is almost totally different than those of HF radars, since:

- Radio wave propagation mechanisms are different at MW and HF. For example, surface waves dominate at HF and radar signal can reach beyond the LOS. On the other hand, only direct and ground-reflected waves occur at MW, therefore, radar performance is limited to LOS.
- As non-coherent, randomly fluctuating signal, noise has different characteristics for HF and MW radars. In an HF radar, external (atmospheric, environmental) noise limits minimum detectable signal, but thermal noise is of interest in MW radars, which is 40 - 70 dB lower than external noise.
- As coherent randomly fluctuating signal, clutter is also different in HF and MW radars. The amplitude distribution characteristics of land clutter is totally different from sea clutter; they also have different characteristics at MW and HF.
- There is no need to say that softwares of MW and HF radars are also quite different. Detection algorithms, tracking routines, etc., depend on total echo signal environment. Therefore, their characteristics are also different.
- Finally, parameters (such as coverage, range/azimuth resolution, beam forming and antenna scanning techniques, etc.) of both radars are totally different.

Nevertheless, simulation of total radar echo signal first requires stochastic modeling of target RCS, noise, clutter, and interference. These are all radar waveform-dependent, and they vary with frequency as well as with the type of radar (mono-static, bi-static, MW or HF radars, acoustic sensors, etc.). Time sequence of the radar total echo or frequency vs. total echo (if it is a Doppler radar) may be obtained via FFT.

Is this enough to model radar correctly? Off course, not. One also needs to estimate path loss precisely if the correct SNR (and/or CNR) is to be simulated. Since detection process mostly relies on the Constant False Alarm Rate (CFAR) principle, the precise calculation of signal strength is crucial in determining the detection threshold at the radar receiver. The requirement of precise path loss prediction at different frequencies, at different ranges, etc., necessitates powerful propagation simulations.

Assume that the CFAR algorithm is to be tested. Then, realistic total signal environment must be simulated. Target RCS fluctuations, noise and clutter are all stochastic processes, which involve randomness. Mean values of target RCS, noise floor as well as clutter distribution must be precisely calculated. This requires powerful RCS codes, propagation simulators and long-term noise/clutter observations. For example, if sea clutter is to be modeled, one needs to understand sea waves/EM waves interaction, surface current, wave height effects, low altitude atmospheric distributions, wind speed and direction, etc., very well.

Now, assume that a kind of tracking algorithm is to be tested. In this case, precise path loss calculations and total echo signal are not necessarily modeled in detail. A kind of scenario where the radar coverage and other parameters are determined and targets with chosen routes (and maneuvers) are defined is played and only detection/no detection decisions are fed to the input of tracker. Moreover, 2D (range/azimuth) pre-calculated coverage (look-up tables) may be used to produce CFAR detector output.

A radar simulation is a multi-level task. At first stage, the developer should decide where and when to use mathematical or numerical models of sub-systems and/or sub-components, or when to switch to pre-prepared look-up tables. For example, propagation path loss at discrete ranges for possible radar operating frequencies may be calculated off-line and stored as propagation look-up tables. Instead of running heavy computations at each simulation step, the simulator may get values directly from these tables. Simple interpolations may be used for intermediate values both in frequency and range. Similarly, RCS vs. aspect angles at variety of operating frequencies may be stored as RCS look-up tables. These tables may be obtained by using powerful RCS simulators once for bulk RCS values and fluctuations (that obey any kind of amplitude distributions) may be added by random number generators and distribution algorithms while running the simulator. Pre-calculated clutter tables may also be used in generating clutter as well as noise, which are both stochastic processes. Long-term seasonal atmospheric records and CCIR recommendations may be used for realistic clutter and noise modeling, respectively.

In this chapter, HF surface wave radars (HFSWR) simulation is discussed. HF radars use both ground and sky waves to cover ranges beyond the LOS. HF sky wave radars use the earth/ionosphere propagation path and can reach up to thousands of nautical miles (nm) and they must be bi-static (to handle skip zone). HFSWR use ground (surface) waves, and operation range may extend up to 200 - 250 nm (1nm $\approx$ 1.86 km). They both are used not only in detection surface and air targets, but also to remotely sensing sea/ocean surface. Since the band covers frequencies between 3 MHz and 30 MHz, radar wavelengths are 100 m to 10 m. HFSWR uses coherent integration of consecutive echoes scattered from a target in response to the transmitted pulse train. Target detection is performed in FD based on the Doppler Effect. The principle of an HFSWR is depicted in Fig. 6.1.

HFSWR are mostly operated from the coast. A broadband transmitter and the receiver array are co-located either on a hill or at sea level. Radar power is coupled to the sea surface (using Brewster angle principle), and vertically polarized waves propagate between radar and target. Some power may also couple to the ionosphere path, because of imperfect overhead nulling property of the antennas. Also, target scattered waves are also coupled to the ionosphere path in return, because of the complexity of the target itself and sea surface fluctuations.

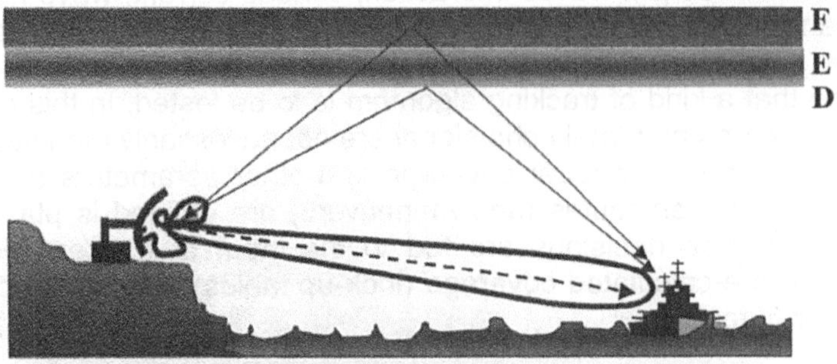

Figure 6.1: *HFSWR, rough sea surface and possible propagation paths (along the surface and via E and F layers of the ionosphere)*

Target discrimination is constrained by range/azimuth resolution. Resolution is the ability to distinguish nearby targets. Equally as important as the resolution is the accuracy. Accuracy indicates uncertainty in a measurement of the absolute value of a parameter (e.g., range, bearing, velocity, etc.). A typical radar pulse train, range bins and target echo are depicted in Fig. 6.2.

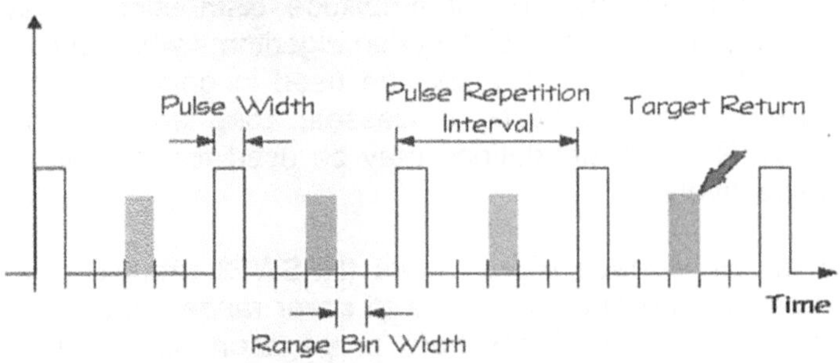

Figure 6.2: *transmitted pulse train and target echos*

Here, a continuous pulse train, with pulse (width) duration, τ, and pulse repetition frequency, PRF, is transmitted. The maximum range, R_{max}, is determined by PRF

$$R_{\text{max}} = \frac{c}{2 \times PRF} \quad \text{[m]} \tag{6.1}$$

and pulse width τ determines the range bin (i.e., range resolution) as

$$\Delta R = \frac{c\tau}{2} \quad \text{[m]} \tag{6.2}$$

which yields number of range bins as ($R_{max}/\Delta R$) - 1. Unfortunately, understanding accuracy is not as straightforward as resolution. Two important parameters that define accuracy are pulse duration (i.e., bandwidth, $B \approx 1/\tau$) and SNR [2]. That is, the range error δR is equal to

$$\delta R \approx \frac{c}{2B\sqrt{(2 \times SNR)}} \quad \text{[m]} \tag{6.3}$$

Another important radar parameter is the *pulse integration*. Theoretically, an echo of a target for a unique transmitted pulse is enough to measure target range. Unfortunately, unlike Fig. 6.1, target return is usually embedded in the clutter and noise and it is almost invisible in TD. That's why the Doppler effect is used to the measure (i.e., to detect) the velocity of a target. Radial velocity (v_r) of a target causes a slight frequency shift in the radar's operating frequency and this is called Doppler shift. This Doppler shift for a radar wavelength λ is equal to

$$f_D = V_r \frac{2}{\lambda} \quad \text{[Hz]} \tag{6.4}$$

In order to measure Doppler shift, one needs to obtain a time series, which contains multi-echoes, and then apply FFT to get Doppler spectrum. The length of time taken to make an observation with a radar set is called *integration time,* and the process is named coherent integration. It is well known from the sampling theorem that maximum Doppler frequency f_{Dmax} and Doppler resolution Δf_d are directly related to integration time and sample interval in time domain as

$$\Delta f_D = \frac{1}{T_{int}}, \quad f_{D\max} = \frac{1}{2\Delta t} \quad \text{[Hz]} \tag{6.5}$$

HFSWR detect targets in three-dimensional space: (a) range, (b) bearing, and (c) velocity (Doppler).

(a) The range resolution is directly proportional to the bandwidth of the transmitted waveform. HFSWR typically operate with a maximum bandwidth of less than 20 kHz, for which the resolution is about 4 nautical miles (approximately 7.5 km).

(b) The azimuthal resolution is directly proportional to the aperture size of the antenna. The aperture is measured in terms of the radar wavelength. For an antenna array with element separation of half a wavelength, a 16-element antenna array produces a beamwidth at boresight of approximately 9°. This translates to a cross range of 30 nm at a distance of 200 nm. As the beam is steered away from boresight, the beamwidth increases.

(c) The velocity resolution is directly proportional to the coherent integration time. HFSWR employ integration intervals corresponding to approximately 20 seconds for air targets, 165 seconds for ship targets and 1,200 seconds for near stationary targets (such as icebergs). These correspond to velocity resolutions of 4 knots (1 m/sec ≈ 1.94 knot), 0.5 knots, and 0.06 knots, respectively.

Even though the resolution capabilities of HFSWR seem at first glance to be rather moderate, in practice it is not a serious issue since two targets can be resolved provided that they are separable in one of the three dimensions. The probability of having two targets in close proximity in all three dimensions is not high. In the event that two targets are not resolvable, the radar will track either the larger or the composite return of the two until such time that they can be resolved.

6.2 HFSWR CHARACTERISTICS

Total radar echo includes target's signature, noise, clutter, and interference signal, which are all stochastic processes. That is, all components have temporal fluctuations. Here, stochastic modeling is discussed in HFSWR. First, important aspects of HFSWR are outlined and then stochastic simulation steps are summarized.

6.2.1 HFSWR Equation

HFSWR for long-range surveillance is a developing technology [3,4] that has already demonstrated excellent capabilities in the long-range detection of ships and aircraft [5] as well as icebergs [6] and remote sensing of the ocean environment [7 - 11]. Operating at the lower end of HF band allows the radar to achieve ranges in excess of 200 nm by means of surface waves, where the high conductivity of seawater results in relatively low attenuation of the vertically polarized radio waves [12,13]. The received power from a target for a mono-static HFSWR is defined as

$$P_r = \frac{P_t G_t G_r \sigma A^4 \lambda^2}{(4\pi)^3 R^4 L_s} \quad [W] \tag{6.6}$$

where

P_r: Receive peak power; P_t: Transmit peak power
G_t: Transmit antenna gain; G_r: Receive antenna gain
σ: Target radar cross section (RCS); λ: Radar wavelength
R: Target range; L_s: System loss
A: Norton SW attenuation factor at range R ($A < 1$)

It can be observed that the range equation for a HFSWR differs from that of a MW radar due to the presence of the surface wave attenuation factor, A. The surface wave attenuation factor is typically combined with the free-space propagation loss, yielding a total one way propagation loss, L_p;

$$L_p = \left(\frac{4\pi R}{\lambda A'}\right)^2 \tag{6.7}$$

where A' is the modified surface wave attenuation that includes the surface roughness loss [13,14].

For a MW radar, the dominant noise source is thermal receiver noise, whereas, for HFSWR, external noise is dominant. The radar range equation is therefore written in terms of SNR after the returned echoes have been coherently integrated over many pulses,

$$\frac{S}{N} = \frac{P_{av} G_t G_r \acute{o} T_{\text{int}} 4\eth}{L_p{}^2 L_s \ddot{e}^2 N_o} \tag{6.8}$$

where P_{av}, T_{int} and N_o are average transmit power, coherent integration time (CIT) and noise density, respectively. It should be noted that peak power and average power of pulsed HFSWR are related via

$$P_{av} = P_t \times \tau \times \text{PRF} \qquad \text{[W]} \tag{6.9}$$

where peak powers may be on the order of MW, but average powers are usually in the orders of kW.

6.2.2 Antenna Gain

HFSWR transmit and receive antenna arrays need hundreds of square meters leveled area in coastal regions. A typical (top view) antenna site of a HFSWR is depicted in Fig. 6.3.

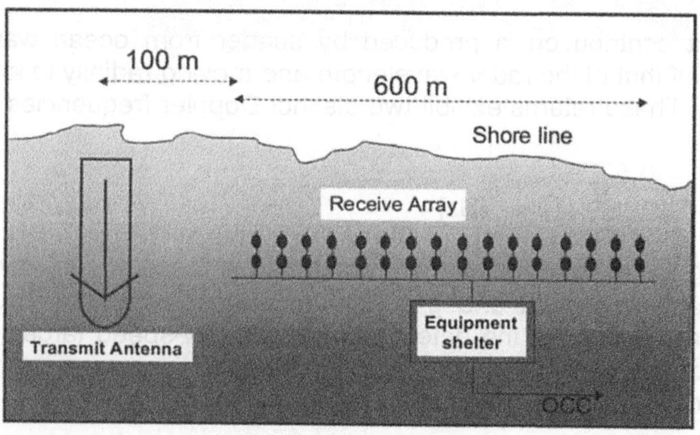

Figure 6.3: *Transmit and receive antenna parks of HFSWR*

A vertically polarized log-periodic-type transmit antennas with quarter-wavelength monopoles are used to cover all the region. A 16- to 24-element-doublet, triplet, or quadlet is used as a receiving array. Here, doublets, triplets or quadlets are used as end-fire array elements to obtain high FBR. Since the ground is a poor conductor (with respect to sea), a ground screen is used under both transmit and receiver arrays to increase antenna gain. On the other hand, using ground screen reduces overhead null, which causes strong ionospheric interference.

Significant confusion has arisen in specifying antenna gains and target RCS when using this radar equation [15]. There are two effects [16] that must be taken into account in the determination of the proper values of G_t, G_r, and σ. The first is the ground-plane effect, which refers to doubling of the field intensity or quadrupling of the power flux density when an antenna is radiating over a PEC plane. This results in a 6-dB increase in the antenna gain. The second effect is the ground proximity effect, which refers to the reduction in the antenna gain (or RCS) due to mutual coupling of the antenna (or target) and its image as its distance from the conducting surface decreases. A reduction of 3 dB in gain (or 6 dB in target RCS) will result when the antenna (or the target) is situated on the conducting surface.

Another point that is often overlooked in HFSWR equation is the proper value of the antenna gain under external noise-limited conditions. Most HFSWR employ a receive array of vertically polarized antenna elements. The element pattern is very broad in azimuth, and the peak gain is at zero elevation. If the level of external noise, which comprises the galactic, atmospheric, and man-made, is substantially higher than the receiver noise, then the element gain of the receive antenna has no effect on the system's SNR. Consequently, for the receive array of an HFSWR only the array gain is relevant.

6.2.3 Ocean Clutter

The clutter return from the ocean surface has a characteristic Doppler spectrum. A typical ocean spectrum is depicted in Fig. 6.4. Ocean parameters and their effects on the spectra are also mentioned in the figure.

The dominant contribution is produced by scatter from ocean waves having a wavelength half that of the radar wavelength and moving radially to and away, from the radar site. These returns exhibit two distinct Doppler frequencies (called Bragg peaks):

$$f_b = \pm\sqrt{\frac{g}{\pi\lambda}} \qquad \text{[Hz]} \tag{6.10}$$

corresponding to the velocities of the propagation of these ocean waves [17]. Here, λ is the radar wavelength and g is the acceleration due to gravity. It is this continuum that often limits the detection of small, low-speed targets. The level of the second-order continuum at a given Doppler increases with both radar frequency and sea-state.

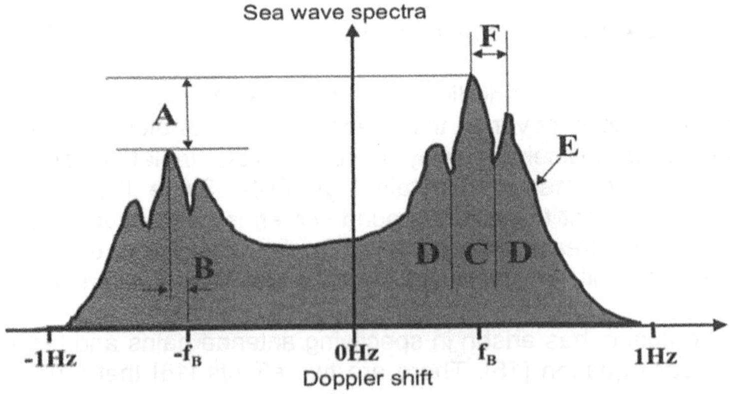

Figure 6.4: *Typical ocean spectra and parameters derived from (A) surface wind direction, (B) radial current, (C,D) wind speed, and (E,F) scalar wave spectrum*

The clutter factor is described in terms of the SCR as

$$\text{SCR} = \frac{\sigma_t}{\sigma_c}; \qquad \sigma_c = \sigma_0 A \qquad [\text{m}^2] \tag{6.11}$$

where σ_c is the average value of the clutter RCS. σ_0 is the cross section per unit area (back-scattering coefficient) and *A* is the area of the surface of the clutter patch at the target range. For the surface wave propagation, σ_0 is assumed to be approximately –32 dB and frequency-independent [16]. At a given range *d*, the area *A* is determined by the azimuth beamwidth of the receiver array $\Delta\varphi$ and by the effective pulse width "τ" as

$$A = d\frac{c\tau}{2}\Delta\varphi \qquad [\text{m}^2] \tag{6.12}$$

It should be noted that the ocean spectra for the lower HF frequencies (e.g., 3 MHz - 15 MHz) is between –1 Hz and +1 Hz. The majority of the surface targets Doppler frequencies (caused by their radial velocities) lie in this region (i.e., between –0.5 Hz and + 0.5 Hz); therefore, instead of SNR in (6.8), SCR in (6.11) is of interest. On the other hand, low-altitude air targets' Doppler frequencies are mostly out of this region (e.g., ± 3 Hz to ± 10 Hz, etc.), where environmental noise floor limits the detection; therefore SNR in (6.8) is the case. Targets with Doppler frequencies that are equal to first-order Bragg peaks are not detected, and these radial velocities are called *blind velocities*. Targets Doppler frequencies are proportional with the radar operating frequency, but first-order Bragg peaks are proportional with the square root of the operating frequency. Therefore, targets blocked by the first-order Bragg peaks can be detected by changing the operating frequency, since then they are separated on the Doppler axis.

Ocean spectra are strongly related to the wind speed, the direction, and the area affected. This relation is given in Fig. 6.5, where wave height vs. wind period and wave height vs. effective wind area are given for different wind speeds. According to these plots, for example, a fully developed sea under 20-knot wind speed requires wind blow of nearly 12 hr on approximately 300 km of area. In this case, wave height will be 2 m, which is stated with Sea State (SS) 4.

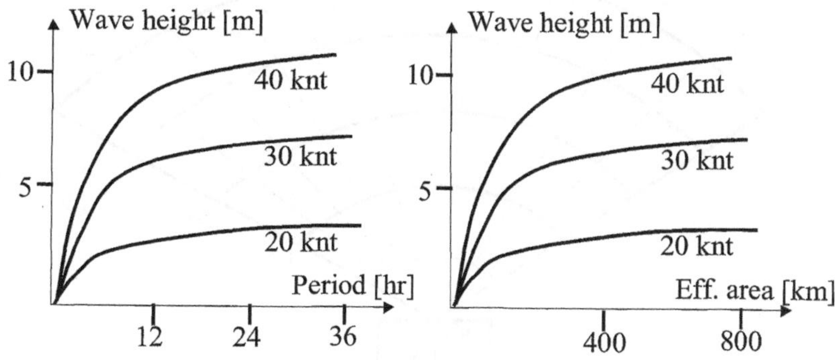

Figure 6.5: *Wave height vs. wind period and effective area*

Fully developed sea is classified in terms of SS and sea wave height, as illustrated in Fig. 6.6. SS 0 corresponds to calm sea (i.e., wave heights are less than 0.5 m). SS 1 and SS 5 correspond to wave heights between 1 m and 3 m. SS 2 and SS 3 are called moderate, where SS 4 and SS 5 represent rough sea. SS 6 to SS 9 are for extremely rough sea, where unexpected non-linear variations may occur.

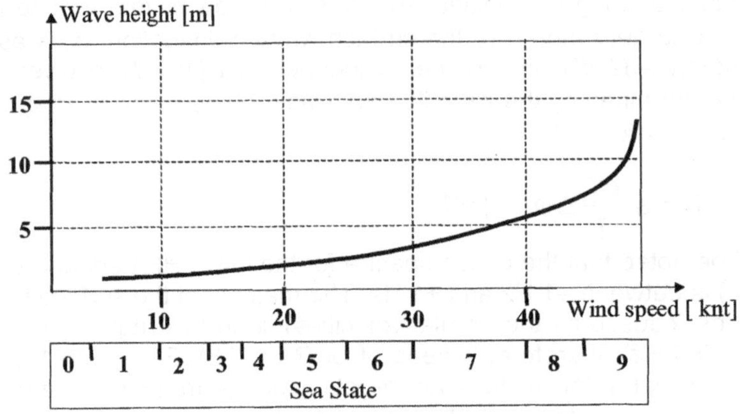

Figure 6.6: *Sea states / sea wave height relation*

6.2.4 Radar Resolution cell

HFSWR is a pulsed Doppler radar which discriminates targets by using their relative radial velocities with respect to radar receiver. Its coverage may extend up to 400 - 500 km in range and 100° to 120° in azimuth. A typical coverage is depicted in Fig. 6.7. Here, range is divided into N range bins (determined by the radar pulse length) and M azimuth bins (determined by receiver array length), yielding $N \times M$ radar resolution cells. Typical values for maximum range of 400 km and 120° azimuth coverage - for example, for 20-kHz bandwidth and 4° azimuth resolution - are more than 1600 radar cells.

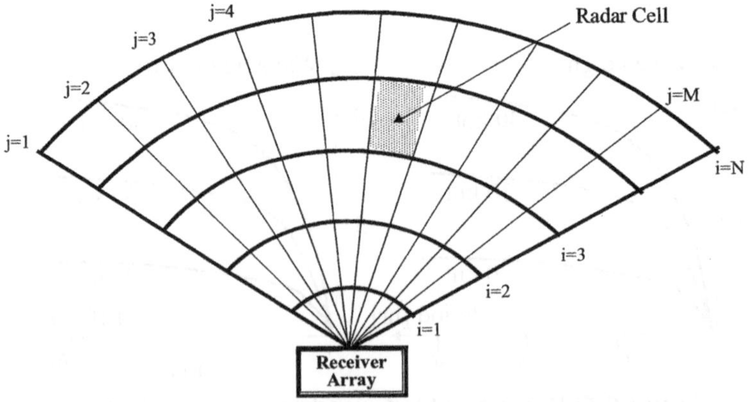

Figure 6.7: *HFSWR coverage and radar (resolution) cell*

HFSWR works as explained below:

- The chosen radar pulsed waveform is transmitted to cover all the radar cells.
- Back-scattered echoes from all possible targets inside the coverage are received by each of the isotropic elements of the receive array.
- Azimuth discrimination is achieved in the receiver array by digital beam synthesis.
- Range bin discrimination is performed by time gating during the receiving phase.

As a result, total echo signal for all the resolution cells are accumulated during the integration period and Doppler spectrum is obtained via FFT.

Range - azimuth resolution and time series accumulation in each radar cell are explained schematically in Fig. 6.8. Each beam (azimuth direction) consists of N range cells, in which $S_{i1}(t)$, $S_{i2}(t)$, .., $S_{iM}(t)$ time series are accumulated. Each series is discretized via ADC (analog-to-digital converter) and $S_{i1}(k)$, $S_{i2}(k)$, .., $S_{iM}(k)$, k = 1,2, .., K_{FFT} are obtained. The accumulation continues during the integration period T_{int} with time intervals Δt (where $\Delta t = T_{int}/K_{FFT}$), and then Doppler spectra at each resolution cell is obtained via FFT. The rest is to apply suitable (constant and/or dynamic) CFAR threshold detection process followed by a tracking routine.

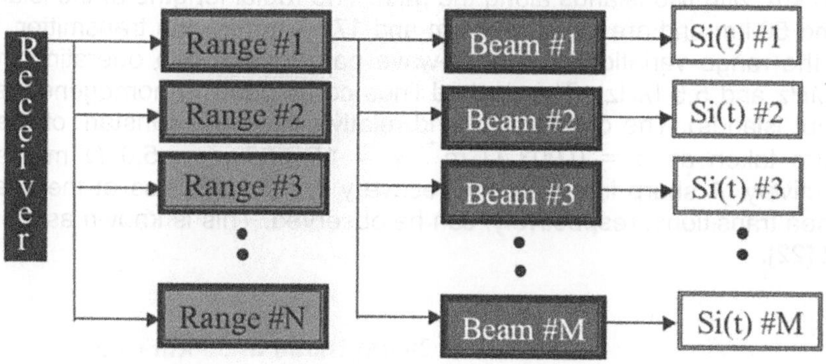

Figure 6.8: *HFSWR signal accumulation*

It should be noted that the radar cell is very much larger than surface targets. There may be many targets inside a resolution cell of typical 15 km by 10 to 30 km in dimension. Therefore, intelligent algorithms should be used in order to discriminate targets inside a resolution cell as well as to track them continuously.

6.2.5 Surface-Wave propagation

One of the main components of the propagation loss, *Lp*, in (6.7) is the surface wave attenuation factor *A'*. This has been computed by Norton [12] for the case of a smooth spherical earth. Barrick [15,16] extended the theory to cover the case of a rough surface that accounts for higher sea states. As an example, Barrick predicts that the added loss at 500 km for the frequency range of interest is between 3 dB and 10 dB up to sea state 5 (SS 5, 20 knot wind speed and 2.5-m to 3.0-m wave height).

To calculate the excess path loss due to surface roughness, either a term with ocean wave height is added in the reflection coefficient [13] or the ocean wave spectrum is directly modeled [18]. Comparisons of the results of these two models with real data show good agreement up to SS 5. Further studies are necessary to find reasonable models for higher sea states (for heavy storms), where sharp increases in excess loss may occur because of high non-linearity.

In summary, surface-wave attenuation is proportional with the carrier frequency, same as the added loss due to surface roughness. Therefore, to minimize attenuation it is required that the radar is operated at the lower end of the HF band. Although statement of the surface-wave attenuation problem is quite old [19 - 21], research is still being undertaken to find numerically efficient programs for accurate surface-wave propagation calculations. A powerful simulator HFMIX is introduced in Chapter 1 to calculate surface-wave propagation characteristics including multi-mixed paths, which hybridizes the ray [12] and mode [20] representations.

6.2.6 Mixed-Path Propagation

The effect of mixed path propagation is better understood by considering the following two examples. First consider the propagation loss over a radial distance of 500 km, with two islands along the path. The radial lengths of the islands are 20 km and 50 km and are located 50 km and 170 km from the transmitter. Figure 6.9 plots the range variation of surface-wave path loss at two operating frequencies (3.5 MHz and 5.5 MHz). The dashed lines correspond to homogeneous sea path (without islands). The conductivity and relative dielectric constant of the land and sea are taken as σ_l = 0.003 $\Omega^{-1}m^{-1}$, ε_{rl} = 15 and σ_s = 5.0 $\Omega^{-1}m^{-1}$, ε_{rs} = 80.0, respectively. A sharp increase and recovery in the path loss at the sea/land and land/sea transitions, respectively, can be observed. This is known as the Millington effect [22].

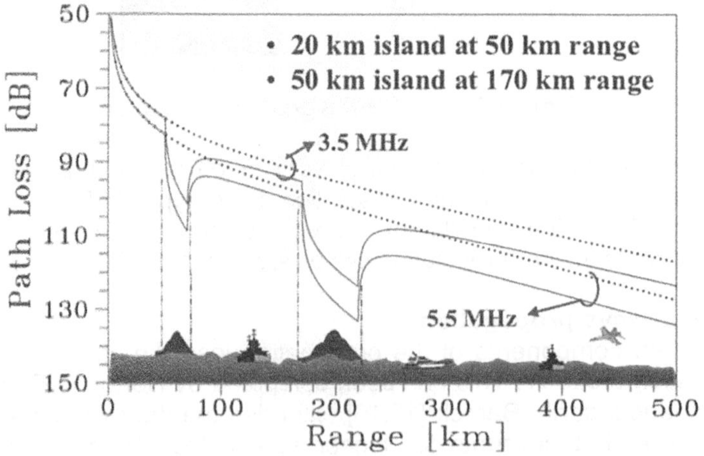

Figure 6.9: *Path loss vs. range over multi-mixed path*

In the second example, the radial propagation path includes five islands with 10-km, 5-km, 20-km, 10-km, and 10-km radial lengths, along with radial distances of 10 km, 25 km, 50 km, 80 km, and I00 km from the source, respectively. The results for the second scenario are plotted in Fig. 6.10.

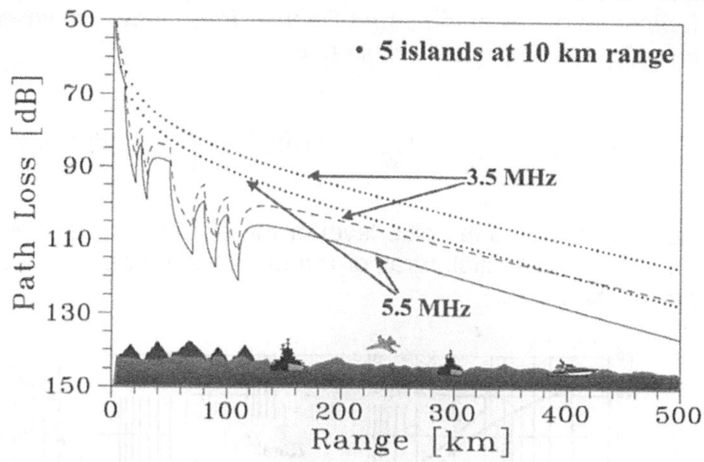

Figure 6.10: *Path loss vs. range over multi-mixed path*

In summary, the following observations can be made regarding multi-mixed path propagation:

- Having an island along the propagation path increases the path loss.
- Excess loss can be minimized by operating the radar at a lower frequency.
- For the given two examples, excess loss caused by the islands is between 5 dB and 15 dB.
- Nearby islands cause more loss than far islands.
- Critical parameters for mixed-path losses are the radial lengths of the islands and the distance in between them.

It should be noted that the two examples presented are almost extreme cases. However, even for these extreme cases, HFSWR can still yield the necessary performance to provide effective surveillance. On the other hand, surface targets beyond a few kilometers of an island may not be detected because of the sharp decrease in target signal, which may cause blind zones inside the radar coverage.

6.2.7 Noise Density

The next parameter that influences the performance of an HFSWR is the noise *N*. Typical noise types and their range of effective frequencies are pictured in Fig. 6.11. As shown in the figure, day- and nighttime atmospheric noise and man-made noise in rural and urban areas determine the noise floor at HF frequencies. Also, night-time atmospheric noise level is much higher than day time, especially at lower HF frequencies where HFSWR operate. Therefore, performance degradation at night-times should be expected in HFSWR.

As given in (6.8), noise density is defined for the case where the target echoes have been coherently integrated over a time interval, T_{int}. Since coherent integration concentrates the (constant velocity) target energy into a given Doppler bin, the value of N (please note that N is also used an integer variable to represent number of radar range bins) refers to the power density of the combined (clutter + noise + interference) process at the target Doppler frequency. Environmental noise can be expressed by the empirical formula [23]

$$N = N_e B \ [dBW]; \quad N_e = N_o - 12.6 \ln[\frac{f_{MHz}}{3}] \ \text{[dBW/Hz]} \tag{6.13}$$

where N_0 is equal to man-made noise level of the selected radar site (typically, –136, –148, –164 dB in residential, rural, or remote sites, respectively) and N_e is the noise density given as dB/Hz.

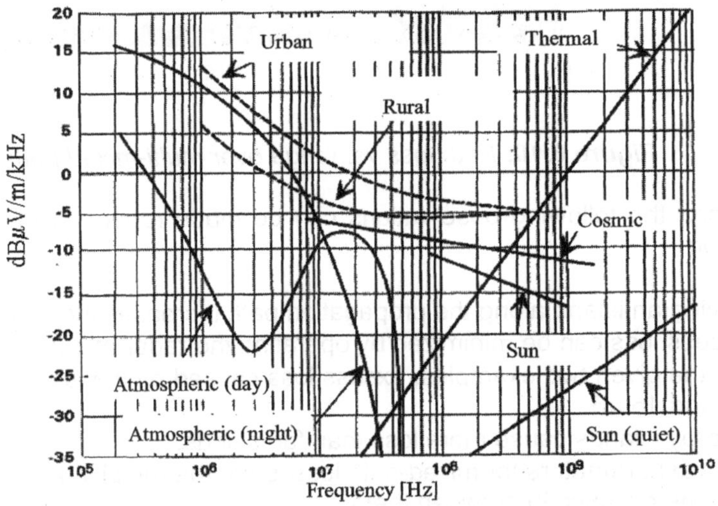

Figure 6.11: *Noise types and effective frequency ranges*

The detection performance of a HFSWR is ultimately limited by external noise [23,26]. External noise is divided into three categories; atmospheric, galactic, and man-made. For frequencies below 10 MHz, atmospheric noise usually dominates. At the lower end of the HF band the external noise level is approximately 40 to 60 dB higher than thermal noise and is subject to diurnal and seasonal variation [23]. Long-distance communication systems use sky-wave propagation paths and occupy the whole HF band (3 - 30 MHz). Increase in activity at lower end of the HF band, during the hours of darkness, is associated with the disappearance of the E layer and the merging of the F1 and F2 layers.

In summary, atmospheric noise is inversely proportional to frequency and is higher at the lower end of the HF band.

6.2.8 Ionospheric Self-Interference

One of the most significant causes of interference to an HFSWR is the ionosphere. Potential interference paths follow both surface/sky, sky/sky and sky/surface paths. One-way propagation delays for both surface and sky waves are on the order of milliseconds for the coverage mentioned above. For example, two-way time delays for a possible target at typical 100-km, 250-km and 500-km ranges are approximately 0.7 msec, 1.7 msec, and 3.3 msec, respectively [25]. It corresponds to sky-wave interference at typical 80- to 120-km and 200- to 250-km ranges. The sky wave interference becomes more complex when propagation is supported by F1, F2, and E layers.

To help mitigate against this ionospheric clutter, it is desirable that the transmit and receive antenna system provide a deep, broad null against signals arriving at near vertical incidence (NVI). In addition, during the daytime the radar can be operated at the critical frequency (CF), where sky-wave radiated energy is absorbed by the D layer. Alternatively, the radar can be operated above the maximum usable frequency (MUF) where energy propagated via the skywave mode propagates through the ionosphere and is not returned to the earth's surface. It appears that one way of reducing this NVI self-generated ionospheric interference good transmit and receive antenna array design with deep over-head nulls are required.

6.2.9 Spectrum Occupancy

A radar designer has no direct control over all these sources of interference; however, indirectly, these interferences can be mitigated, to a certain extent, by a judicial use of the radar frequency. The challenge is finding useful unoccupied frequency in an already crowded spectrum. Hence an important issue for HFSWR operation is spectrum management. Typical HF spectrum measurement results [24] are given in Fig. 6.12, which belongs to mid night (on the left) and mid day (on the right) times.

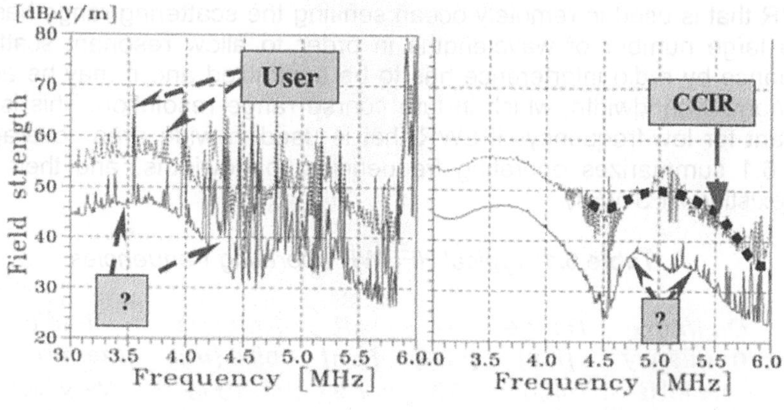

Figure 6.12: *Night- and daytime spectrum occupancies*

The solid and dashed lines correspond to measurements with average and peak detectors, respectively, with a 10 kHz bandwidth. The predicted CCIR noise levels [23] are also plotted (dotted line). Question marks are put in regions where broadband peculiarities are observed, which need to be investigated until their reason is found. Are they local effects? Do they appear in the spectrum throughout the year? These kinds of questions should be answered before the installation of the HFSWR in the region. As observed from the figure, most of the long range communication systems operate during nighttime, which means that a reduced HFSWR performance should be expected at nights.

Existing HFSWR, which are used either for ocean remote sensing or for continuously monitor surface and/or low-altitude air activity in wide areas, have limitations because of both ionospheric interference and spectrum occupancy. A compromise among range resolution, operating frequency, and signal bandwidth has to be found. For example, ionospheric reflections limit continuously monitoring surface and low-altitude air activity in wide areas, which may be overcome by choosing the operating frequency higher than the MUF. The MUF is highest at noon or in early afternoon and is highest during periods of greatest sunspot activity [23]. The MUF can drop to 4/5 MHz and go up to 30 MHz. HFSWR operates with frequencies below the MUF are strongly affected by the ionospheric interference (with HF radio transmission from all over the world). The signal bandwidth should be narrow in order to realize a sufficient SNR. For example, the signal bandwidth should not exceed 15 - 20 kHz for the operating frequencies between 3 and 7 MHz, which corresponds to a range resolution of 10 km to 7.5 km. Increasing the frequency also allows us to use higher signal bandwidth, but decreases the operating range because of the higher signal path loss at higher frequencies. For example, HFSWR that use upper HF band (i.e., 25 - 30 MHz) may use a bandwidth of up to 500 kHz, which corresponds to a range resolution of 300 m. On the other hand, surface-wave path loss drastically increases at these frequencies and limit the maximum range (for example, a 3-MHz HFSWR may reach up to 500 km in range, but this is reduced to 40/50 km for 30 MHz, with the same transmitter power).

HFSWR that is used in remotely ocean sensing the scattering range has to extend over a large number of wavelengths in order to allow resonant scattering. The disturbance by radio interference has to be minimized and it may be achieved by using narrow bandwidth, which in turn coarse range resolution. This is especially important for low-frequency HFSWR that is used in wide area coverage [27,28]. Table 6.1 summarizes operating frequencies, bandwidths, and their effects (in some existing HFSWRs).

***Table 6.1**: Typical HFSWR operating frequencies*

Operating frequency	*Range [km]*	*B [kHz]*	*ΔR [km]*	*Antenna size [m]*	*Radio interference*
3.4 MHz	400	20	7.5	≈ 1000	Very high
6.75 MHz	200	20	7.5	≈ 870	Very high
7 MHz	200	20	7.5	≈ 200	Very high
25 MHz	60	125	1.2	≈ 90	Low
27 MHz	55	500	0.3	≈ 83	Low
30 MHz	50	125	1.2	≈ 75	Low
55 MHz	15	600	.25	≈ 83	Low

6.2.10 Target Reflectivity

Target Reflectivity is measured in terms of radar cross section (RCS). The RCS of a target is highly dependent on the size of the target (*l*) measured in terms of the radar wavelength (λ). There are three distinct RCS regions known as Rayleigh, resonance, and optical regions corresponding to $\lambda << l$, $\lambda \approx l$, and $\lambda >> l$, respectively [29]. Targets that have a vertical extent that is comparable to the wavelength of the radar carrier frequency will be in the resonance region. When the wavelength is large compared to the target vertical extent, the RCS will be in the Rayleigh region where the RCS of a target decreases rapidly with increasing radar wavelength. The RCS of ships and aircraft have been extensively studied in the optical region. However, at HF the dimension of these targets are such that the RCS must be determined in both the Rayleigh and resonance regions.

In the optical region, individual components of a complex target may be treated in isolation, but at resonance the target must be modeled as a whole. In the optical region analytical asymptotic techniques, such as geometrical optics (GO), physical optics (PO), and the uniform theory of diffraction (UTD), are used to predict RCS behavior. At resonance, numerical techniques such as Method of Moments (MoM) [30], finite-difference time domain (FDTD) [31], or transmission line matrix (TLM) [32] analysis must be employed.

A typical example is presented in Fig. 6.13, where a 45-m navy frigate model is used as a target and its RCS is calculated via FDTD [33].

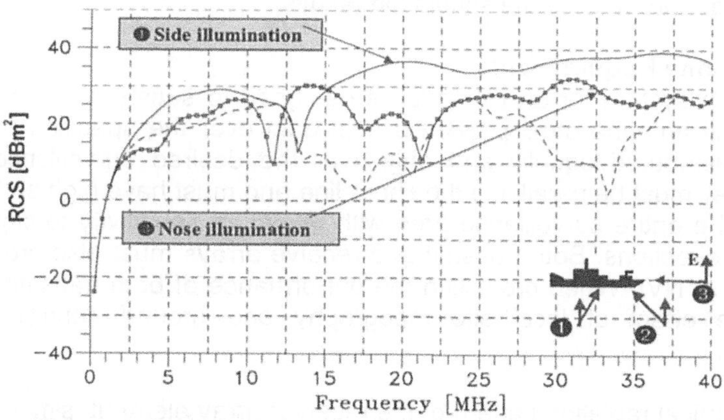

Figure 6.13: *RCS vs. frequency of a 45-m navy frigate*

As shown in the figure, RCS of this 45-m surface target varies between 10 dB and 40 dB in the HF band. HFSWR targets may be as much as 300 m in length and 20 - 30 m in height (such as oil tankers). Therefore, RCS of HFSWR may usually be on the order of 20 dB to 40 dB [34-36].

In the radar simulator, RCS values of targets may be obtained via different ways. It can be calculated via powerful simulators, such as FDTD and TLM, which requires realistic discretization of the target and is computationally expensive. It can also be obtained via high-frequency asymptotics. In this case the target is assumed as the superposition of (may be hundreds of) scattering centers with different scattering amplitudes and sectors (see Fig. 6.14). The latter requires validation against measurements at different operating frequencies with illuminations from all aspect angles.

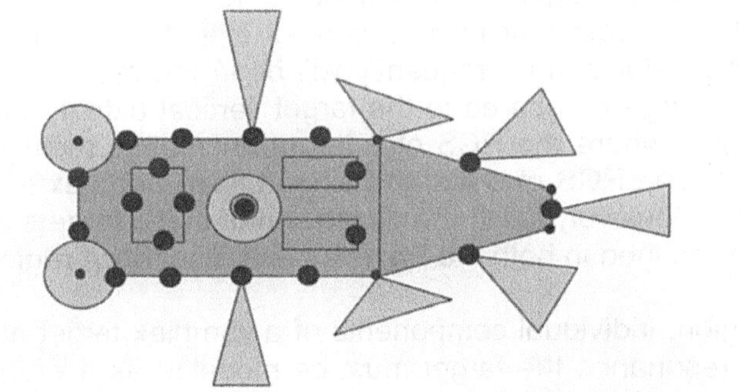

Figure 6.14: *Top view of a surface target and its scattering centers with different scattering properties*

In either case, the simulator requires an RCS database of all possible targets that are going to be played in the radar simulator. The more realistic the target RCS models, the more reliable the simulation results.

6.2.11 Antenna Requirements

The antenna system for an HF radar is designed to satisfy a number of criteria. The transmit antenna must provide a high gain over the specified band. Energy must be distributed equally and only over the desired surveillance area. The receive array must be parallel to the shore line and must have high and equal array gain over the entire surveillance area with minimum sensitivity to signals arriving from other directions. Both transmit and receive arrays must also provide a deep, broad, null at NVI. When predicting the performance of both transmit and receive arrays, the effect of local site topography and ground conditions must be considered.

A typical vertical radiation pattern of the receiver array element, simulated via NEC package, is given in Fig. 6.15. Here four-quarter-wavelength-high, vertical monopoles, separated by quarter wavelength, are used as array endfire elements. The pattern along the antenna axis is plotted over poor ground (σ = .003 $\Omega^{-1}m^{-1}$, ε_r = 4.0) for two cases, with and without ground screen. The gain with ground screen, is 2.1 dBi and points out $\theta = 60^o$ (i.e., 30° from the ground). Without ground screen this gain is almost 10 dB lower. However, employment of a ground screen can be seen to have a negative effect on direct overhead nulling. The solid and dashed lines correspond to sky-wave radiation patterns without and with ground screen, respectively. In this figure, surface-wave coupling effect is also plotted (see dotted line). A quadlet with the ground screen is assumed to be 50 m away from the ocean with electrical parameters σ = 5.0 $\Omega^{-1}m^{-1}$ and ε_r = 80.0. As can be seen from this figure, because of the surface-wave coupling, maximum radiation points to the ocean surface with a gain of 2.7 dB.

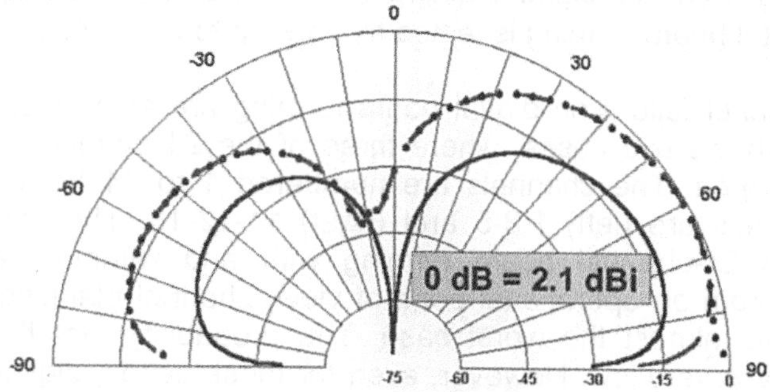

Figure 6.15: *A typical vertical radiation pattern of the receiver*

Operating at the low end of the HF band requires that the receiver array occupy a significant shoreline area, with the aperture of the array inversely proportional to frequency. For example, at 3 MHz, a 5° azimuthal beamwidth requires an array aperture of approximately one kilometer. As mentioned before, powerful algorithms may be used to cancel undesired signals. However, for successful unwanted signal elimination, the first step must be a well-designed antenna array.

6.2.12 Digital Beam Synthesis

Target detection is based on range and azimuth information, which is supplied by time gating and electronic beam synthesis in the receive system. Therefore, the limitations of the receive array to the beam steering capabilities must be understood not only from theoretical, but also from practical, points of view. For example, what is the influence of site topography? What do the beams look like when the array is located over lossy ground? How do the beam patterns change as they are steered? Is the system design resilient to channel failure? These questions must be answered if reliable tracking following the detection is to occur.

These problems are investigated via NEC simulation package. To illustrate digital beam steering capability, some typical simulation results are given in Fig. 6.16.

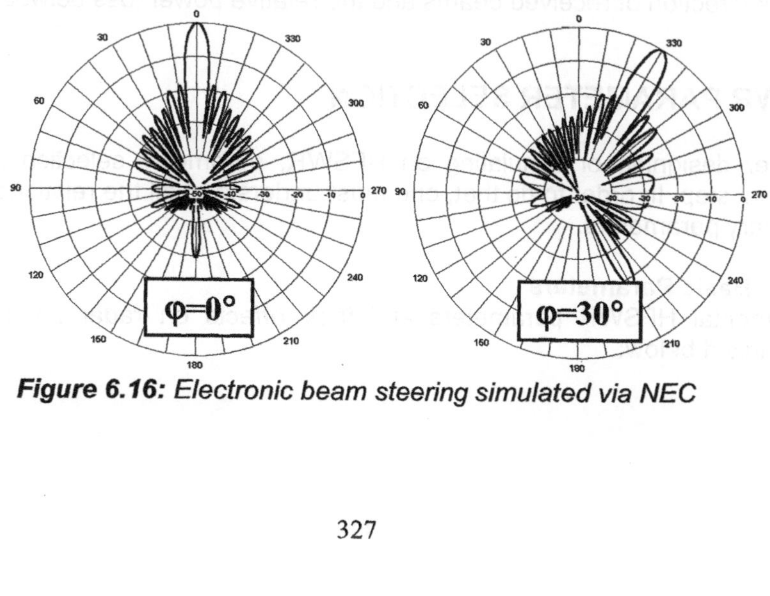

Figure 6.16: *Electronic beam steering simulated via NEC*

Here, digital beam forming via electrical phasing is illustrated for a 24-element quadlet array (24 × 4). The most pure beam is formed at boresight. As the beam angle diverges from boresight, unwanted side- and backlobes may appear. With this array, digital beam forming is tested to work up to $\varphi = \pm 45°$.

Effects of channel failure on digital beam forming are also investigated via NEC. Figure 6.17, shows two cases, where three of the 24 channels are blanked and beams are formed. The channels are numbered 1 to 24 from left to right. The blanked channels are (left) 1-3-5 and (right) 11-12-13. The thin and thick lines correspond to 24-channel beam forming with and without blanked channels, respectively. From an operational point of view, channel blanking modeled in Fig. 6.17 represents almost the worst case. The probability that three channels fail simultaneously is very low. However, even for these cases, digital beam synthesis is still possible with only minor degradations at the sides of the radar coverage.

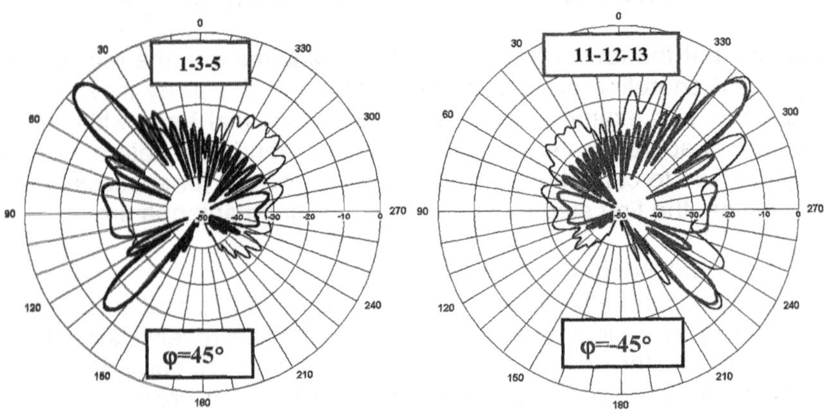

Figure 6.17: *Beam forming, under channel failure*

Land/sea transition, ground irregularities and the height of the site above sea level also strongly affect the shapes of the electronically formed beams. Therefore, receiver antenna measurements at the site must be undertaken to calibrate the receive arrays. The calibration is composite and should take both the transmitter and the receiver arrays into account simultaneously. The calibration establishes the look direction of received beams and the relative power loss between beams.

6.3 HFSWR PARAMETER SELECTION

In either designing or simulating an HFSWR, parameter selection is the most important step. In order to do that, one must understand all the relations among the necessary parameters.

6.3.1 Basic Parameters

Fundamental HFSWR parameters and their effects on radar performance are summarized below:

Radar frequency:
- affects the man-made, atmospheric, and galactic components of the total external background noise
- affects attenuation with range of the target and clutter signals
- affects the shape of the normalized clutter signal
- target Doppler frequency is proportional to it
- square of the Bragg frequency is proportional to it

Receiver or system gain (excluding the antennas): This is the overall power gain of the ground wave radar system, including gains in signal processing and losses in the antenna cables, but excluding the antenna gains.
- the target, clutter and observed noise signals are proportional to it
- it affects signal-to-noise and clutter-to-noise ratios
- signal-to-clutter ratios are independent of it

Effective transmitter peak power:
- the target and clutter signals are proportional to it
- the observed noise signal is independent of it
- the signal-to-noise and clutter-to-noise ratios are proportional to it
- the signal-to-clutter ratio is independent of it

Coherent integration time:
- the target signal is proportional to it
- the clutter and observed noise signals are independent of it
- the inverse of the Doppler resolution (in Hz) is equal to it, i.e. the spacing between the points of the processed radar signal in the Doppler frequency domain
- the signal-to-noise and signal-to-clutter ratios are proportional to it
- the clutter-to-noise ratio is independent of it

PRF (number of radar pulses transmitted per second):
- the observed noise signal is inversely proportional to it
- the target and clutter signals are independent of it
- the Doppler bandwidth (in Hz) is equal to it; i.e., the final processed bandwidth of the radar signal in the Doppler frequency domain
- the signal-to-noise and clutter-to-noise ratios are proportional to it
- the signal-to-clutter ratio is independent of it

Pulse length (τ):
- the clutter signal is proportional to the area of the ocean clutter patch, which is proportional to the ocean clutter patch width, which is proportional to the effective length of the radar pulse
- it affects the shape of the first-order normalized clutter signals (due to the change in the patch width)
- the observed noise signal is proportional to the receiver band width, which is equal to the inverse of the effective length of the radar pulse
- the target signal is independent of it
- the signal-to-noise ratio is proportional to it
- the signal-to-clutter ratio is inversely proportional to it

- the square root of the clutter-to-noise ratio is proportional to it

Wind speed:

- affects the attenuation with range of the target and clutter signals
- affects the shape of the normalized clutter signal
- affects the signal-to-noise and clutter-to-noise ratios
- signal-to-clutter ratio is independent of it

Wind direction:

- it affects the shape of the normalized clutter signal
- the signal-to-noise, clutter-to-noise and signal-to-clutter ratios are independent of it
- a cross wind (i.e., 90 degrees) produces a clutter spectrum that is symmetrical about 0 Hz
- a 0-degree wind produces a clutter spectrum that is higher on the negative Doppler side and is much lower on the positive Doppler side
- wind directions of X and –X degrees produce identical results
- wind directions of 90+X and 90–X degrees produce clutter spectra that are mirror images about 0 Hz

Target speed:

- it determines what point on the clutter spectrum is obscuring the target signal
- the signal-to-noise ratio is independent of it
- it effects the signal-to-clutter and clutter-to-noise ratios

Target direction:

- it determines what point on the clutter spectrum is obscuring the target signal
- the signal-to-noise ratio is independent of it
- it effects the signal-to-clutter and clutter-to-noise ratios

Target range:

- the target and clutter signals attenuate with increasing range
- the observed noise signal is independent of it
- the signal-to-clutter ratio is inversely proportional to it
- it effects the signal-to-noise and clutter-to-noise ratios

Target RCS:

- the target signal is proportional to it
- the clutter and observed noise signals are independent of it
- the signal-to-noise and signal-to-clutter ratios are proportional to it
- the clutter-to-noise ratio is independent of it

Target height:

- the net target RCS changes (the user must manually make this change)
- a (usually) small change occurs in the attenuation factor

Transmitter antenna gain:

- the target and clutter signals are proportional to it
- the observed noise signal is independent of it

- the signal-to-noise and clutter-to-noise ratios are proportional to it
- the signal-to-clutter ratio is independent of it

Receiver antenna gain:
- the target and clutter signals are proportional to it
- the observed noise signal is independent of it
- the signal-to-noise and clutter-to-noise ratios are proportional to it
- the signal-to-clutter ratio is independent of it

Receiver beam:
- the clutter signal is proportional to the area of the ocean clutter patch, which is proportional to the receiver antenna beam width
- it affects the shape of the normalized clutter signals
- the signal-to-noise ratio is independent of it
- the clutter-to-noise ratio is proportional to it
- the signal-to-clutter ratio is inversely proportional to it

Environmental noise:
- the target and clutter signals are independent of it
- the observed noise signal is proportional to it (the program assumes that the receiver system contributes no significant noise)
- the signal-to-noise and clutter-to-noise ratios are inversely proportional to it
- the signal-to-clutter ratio is independent of it

Radar signal bandwidth:
- the observed noise signal is proportional to it
- range resolution is inversely proportional to it

Clutter patch area:
- the clutter signal is proportional to the area of the ocean clutter patch, which is proportional to the width of the patch
- it affects the shape of the first-order normalized clutter signals

These physical parameters, their interrelations, range of validities, etc., must be well-understood, if a realistic HFSWR is desired to be simulated, which is more difficult than physical realization of the HFSWR system. For example, you do not need to understand and simulate mixed-path propagation effect in physical realization.

6.3.2 Numerical Design Example

The design of an HFSWR starts by specifying the requirements.

Requirement:

An HFSWR shall operate to cover a coastal region up to 200 nm in range and 100° in azimuth. The radar shall be capable of detecting surface targets with radial velocities up to 50 knots (90 km/hr) and discriminating targets with radial velocity difference of $\Delta v_r = 1$ knot, and air targets with radial velocities up to Mach 1 (1200 km/hr) and $\Delta v_r = 10$ knots.

A typical design:
First of all, power budget of the HFSWR shall be determined from the radar equation. Giving maximum range as 200 nm (nearly 400 km) needs low-frequency HFSWR and a high-power transmitter, since path loss drastically increases with frequency. For example, two-way path loss of a 5-MHz signal at 400-km range is nearly 220 dB. This becomes nearly 360 dB at 20 MHz. In other words, HFSWR signal that reaches 400 km at 5 MHz can only reach up to 60/70 km with the same power at 20 MHz. Typical transmitter average power is on the order of kilowatts (between 5 and 20 kW) for the HFSWR that operate in 3 - 6 MHz region. Nominal signal bandwidth is usually 20 kHz, which restricts range resolution to 7.5 km. A receiver array of 16 to 24 quarter-wavelength monopoles (separated by quarter wavelength) requires nearly 1-km coastal region. This array allows digital beam steering capabilities with azimuthal beamwidth of 4° - 6°. This corresponds to a cross-range resolution of nearly 40 km at 400-km range.

Typical antenna gains are 13 dBi to 15 dBi for the transmitter and 18 dBi to 20 dBi for the receiver. The thermal noise floor for the 20-kHz bandwidth is approximately –144 dBW. On the other hand, environmental noise floor for these operating frequencies and signal bandwidth is approximately 60 dB higher than thermal noise, which means the noise floor is nearly –84 dBW.

As a summary, power budget may be listed as in Table 6.2.

Table 6.2: *Typical long-range HFSWR power budget*

Parameter	Value	Value [dB]
Average Transmitted power	10 kW	40
Transmitter antenna gain	30	15
Receiver antenna gain	100	20
Noise floor	-84 dBW	84
Signal processing gain		40
Target RCS	100 m^2	20
TOTAL	-	219

Typical two-way path loss is on the order of 210 to 220 dB, which is almost equalized with the total power budget given in the table. On the other hand, other parameters shall be chosen as follows:

- Maximum range of 400 km determines the PRF as 375.
- Operating frequencies between 3 MHz and 6 MHz results in first order Bragg peaks in the range of ±0.17 Hz to ±0.3 Hz in Doppler domain.
- Maximum surface targets with velocities of 50 knot result in 0.5 Hz to 1.0 Hz Doppler frequencies. Target discrimination based on 1 knot radial velocity difference requires a Doppler resolution of 0.01 Hz to 0.02 Hz.
- Because of these Doppler requirements, coherent integration time shall not be less than 100 sec for surface targets.

Air targets with Mach 1 (1200 km/hr) radial speed cause Doppler shifts of ±6.6 Hz to ±13.2 Hz, and target discrimination with 10 knot requires a Doppler resolution of 0.1 Hz to 0.2 Hz, which necessitates an integration time of 5 sec to 10 sec.

Finally, depending on the FFT size chosen, number of pre-detection integration n shall be determined form PRF $\times T_{int} = n \times K_{FFT}$ equation.

6.4 STOCHASTIC MODELING OF HFSWR SIGNAL ENVIRONMENT

Total radar echo consists of target signal, noise, and clutter. Long-term observations [26] indicate that environmental noise amplitude is almost Gaussian distributed with a variance which can be expressed by (6.13). Noise samples may be produced as follows:

- An array of K_{FFT} real numbers are generated with a random number generator which have Gaussian amplitude distribution.
- The array is normalized to zero mean and unit variance.
- The array is then re-normalized to the average noise power calculated from (6.13)

Following this procedure yields sample-to-sample uncorrelated noise pulse train.

The target is modeled based on its RCS and Doppler shift, caused by the radial component of the target velocity with respect to radar. The RCS of the target is estimated as [29,37]

$$\acute{o}_t = 52 f_{MHz} D^{3/2} \quad [\mathrm{m}^2] \tag{6.14}$$

where D is the full-load displacement of the vessel in [ktons]. The fading characteristics of the target are assumed to be Swerling II [1].

Ocean clutter is modeled as an ensemble of M complex elements that have Gaussian amplitude distribution and are pulse-to-pulse correlated. This correlation is established through imposition of a Gaussian shaped power spectrum [38]. The clutter spectrum is modeled as follows:

- Two Gaussian-distributed unit amplitude phasor arrays g_{1k} and g_{2k} are generated as $\xi_k = g_{1k} + i\, g_{2k}$, $|\xi| = 1$.
- The Gaussian-shaped power spectrum $S(f_k)$ of the sea clutter around dominant Bragg frequencies are generated at K_{FFT} points (at f_1, f_2, .. , f_{max} where f_{max} is the maximum frequency in the Doppler spectrum).
- The width of the Gaussian-shaped power spectrum is related to the ocean wave-height.
- The ensemble of M ($M = K_{FFT}$) elements X_k are generated via $X_k = \xi_k \sqrt{S(f_k)}$ ($k = 1,2, .., M$).
- Inverse Fourier transform is utilized via FFT to obtain the complex elements Y_n as

$$Y_n = \sum_{k=0}^{K_{FFT}-1} X_k e^{i2\pi kn / K_{FFT}}, \quad n = 0, 1, 2, ..., K_{FFT}-1 \tag{6.15}$$

This procedure gives TD samples, which have a Rayleigh-distributed amplitude envelope and a Gaussian-shaped power spectrum.

Addition of all these synthetically produced components together yields the total echo signal. The radar PRF and the integration time T_{int} determine the number of samples in the total echo array. K_{FFT} is typically chosen between 128 and 2048, which also determines the number of echoes averaged incoherently before FFT, which is equal to PRF $\times$ T_{int}/K_{FFT}. A typical synthetically produced Doppler spectrum is plotted in Fig. 6.18. Here, the operating frequency is 5 MHz. This yields the dominant Bragg frequencies of $\pm$0.228 Hz that are normalized to $\pm$1 Hz. For this operating frequency the blind velocities are $\pm$13.7 knots. The CNR is chosen to be 30 dB. Two surface targets with 20 dB and 15 dB SNR ratios and 14-knots and 20-knots radial velocities, respectively, are included in the model. Dashed and solid lines correspond to one spectrum and average of consecutive 20 spectra, respectively. In Fig. 6.18, dominant Bragg returns at $\pm$1-Hz, second-order continuum and the target with-20 knots velocity are clearly seen. The other target with 14-knots radial velocity is obscured by the dominant Bragg return at 1 Hz.

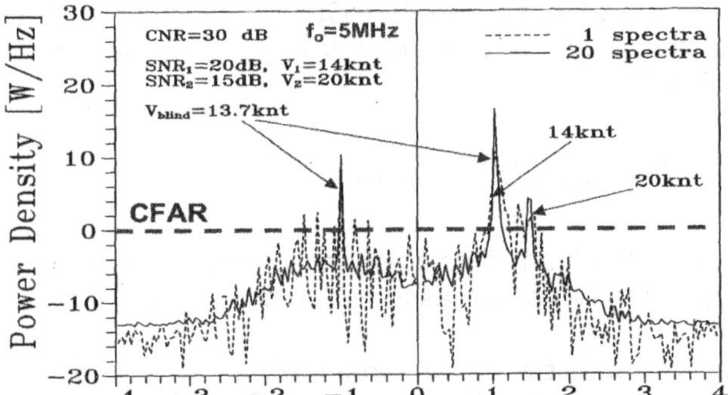

Figure 6.18: *Doppler spectra of the total radar echo (f = 5MHz)*

First-order Bragg peaks are proportional with the square root of the radar operating frequency, while targets' radial velocities are proportional with the operating frequency. This means that targets that are obscured by the first order Bragg peaks (i.e., targets with radial velocities in the order of blind velocities) may be distinguished by changing radar's operating frequency. This is clearly illustrated in Fig. 6.19. Here, all the parameters are same as in Fig. 6.18, except the operating frequency is moved from 5 MHz to 8 MHz. In this case, blind velocities move to $\pm$ 10.8- knots and target with 14-knots radial velocity is clearly seen in the spectrum.

In real HFSWR environment the Doppler spectrum of the return echo is not as clear as synthetic spectra given in Figs. 6.18 and 6.19. Two real spectra are given in Fig. 6.20 observed with Canadian HFSWR at the east coast [39,40]. As observed in the figure, first-order Bragg peaks and second order continuum around are not as clear as in synthetically produced data.

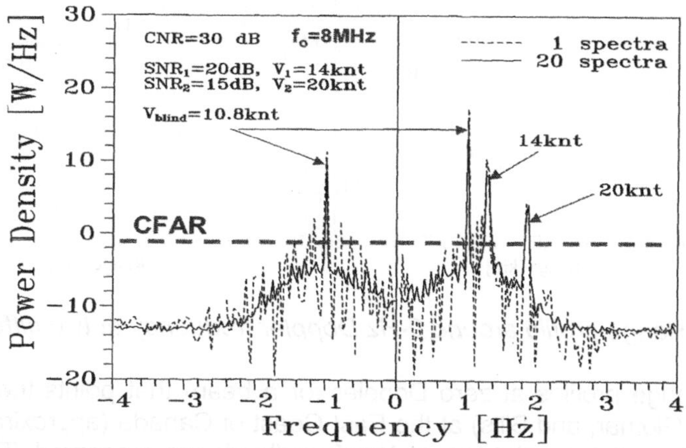

***Figure 6.19:** Doppler spectra of the total radar echo (f = 8 MHz)*

There are many ways of simulating a complex system like a HFSWR. One may stochastically model the real radar signal environment and simulate the detection process, while the others may do it in a simple (but reliable) way and focus on the tracking and classification algorithms to play on the performances of different tactical scenarios. Here, both are taken into account. The total signal environment is stochastically modeled (including uncorrelated noise, correlated sea clutter, and target echoes), path losses are calculated via HFMIX and extended Kalman filters with pre-assigned gates are used to simulate not only detection but also tracking of targets [38].

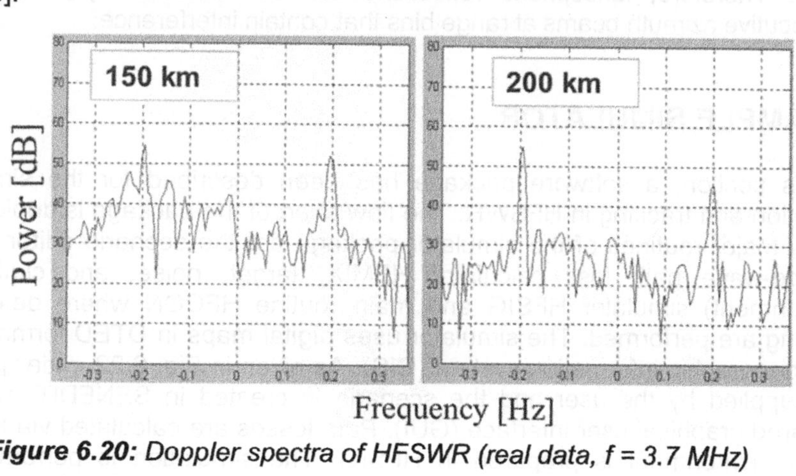

***Figure 6.20:** Doppler spectra of HFSWR (real data, f = 3.7 MHz)*

To show complexity and dynamic behavior of the HFSWR environment, real data obtained via one of the Canadian East Coast system [40] is presented in Fig. 6.21.

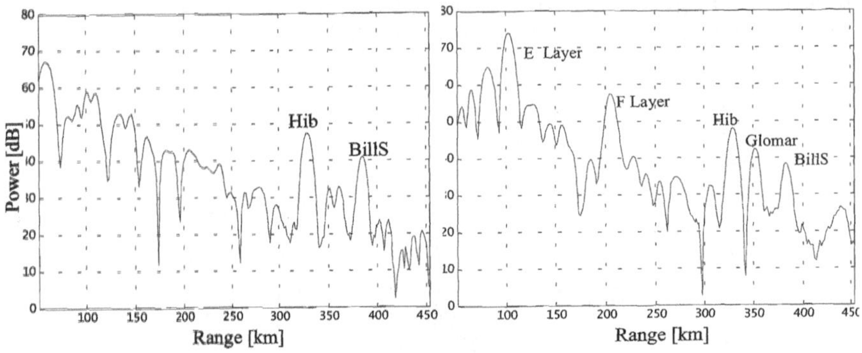

Figure 6.21: *Range profile around 0-Hz Doppler frequency at two different times*

In the figure, range profiles at zero Doppler for a beam that points the offshore oil platforms (Hib, Glomar, and Bills) at the East Coast of Canada (approximately 300 to 350 km away from the coastal line at Newfoundland) are presented. The two plots correspond to two different times in mid-April 1999. The data on the left are taken at 13:30 and clearly shows the three stationary offshore platforms, it can also be observed that the ocean clutter at 0 Hz falls off at approximately an R^3 rate (where R is the radial distance from the radar) and approaches the noise floor of 15 dB at 400 km. The data on the right, taken approximately 1 hour later, show the presence of two additional peaks corresponding to E and F layer reflections, at ranges of 90 to 110 km and 200 to 250 km, respectively, which is consistent with the theoretical results obtained from the computer simulations [39]. It should be noted that although these ionospheric peaks have similar range profiles to a point target, their Doppler profile is usually broadband. The Doppler smearing occurs as a consequence of ionospheric winds. Therefore, ionospheric reflections can be removed by cross-correlating consecutive azimuth beams at range bins that contain interference.

6.5 A SAMPLE SIMULATOR

In this section, a software package has been described for the simulation of detection and tracking in HFSWR. The flow chart of the package is depicted in Fig. 6. 22. Major routines of the simulator package are the scenario editor SENEDIT, surface-wave path loss calculator HFMIX, target, noise, and clutter (signal environment) simulator HFSIG and main routine HFCON where detection and tracking are performed. The simulator uses digital maps in DTED format, supplied by a geographic information system (GIS). As given in Fig. 6.22, radar parameters are supplied by the user and the scenario is created in SENEDIT via specially prepared graphical user interface (GUI). Path losses are calculated via HFMIX and signal environment is prepared in HFSIG. The simulation is performed in the simulator control center HFCONT. The radar parameters used in the simulator are listed in Table 6.3.

The path loss calculator HFMIX is discussed in detail in Chapter 1. The contents HFSIG routine is explained with typical figures in the previous section. Therefore they are not repeated here. The other sub-package SENEDIT and the main simulator HFCON are given below.

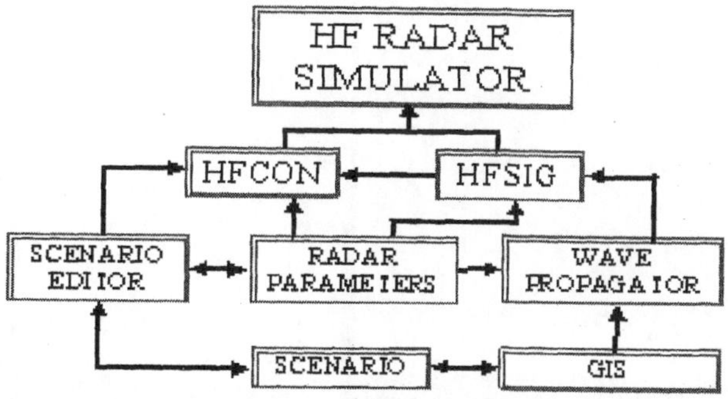

Figure 6.22: *Flow chart of the SWHFR simulator*

6.5.1 Scenario Editor

The third basic routine in the HFSWR simulator is the scenario editor, SENEDIT. A scenario consists of a HFSWR with its location and coverage area and a number of different kinds of surface targets (identified with their average bulk RCS values) which are moving along the chosen sailing routes. The scenario is prepared over a simulation map (terrain) which consists of a bitmap saved digital data. In order to prepare a scenario in SENEDIT, first the radar is located. This can be done both by giving its latitude/longitude coordinates, or by dragging the radar icon over the terrain data where the simulation will be performed.

Table 6.3: *Radar parameters*

Radar Parameters	Symbol	Unit
Frequency	f	Hz
Pulse rep. frequency	PRF	Hz
Signal bandwidth	B	kHz
Azimuth beamwidth	$\Delta\varphi$	Rad
Azimuth coverage	φ_{max}	Rad
Transmitter power	P_t	dBW
System loss	L_s	dBW
Antenna gains	G	dB_i
Detection margin	M_{det}	dB
Integration time	T_{int}	Sec
Clutter RCS coefficient	σ_0	dBm^2/m^2

Two sub-windows, the radar and target windows, are used to supply the necessary HFSWR parameters (see Table 6.3) and to define sailing routes of the targets which are included in the scenario. The supplied HFSWR's parameters are used in path loss calculations and in stochastic modeling of the total echo. Supplying the radar parameters also defines the operational area (radar coverage). The surface targets and their sailing routes are prepared in the target window. In Fig. 6.23, preparation of a typical surface target's sailing route is given.

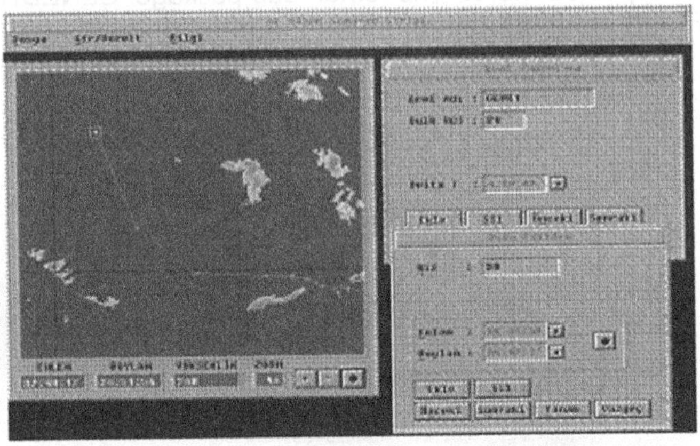

Figure 6.23: *The SENEDIT window*

On the left, the terrain data, which are loaded automatically when SENEDIT is called, are shown. From the other two sub-windows on the right, the name, bulk RCS, and initial delay after which the target is included in the scenario are supplied. Also, the sailing routes are drawn as broken lines ,each of which may have different velocity and anchor time. Different magnification levels can also be used to select accurate sailing routes for each surface target. From these targets, routes prepared by GUI-supported SENEDIT window tables are prepared for each target. For example, a typical target movement table is given in Table 6.4. The target number and its bulk RCS are given first. Then locations at different time instants are listed as latitude/longitude pairs, together with anchor times at these locations. For example, if the simulation starts at time t = 0, then, as mentioned in the table, Ship #1 will enter the simulation at time t = 06:45:00 (6 hours 45 min after the start) at the location given with latitude/longitude pair (25:34:00, 35:44:00). Ship #1 will follow a straight line up to the coordinates (25:34:35, 35:45:00) with a speed defined the path divided by the time interval. Then, it will wait there for an hour and start to move the third coordinate (25:35:21, 35:46:14) another speed defined by the second straight path divided by second time interval (after subtracting the anchor time).

Table 6.4: *A target and its movement in a scenario*

Ship #1	Bulk RCS = 30 dB	
Time	**Lattitude, Longitude**	**Anchor time**
t_1 = 06:45:00	25:34:00, 35:44:00	Δt_1=00:00
t_2 = 11:12:15	25:34:35, 35:45:00	Δt_2=01:00
t_3 = 16:36:00	25:35:21, 35:46:14	Δt_3=00:00
t_4 = 21:00:00	25:36:00, 35:47:16	Δt_4=03:00
t_5 = 23:00:45	25:37:00, 35:48:22	Δt_5=00:00

A typical scenario with an HFSWR radar coverage and four targets with their sailing routes are given in Fig. 6.24. The scenario prepared in this way is saved as a given name (e.g., scen #1.dat, scen #2.dat, etc.). A scenario includes:

- The name of the digital map (on which the scenario is going to be played)
- HFSWR name (a radar parameter file with this name is going to be included in the scenario)
- HFSWR location and coverage (maximum range, minimum/maximum azimuth bearings)
- Number of targets and their sailing route tables (such as given in Table 6.4)

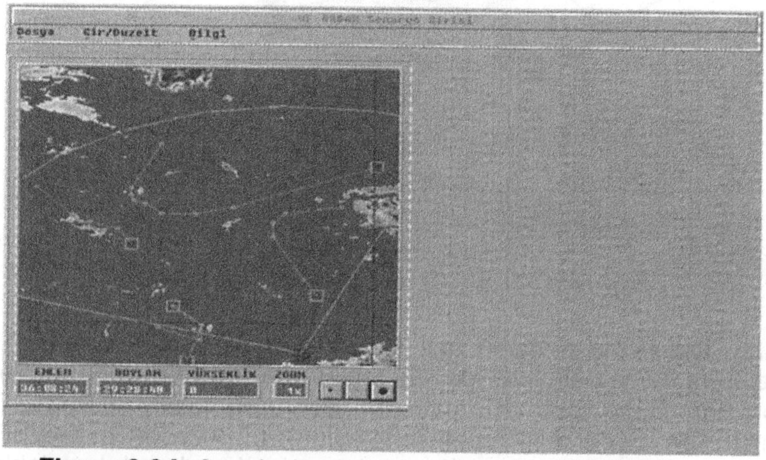

Figure 6.24: *A typical scenario prepared in SENEDIT*

6.5.2 HFSWR Simulator

The main part of the HFSWR simulation software package is HFCONT, where the detection and tracking of the targets are simulated. This routine is called together with

- a terrain data (digital map) and
- a scenario, which is created in the SENEDIT routine

Creating or loading a scenario determines the operating frequency, the radar coverage and the range and azimuth resolutions. The HFCONT first obtains the radial path of each azimuth resolution cell (with possible islands along it) from the terrain data. It then links to HFMIX, by which path-loss variation for each radial section is calculated. From these calculations the radial-range/path loss variations for all the angular sections are prepared as a look-up table to speed-up the simulation. These tables must be re-prepared when the location or the operating frequency of the HFSWR is changed. For example, a typical HFSWR location and its coverage on a chosen map is given in Fig. 6.25. When this scenario is started to play, radial propagation paths, some of which include multi-mixed paths (island transitions), are determined first. These paths are chosen radially at cell center and labeled as path #1, path #2, etc. from leftmost radial to rightmost. Surface-wave path-loss table is prepared at a given frequency along these propagation paths. On the other hand, it may be desirable to use the same radar with exactly the same parameters (including max range and azimuth coverages) may desired to be used at different location on the same region.

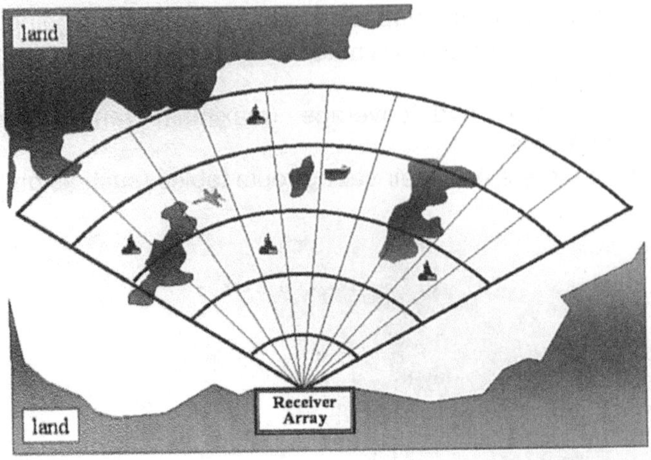

Figure 6.25: *A HFSWR location and coverage*

Figure 6.26, depicts this scenario. In this case, although operation region and radar parameters are same path loss look-up tables are required to be re-prepared. As seen from the figure, path #1, path #2, etc., are totally different. Therefore, whenever (i) the digital map (operation region), (ii) location of the HFSWR are changed, propagation paths must be re-determined and path-loss tables are re-prepared.

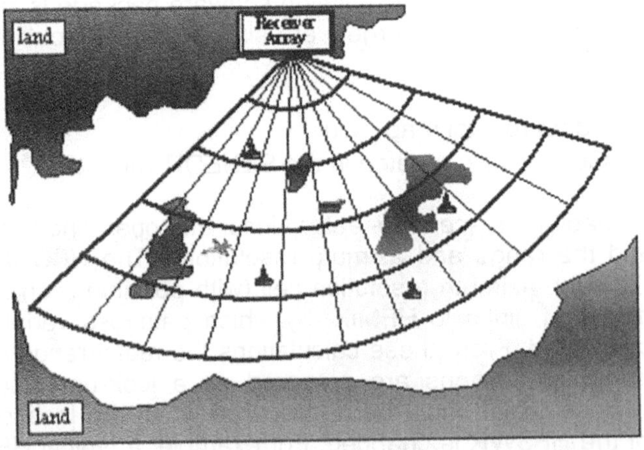

Figure 6.26: *Another HFSWR location and coverage*

After preparing path-loss look-up tables, the average environmental noise and the sea clutter powers are calculated. Since patch area A of the clutter depends on the range gate, clutter average power increases as the range increases. If there is an island in the patch or part of a land falls in that patch, sea and land patch areas in that cell are calculated separately and multiplied by different backscatter coefficient σ_0 to obtain the average clutter power. Finally, SNR of each target and CNR for each range gate are calculated.

6.5.3 Detection Process

The target signal in a HFSWR is generally embedded in the strong sea clutter returns. Only large surface targets at close ranges occasionally exceed the clutter level. Therefore, the detection of a target in the simulator package is performed in Doppler domain. The simulation is built according to the self-informing strategy of the targets. Each target knows its initial position, sailing route, velocity, and inclusion time. When the simulation starts at the initial time, targets in the chosen scenario are numbered as ξ_i, (i=1,2, .., N). The detection process is based on the following calculations, which are carried out at each time step during the simulation:

- Each target declares its sailing status (i.e., range, azimuth and velocity).
- Using this status information, the radial range of each target is calculated and its azimuth sector is found.
- The path loss value is read from the look-up path-loss table for each target.
- SNR is calculated for each target.
- CNR and SCR are calculated for the cells which contain targets at that moment.
- The total echo array of length K_{FFT} is generated as above.
- FFT of each total echo array is taken and a Doppler spectrum of the HFSWR is obtained for each cell which contains targets.
- Clutter suppression is performed for the dominant first-order Bragg peaks in each spectrum.
- For a given detection margin M_{det} = SNR_{min} the peaks above threshold are found and their Doppler frequencies are marked.

The main simulation window of HFCON is given in Fig.6.27.

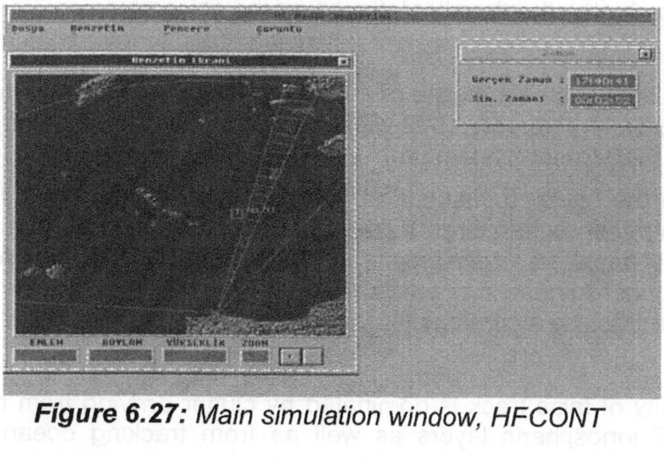

Figure 6.27: *Main simulation window, HFCONT*

The location and coverage of the HFSWR and the detected target is shown in this figure. The real and the simulation times are presented at right top of the main window. During the simulation, both detected and undetected targets can be displayed on the same terrain data. Since the detection in an HFSWR is performed in the Doppler domain, the Doppler spectrum of any target may also be displayed. Any target on the simulation terrain data can be selected with a mouse by double-clicking, and then its Doppler spectrum is displayed. In Fig. 6.28, the simulation window and the Doppler spectrum of a chosen target is shown. Here, the detected targets and the ground-truth scenario are presented in different colors. The two dominant Bragg peaks, the environmental noise floor, the threshold, and the target peak in the Doppler spectrum are all observed in this figure.

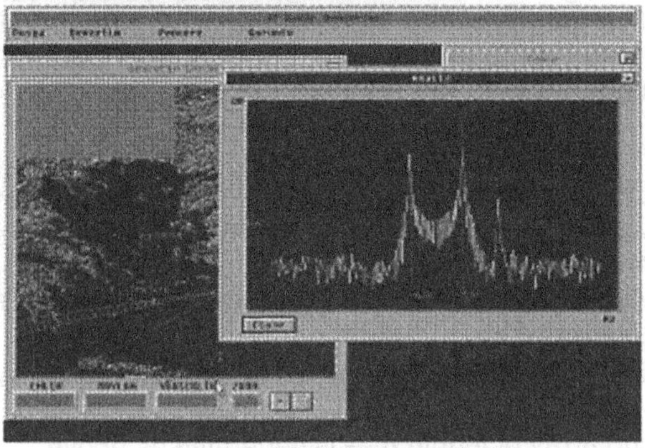

Figure 6.28: *Main simulation window and target's Doppler spectrum*

It should be noted that HFSWR operate with a very high false alarm rate. The probability of detection (P_D) is a measure of the likelihood that a target with a given SNR shall be detected. Variations in both signal strength and the contending noise or clutter level determine the probability that a signal on the threshold boundary will register as a detection or not. The probability of a false alarm (P_{FA}) is determined by the likelihood that the noise or clutter exceeds a given threshold level. Obviously, the higher the threshold, the lower the P_{FA}.

To show the high false alarm rate of the HFSWR, total detections in approximately 75 min, recorded in February 1999 with one of the HFSWR (located in Cape Race) in Canadian East Coast System [40], is plotted in Fig. 6.29.

As shown in the figure, thousands of detections are made in 30 T_{int} (CIT), when nearly ten to fifteen actual targets are present inside the coverage. This does not cause severe problems or limitations in the performance of HFSWR, since false detections (ghost targets) can easily be removed with intelligent multi-hypothesis detection and tracking algorithms [40].

The probability of false track is dominated by clutter arriving from reflections from the E and F ionospheric layers as well as from tracking ocean clutter peaks.

Ionospheric returns accumulate in an annular band spanning several range bins, all azimuths, and most of the ship Doppler band.

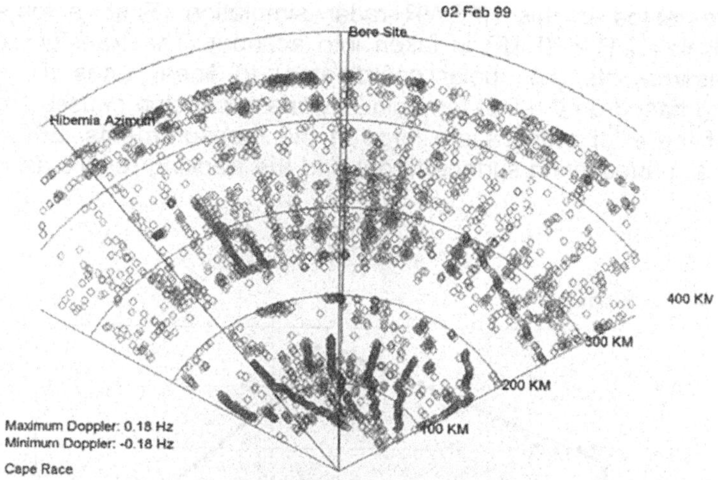

Figure 6.29: *Real detections of HFSWR in 30 CIT (≈ 100 min)*

HFSWR are coherent radars and detection is based on the SNR at a given Doppler, where the noise bandwidth is determined by the inverse of the CIT. A consequence of the much reduced noise bandwidth is that the P_D, even at a relatively low SNR, is extremely high, and for the noise-limited case the P_{FA} at a given Doppler bin is very low. The number of false alarms for a given CIT is equal to the probability of false alarm multiplied by the number of independent range/azimuth Doppler cells (approximately 2.5 million). Consequently even a moderate probability of false alarm translates into a high number of false detections. These false detections are dominated by ionospheric clutter and are characterized as rings of detections corresponding to the vertical range of the various ionospheric layers.

6.5.4 Tracking Process

Target tracking involves association of the tracks obtained from detection process at regular time intervals. Most of the work that has been done about tracking algorithms are for the microwave radars. Kalman filters are the most common and easily implementable algorithms, which are used in target tracking with MW radars. There are three important aspects of MW radars, which prevent the extension of these techniques to HFSWR:

- Data flow of the detection is fast enough that the target parameters do not vary appreciably,
- Range, range rate, and azimuth information is accurate enough to track the targets,
- The target return signal in any resolution cell is generally greater than the noise and the clutter return.

The HFSWR gives poor range and azimuth but very high velocity accuracy as compared with the MW radars. Therefore, different tracking logic based on Kalman filtering are tested in this HFSWR radar simulation. First, a four-dimensional extended Kalman filter (EKF) is taken into account. The transformation of non-linear measurements (i.e., polar coordinates) to linear ones (i.e., rectangular coordinates) based on the first terms in Taylor's expansion causes a convergence problems if the error margins are large in the measurements. EKF is tested for tracking of a maneuvering surface target and the tracking results are given in Fig. 6.30.

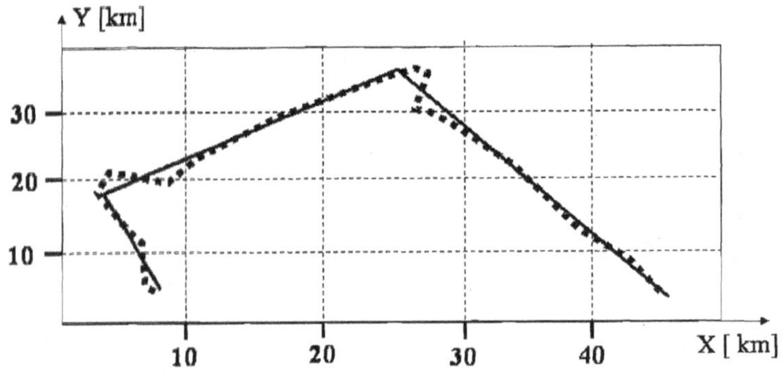

Figure 6.30: *Target tracking while sharp maneuvering [42]*

The HFSWR gives accurate velocity measurement, which is directly used in the mentioned correlation process. As the outcome of these tests, a two-stage tracking algorithm is used in the simulator. In the first stage, Doppler, range, and angle trackers (where position, velocity, and acceleration are included) process the measurement data. The estimate of Doppler tracker is fed into the range tracker in order to increase the performance of the range tracker with accurate velocity information. Because of the same reason, angle tracker is also fed by the estimate of range tracker. The second stage is EKF, which is supported by the estimates of the first-stage trackers. By this way, the simulator gains the ability to track multiple targets with various maneuvers. The example for these improvements is for multi-target tracking and the scenario is shown in Fig.6.31. Here, three surface targets approach each other and fall into the same resolution cell at the same time.

The output of the tracker in HFSWR is presented in Fig. 6.32. Initially, all three targets are detected and the tracks are established with track numbers 1, 2, and 3. In the resolution cell, which is mentioned with the dashed rectangle, the simulator continues to detect the targets but the tracks are lost. In the same resolution cell, HFSWR can only give detection information of the targets (if their radial velocity differences are greater than the Doppler resolution). After targets left the cell, the tracks are re-established with the track numbers 4, 5, and 6.

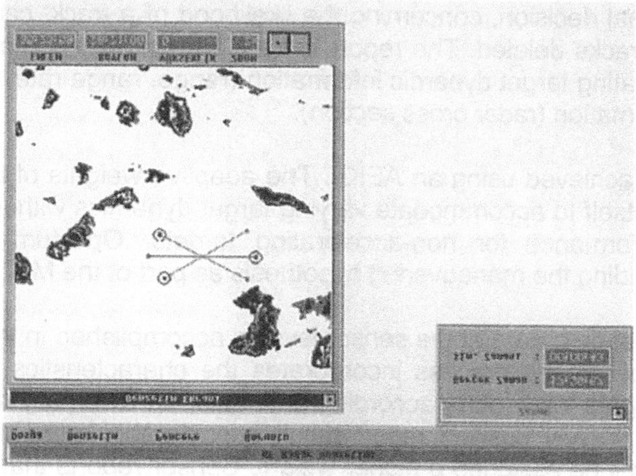

Figure 6.31: *Three surface targets and their routes*

As mentioned previously, multiple targets may be resolved from within a single radar range cell. In addition to these expected target returns there may also be unwanted clutter and interference signal that result in both false alarms and missed detections. The role of the track algorithm is to associate consecutive detections into tracks. For example, in the Canadian East Coast System's HFSWR, this task is accomplished within the Multi-sensor Association Processor (MSAP) [40].

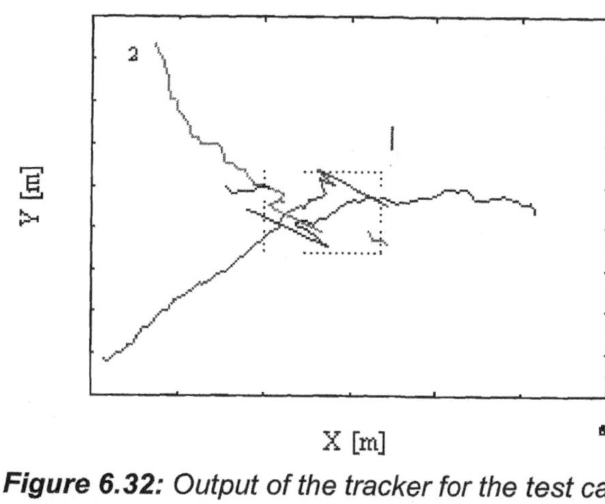

Figure 6.32: *Output of the tracker for the test case*

The MSAP comprises of three functional blocks: multi-sensor fusion processor, multiple hypotheses tracker (MHT), and track filtering based on an adaptive extended Kalman filter (AEKF). It undertakes the combined tasks of sensor fusion, tracking, track fusion and calibration.

The MHT is a deferred decision tracker that propagates multiple hypotheses at the report-to-track assignment stage, e.g.; is it a true detection, a false detection, or a missed detection? The multiple track options are maintained over several update periods until a final decision, concerning the likelihood of a track, can be established and competing tracks deleted. The report-to-track assignment is a multi-dimensional process incorporating target dynamic information (range, range rate, and azimuth) as well as rank information (radar cross section).

Track filtering is achieved using an AEKF. The adaptive weights of the AEKF permit the filter to tune itself to accommodate varying target dynamics without compromising the tracker performance for non-accelerating targets. Optimum performance is achieved by including the maneuvering hypothesis as part of the MHT.

Association of real-time data at the sensor level is accomplished in the report-to-track assignment process. The process incorporates the characteristics of the individual sensors and weights the reports accordingly. That is, an externally reported (e.g., by an independent server) position report being considerably more accurate than an HFSWR report will be assigned a higher weight. Sensor reports that occur on a non-timely basis are correlated at the track level and are used to provide track identification and track management.

Either simulated or established, HFSWR shall be capable of continuously monitor the region and shall be capable of picturing activities as shown in Fig. 6.33. Here, Cape Race HFSWR's picture of the Canadian East Coast System is presented.

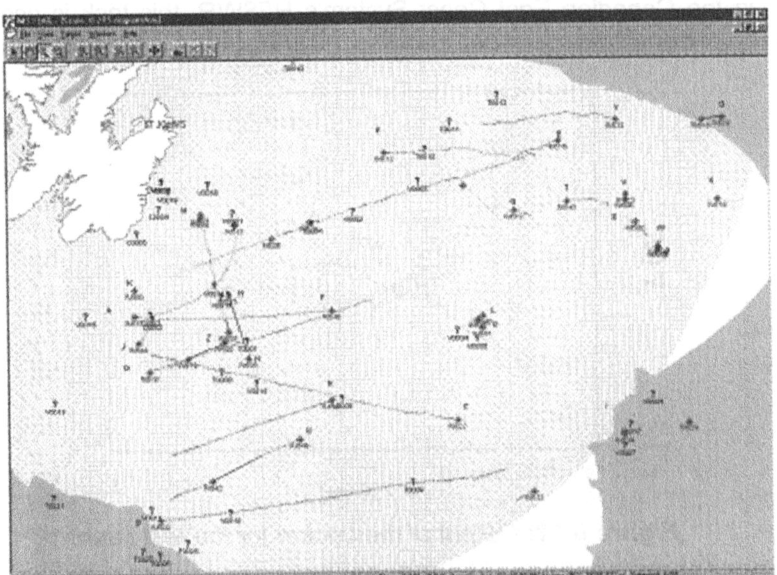

Figure 6.33: *Real HFSWR picture (Cape Race, June 9, 2000)*

6.6 INTEGRATED SURVEILLANCE, MONTE CARLO SIMULATION

Simulation is experimentation with models for design, research, training, education, test, etc. The primary reason in building and using an integrated surveillance system is that simulations allow to investigate and/or understand systems that do not exist (yet) or cannot be conveniently used for experimentation. The simulation requirements for an integrated surveillance system are [4]

- Construction and maintenance of a recognized activity (i.e., the positions, speeds, bearings, classes, etc.) of all targets of interest in a given area

- Flow of traffic, considering the effects of the environment, density, rules and regulations along a waterway, along a highway, in an urban area, etc., towards control and management or safety and efficiency

System level simulation of an integrated surveillance system requires

- Understanding and modeling (either deterministic or stochastic, mathematical or numerical) of different types of sensors with different sets of parameters (such as power, frequency, coverage, capacity, accuracy, etc.)

- Repetitious runs in order to gain confidence and/or establish the sensitivity of the simulator output data on variations in each of the assumed parameters or conditions

- A trade-off between simulation run time and model accuracy in an intelligent way to balance model fidelity and intended applications

- A design decision making on a choice between on-line or off-line simulations partially as stand-alone sensors or sub systems

In the simulation of an integrated surveillance system, sensors and software are generally developed for specific uses and under variety of assumptions, approximations, and limitations. Therefore, it is a challenge to integrate them within domains of their applicability.

Consider a coastal region that extends hundreds of kilometers in range and cross-range. Hundreds of targets may be present in this region at any time. The targets may be military or commercial, small or large, slow or fast, isolated or grouped, moving along a fixed route or having a random route. The mission is to monitor the activity within this region continuously under all-whether conditions. The solution is integrated maritime surveillance (IMS). A typical IMS system fictitiously designed for Turkish maritime surveillance based on HFSWR is pictured in Fig. 6.34. Here, coastal regions in Black, Agean, and Mediterranean seas are covered up to 200 nm ranges with six HFSWR. The surveillance scenario may even be more complex when other types of sensors (MW radars, IR sensor, daytime cameras, etc.) are used. The sensors may be fixed or mobile on land-based, airborne, or maritime vehicles. When we integrate all these sensors with different operating parameters, environmental conditions are a real challenge. The problems may be listed as follows:

- Design (available sensors, optimum number of each type sensors, optimum sensor locations, etc.)

- Software (detection and tracking algorithms, level of sensor or data fusion, etc.)
- System integration (communication between sub-systems, input/output relations, etc.)
- Communication (types of communication, data flow over communication lines, etc.)
- Test (calibration, verification and validation)
- Performance evaluation (performance, success/failure criteria)

Suppose all these challenging problems are handled and the integrated surveillance simulator has been established. Now, how is it going to be played? How are typical test scenarios, realistic run time scenarios going to be prepared? More importantly, how is the system going to be tested and how is its performance going to be evaluated? What will the performance criteria be?

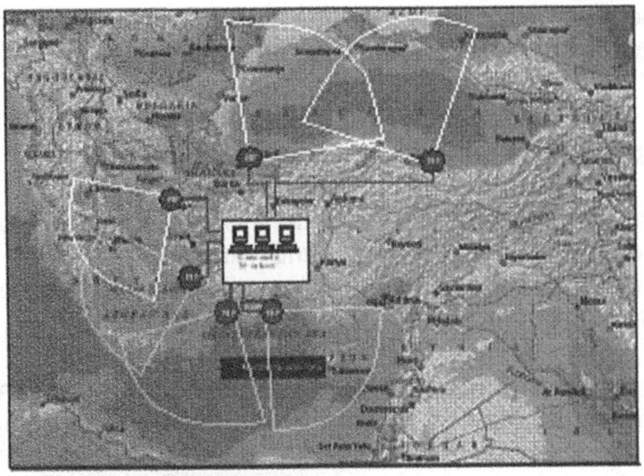

Figure 6.34: *A typical scenario for an integrated surveillance system*

The way to answer all these questions and satisfy the requirements is the Monte Carlo simulation. Monte Carlo simulations have been around for more than fifty years, but only with advances in computer technology have their applications been expanded. Monte Carlo is a mathematical model for computing the odds or probability of an outcome by testing thousands upon thousands of possible results. The mathematics behind Monte Carlo simulation came out of the Manhattan Project to build the atomic bomb during World War II. Monte Carlo simulation is based on generating millions of different scenarios by randomly changing each parameter in each run, and then see what happens.

It should be remembered that Monte Carlo is not a certainty test. It is a probability test. It gives *a feeling* about a possible result. Thinking about probability is not that easy. For example, what is the difference between a 70 % and a 80 % chance of achieving the goal of the tests? Which is acceptable and why?

Monte Carlo simulations in IMS systems is to prepare hundreds of different scenarios, where hundreds of targets present with variety of different locations, speed, routes, etc., run them thousands upon thousands of times with variety of different parameter sets, and see detection, tracking, classification percentages both for individual case and for all in an average sense.

6.7 CONCLUSIONS AND DISCUSSION

HFSWR simulation is one of the most difficult physical and computation problem in EMs. As discussed in detail above, one needs to understand

- Propagation over spherical imperfect earth's surface including multi-mixed paths
- Antenna physics, realistic antenna arrays located on complex ground screen and digital beam forming capabilities, when the array is located along a coastal region over lossy ground
- Near- and long-term (daily, monthly, seasonal, and year-based) variations in environmental (galactic, atmospheric and man made) noise
- Sea clutter, under different atmospheric conditions
- Target reflectivity and fluctuations (RCS analysis)

which are all major subjects of EM. Moreover, signal environment, Doppler spectra must be carefully investigated in order to apply some CFAR techniques in target detection as well as target maneuvers to apply reliable tracking algorithms.

In this chapter, radar simulation is discussed. A stand-alone HFSWR simulation package related to the detection and tracking of surface targets is introduced. The software package includes

- modeling of the surface wave propagation in an efficient hybrid from and the path-loss calculation over the smooth spherical earth with finite conductivity including multi-section mixed paths and surface roughness
- stochastic modeling of the environmental noise, sea clutter and surface targets
- a scenario editor supported by a specially designed graphical user interface

The focus in this simulation is target detection and tracking in HFSWR. To reduce long simulation times, practical considerations are taken into account without loosing the reliability of analytical calculations as well as stochastic modeling. Empirical formulas based on long observations are used in the simulation package to speed up the simulation.

There are many existing HFSWR, which are used from ocean remote sensing to continuously monitoring wide areas. In Table 6.5, most of the existing HFSWR are listed together with their main parameters [27].

Table 6.5: *Current HFSWR and their typical properties*

SYSTEM	*Pulse*	*FM(I)CW*	*Transmit wide/beam*	*Direction finding*	*Beam forming*
CODAR	♦		w	♦	
OSCR	♦		w		♦
PISCES		(I)	w		♦
C-CORE		(I)	w		♦
COSRAD	♦		b		♦
SeaSonde		(I)	w	♦	
WERA		♦	w	♦	♦

References

[1] J. L. Eaves, E. K. Reedy, *Principles of Modern Radars*, Van Nostrand Reinhold, NY, 1987

[2] S. Kingsley, S. Quegan, *Understanding Radar Systems*, McGraw-Hill, London, UK, 1992

[3] A. M. Ponsford and R. E. Moutrays, "Integrated maritime surveillance of Canada's East Coast", *Proc. of WMO/IOC Workshop on Operational Ocean Monitoring Using Surface based Radars*, Rep. No. 32, pp. 120 - 127, 1995

[4] N. Ince, E. Topuz, E. Panayirci, C. Isik, *Principles of Integrated Maritime Surveillance Systems*, Kluwer Academic, Boston, 2000

[5] A. M. Ponsford, S. K. Srivastava, "A Comparison Between Measured and Modeled Radar Cross Section from a Ground Wave Radar", *Proc. of IEE 5th Int. Conf. on HF Systems and Techniques,* IEE339, pp. 61 - 65, Edinburgh, 1991

[6] S. K. Srivastava, A. M. Ponsford, "Long Range Detection of Icebergs using Ground Wave Radar", *Proc. of Int. Offshore and Polar Eng. Conf.*, Vol. II, pp. 471 - 476, Edinburgh, 1991

[7] R. Howell, J. Walsh, "Measurement of Ocean Wave Spectra Using Narrow Beam HF Radar", *Proc. of 12th Can. Symp. on Remote Sensing, IGARSS'89*, Vancouver, BC, Vol. 5, pp. 2969 – 2972, 1989

[8] B. J. Lipa, D. F. Barrick, "Extraction of Sea State from HF Radar Sea Echo", *Radio Sci.,* Vol. 21, pp. 81 -100, 1986

[9] L. R. Wyatt, "The Measurement of the Ocean Wave Directional Spectrum from HF Radar Doppler Spectra", *Radio Sci.,* Vol. 21, pp. 473 - 485, 1986

[10] K. Hasselman, "Determination of Ocean Wave Spectra from Doppler Radio Return from the Sea Surface", *Nature Phys. Sci.*, Vol. 229, pp. 16 - 17, 1971

[11] T. M. Georges, *et al.*, "Large-scale Mapping of Ocean Surface Currents with Dual over the Horizon Radars", *Nature,* Vol. *279* No. 6564, pp. 434 - 436, 1996

[12] N. A. Norton, "The Propagation of Radio Waves over the Surface of the Earth and in the Upper Atmosphere", *Proc. IRE*, Vol. 24, pp. 1367 - 1387, 1936
[13] L. Sevgi, L.B. Felsen, "A New Algorithm for Ground Wave Propagation Problems Based on a Hybrid Ray-Mode Approach", *Int. Journal of Numerical Modeling*, Vol.11 No. 2, pp. 8 - 103, March/April 1998
[14] J. Walsh, "On the Theory of Electromagnetic Propagation Across a Rough Surface and Calculations in the VHF Region", *Report N00232, OEIC, Memorial University*, Canada, 1980
[15] D. Barrick, "First Order Theory and Analysis of MF/HF/VHF Scatter from the Sea", *IEEE Transactions on Antennas and Propagat.*, Vol. 20, pp. 2 - 10, 1972
[16] E.D.R. Shearman, "Propagation and Scattering in MF/HF Groundwave Radar", *IEE Proc. F*, Vol. 130, pp. 579 - 590, 1983
[17] D.D. Crombie, "Doppler Spectrum of Sea Echo at 13.56 MCs", *Nature*, Vol.175, pp. 681 - 682, 1955
[18] M. Fuks, V. I. Tatarski, D. E. Barrick, "Behavior of Scattering from a Rough Surface at Small Grazing Angles", *Waves in Random Media*, Vol. 9 No. 3, pp. 295 - 305, July 1999
[19] D. E. Kerr (Ed.), "The Propagation of Short Radio Waves", Vol. 1.S, *Radiation Lab. Series*, McGraw-Hill, New York, 1951
[20] J. R. Wait, "On the Theory of Mixed-Path Groundwave Propagation on a Spherical Earth", *J. Res. Nat. Bur. Stand.*, 65D, pp. 681 - 691, 1961
[21] V. A. Fock, *Electromagnetic Diffraction and Propagation Problems*, Ch.10. pp. I91, and Ch.12, pp. 235. Pergamon, Oxford, 1965
[22] G. Millington, "Ground Wave Propagation over an Inhomogeneous Smooth Earth", *Proc. IRE*, Part III, Vol. 96 No.39, pp. 53 - 64, January, 1949
[23] CCIR Rec., "Characteristics and Applications of Atmospheric Radio Noise Data", *CCIR Rec. 322-3*, 1988
[24] C. Isik, L. Sevgi, *Atmospheric Noise and Spectrum Occupancy Measurements*, Technical Report, *UZUN UFUK Project Office*, Jul/Aug 1996
[25] ICEPAC, *HF Propagation Prediction Software Package*, NTIA/ITS, National Telecomm. and Information Adm., *Institute for Telecomm. Sciences*, Boulder, Colorado, Oct. 1998 release
[26] G. F. Gott, N. F. Wong, S. Dutta, "Occupancy Measurements Across the Entire HF Spectrum", *AGARD Conf. Publ.* 332, Paris 1982
[27] K. W. Gurgel, G. Antonischki, H.-H. Essen, T. Schlick, "Wellen radar (WERA), A New Groundwave Based HF Radar for Ocean Remote Sensing," *Coastal Engineering*, Vol. 37, pp. 219 - 234, 1999
[28] K. W. Gurgel, H.-H. Essen, S. P. Kingsley, "HF Radars: Physical Limitations and Recent Developments," *Coastal Engineering*, vol. 37, pp. 201-218, 1999
[29] E. F. Knot J. F. Shaeffer, M. T. Tuley, *Radar Cross Section*, Artech House. Norwood, MA, 1985
[30] R. F. Harrington, *Field Computation by Moment Methods*, Macmillan, New York, 1968
[31] K. S. Yee, "Numerical Solution of Initial Boundary Value Problems Involving Maxwell's Equations in Isotropic Media", *IEEE Transactions on Antennas and Propagat.*, Vol. 4, pp. 302 - 307, 1966

[32] P. B. Johns and R. L. Beurle, "Numerical Solution of Two-Dimensional Scattering Problems using TLM", Proc. IEE, Vol. 118, pp. 1203 - 1208, September 1971

[33] F. Akleman, L. Sevgi, "Radar Cross Section and Antenna Modeling with FDTD Method", *Proc. of JINA'98, 10th Int. Symposium on Antennas*, November 17 -19, 1998, France

[34] L. Sevgi, S. Paker, "FDTD Based RCS Calculations and Antenna Simulations", *AEU, International J. of Electronics and Communication*, Vol. 52 No.2, pp. 65 - 75, March 1998

[35] W. Truman, S. J. Kubina, S. R. Mishra, C. Larose,"Radar Cross-section of a Generic Aircraft at HF Frequencies", *Canadian J. Elect. Comp. Eng.*, Vol.18 No.2, pp. 59 - 62, 1993

[36] W. Truman, S. R. Mishra, S. J. Kubina, C. Larose, "RCS of Resonant Scatterers with Attached Wires", *IEEE Transactions on Antennas and Propagat.*, Vol. 41 No. 3, pp. 351 - 354, March 1993

[37] L. Sevgi, "Target Reflectivity and RCS Interactions in Integrated Maritime Surveillance Systems Based on HF Surface-wave Radars", *IEEE Antennas and Propagation Magazine,* Vol. 43 No.1, pp. 36 – 51, Feb. 2001

[38] L. Sevgi, "Stochastic Modeling of Target Detection and Tracking in Surface Wave High Frequency Radars", *Int. J. of Numerical Modeling*, Vol.11 No.3, pp. 167 - l81, May/June 1998

[39] L. Sevgi, A. M. Ponsford, H.C. Chan, "An Integrated Maritime Surveillance System Based on Surface Wave HF Radars, Part I – Theoretical Background and Numerical Simulations", *IEEE Antennas and Propagation Magazine*, Vol. 43 No.4, pp. 28 - 43, August 2001

[40] A. M. Ponsford, L. Sevgi, H.C. Chan, "An Integrated Maritime Surveillance System Based on Surface Wave HF Radars, Part II – Operational Status and System Performance", *IEEE Antennas and Propagation Magazine*, Vol.43 No.5, pp. 52 - 63, October 2001

[41] Hibernia, An oil exploration platform with approximately 213 m height and 110 m radius, located 315 km away from St. John's, Newfoundland, (http://www.hibernia.ca)

[42] L. Sevgi, A. S. Sanal, "Surface Wave HF Radar Simulator", *Proc. of the IEE Radar 97 Conference*, pp. 181 - 184, October 14 - 16, 1997, Edinburg, UK

A

Electromagnetic Fundamentals[1-3]

Some fundamental information related to the numerical methods included in the book is given in this Appendix.

A.1 Maxwell's Equations

The theory behind electromagnetic fields are well-established by Maxwell's equations. The vectors $\vec{E}$, $\vec{D}$, $\vec{H}$, and $\vec{B}$ are used for electric field [V/m], electric flux density [C/m^2], magnetic field [A/m], and magnetic flux density [Weber/m^2], respectively. The parameters σ, ε and μ (non-negative real numbers) are used as constitutive parameters of the medium under interest, and they are, respectively, the conductivity [S/m], permittivity (dielectric constant) [F/m], and permeability (magnetic constant) [H/m]. A medium is called to be "simple" when it is (i) homogeneous, (ii) linear, and (iii) isotropic, and in a simple medium, EM vectors are related to each other and also to the excitation current density as

$$\vec{D} = \varepsilon\vec{E}\ , \qquad \vec{B} = \mu\vec{H}\ , \qquad \vec{J} = \sigma\vec{E} + \vec{J}_v \tag{A.1}$$

The excitation of EM fields are expressed via electric current density $\vec{J}$ [A/m^2] and electric charge density ρ [C/m^3], and electric current density has conduction and convection components as $\vec{J} = \vec{J}_c + \vec{J}_v$. Conduction and convection current densities are also expressed as $\vec{J}_c = \sigma\vec{E}$ and $\vec{J}_v = \rho\vec{v}$, where $\vec{v}$ is the velocity of the charge.

In simple media, Maxwell's equations are given as

$$\nabla \times \vec{E} = -\mu \frac{\partial}{\partial t}\vec{H} \tag{A.2}$$

$$\nabla \times \vec{H} = \vec{J}_v + \sigma\vec{E} + \varepsilon\frac{\partial}{\partial t}\vec{E} \tag{A.3}$$

$$\nabla \bullet \vec{E} = \frac{\rho}{\varepsilon} \tag{A.4}$$

$$\nabla \bullet \vec{H} = 0 \tag{A.5}$$

Maxwell's equations can be summarized in words as follows:

- *An electric field is produced by a time-varying magnetic field.*
- *A magnetic field is produced by a time-varying electric field or by an electric current.*
- *Electric field lines start and end on charges and are continuous.*
- *Magnetic field lines are continuous.*

The rotational equations (A.2) and (A.3) are first order coupled equations. They relate spatial behavior of a field component (e.g., $\vec{E}$) to the time behavior of the other component (e.g., $\vec{H}$) in a medium specified by its parameters (σ, μ, ε) in terms of sources (e.g., $\vec{J}$). Here, in general, vector fields are four-dimensional; that is, $\vec{E}=\vec{E}(\vec{r},t)$ and $\vec{H}=\vec{H}(\vec{r},t)$ ($\vec{r}$=(x,y,z) in rectangular coordinate system).

Maxwell's equations can be re-arranged, and de-coupled equations can be obtained where only $\vec{E}$ or $\vec{H}$ appears. By applying $\mu\partial/\partial t$ to (A.2) first and then adding (A.3) to the result (with the help (A.4)), one obtains

$$\nabla\times\nabla\times\vec{E}-\varepsilon\mu\frac{\partial^2\vec{E}}{\partial t^2}-\sigma\mu\frac{\partial\vec{E}}{\partial t}=\frac{1}{\varepsilon}\nabla\rho+\mu\frac{\partial\vec{J}}{\partial t} \quad \text{(A.6)}$$

Similarly, applying $\varepsilon\partial/\partial t$ to (A.3) and adding with (A.2) (also by using (A.5)) results in

$$\nabla\times\nabla\times\vec{H}-\varepsilon\mu\frac{\partial^2\vec{H}}{\partial t^2}-\sigma\mu\frac{\partial\vec{H}}{\partial t}=-\nabla\times\vec{J}_v \quad \text{(A.7)}$$

Either eqs. (A.2) and (A.3) or (A.6) and (A.7) are used to determine $\vec{E}$ and/or $\vec{H}$ in terms of the sources ($\vec{J}_v$ and/or ρ) when boundary conditions of the chosen problem geometry in spatial domain and initial conditions in time domain are specified.

In free space (A.6) and (A.7) reduce to classical wave equations

$$\nabla^2\vec{E}-\varepsilon_0\mu_0\frac{\partial^2\vec{E}}{\partial t^2}=0 \quad \text{(A.8)}$$

$$\nabla^2\vec{H}-\varepsilon_0\mu_0\frac{\partial^2\vec{H}}{\partial t^2}=0, \quad \text{(A.9)}$$

where

$$\varepsilon_0=\frac{1}{36\pi\times10^9} \quad \text{[Farad/meter]} \quad \text{(A.10)}$$

$$\mu_0=4\pi\times10^{-7} \quad \text{[Henry/meter]} \quad \text{(A.11)}$$

and represent waves progressing with a wave speed of

$$c=\frac{1}{\sqrt{\varepsilon_0\mu_0}} \quad \text{[m/s]} \quad \text{(A.12)}$$

A.2 Boundary Conditions

On the interface separating two different media, none of which is a perfect electrical conductor (PEC), the tangential components of the electric and of the magnetic fields are continuous, as expressed by

$$\vec{n}\times\left(\vec{E}_1-\vec{E}_2\right)=0 \qquad \text{[V/m]} \tag{A.13}$$

$$\vec{n}\times\left(\vec{H}_1-\vec{H}_2\right)=0 \qquad \text{[A/m]} \tag{A.14}$$

where $\vec{n}$ is the unit vector normal to the interface, directed from medium 2 towards medium 1, and the indices 1 and 2 specify the medium within which the fields are defined, respectively. On the surface of a PEC, electric field vector must meet the condition

$$\vec{n}\times\vec{E}=0 \qquad \text{[V/m]} \tag{A.15}$$

Similarly, on the surface of a perfect magnetic conductor (PMC) (without surface current), the magnetic field vector must satisfy the condition

$$\vec{n}\times\vec{H}=0 \qquad \text{[A/m]} \tag{A.16}$$

Boundary conditions on the normal components of the electric and magnetic fields yield

$$\vec{n}\bullet\left(\vec{D}_1-\vec{D}_2\right)=\rho_s \qquad \text{[V/m]} \tag{A.17}$$

$$\vec{n}\bullet\left(\vec{B}_1-\vec{B}_2\right)=0 \qquad \text{[A/m]} \tag{A.18}$$

Surface charges ρ_s can only be found on the interface between two media when at least one of the two has a non-zero conductivity.

A.3 Radiation Condition and Edge Condition

In an open (unbounded) space with all EM sources confined in a finite region, the constraint that governs the behavior of the fields at infinity is stated in terms of *radiation boundary condition* (RBC). RBC can be stated in one of two ways: (i) If the medium is lossy, then fields must vanish as the observer goes to infinity. (ii) If the medium is lossless and isotropic, then the Sommerfeld radiation condition is used as RBC. In this case, as the observer moves away the source of the field has a phase progressing outward and has an amplitude that decreases at least as rapidly as r^{-1}. That is, any transverse component ψ of the field (with respect to r direction) must satisfy the condition

$$\lim_{r\to\infty} r\left\{\frac{\partial\psi}{\partial r}-ik\psi\right\}=0 \tag{A.19}$$

where $k=\omega\sqrt{\varepsilon\mu}$ is the propagation constant of the medium.

Another situation where Maxwell's equations may not be unique arises when the configuration of the problem contains geometrical singularities, such as sharp edges and tips. In this case an additional physical condition is used, known *as the edge condition*. It is the requirement that the electric and magnetic energy stored in a close proximity of the singularity must be finite; that is,

$$\int_{v}\left(\varepsilon\left|\vec{E}\right|^2+\mu\left|\vec{H}\right|^2\right)dv\rightarrow 0 \tag{A.20}$$

as the volume v contracts to the neighborhood of the edge or tip.

A.4 Duality Relations

Both first-order, coupled and second-order, de-coupled Maxwell's equations have symmetry in terms of parameters and variables. Any equation between the fields and sources can be used to directly obtain equations describing the complementary fields and sources. This is known as the duality principle and dual quantities are listed in Table A.1. $\vec{K}$, ρ_e, and ρ_m are magnetic current density, electric charge density and magnetic charge density, respectively.

Table A.1: Dual quantities in EM (second row is the dual of the first one)

$\vec{E}$	$\vec{H}$	$\vec{J}$	$\vec{K}$	ε	μ	ρ_e	ρ_m
$\vec{H}$	$-\vec{E}$	$\vec{K}$	$-\vec{J}$	μ	ε	ρ_m	$-\rho_e$

A.5 Time-Harmonic Case

General behavior of EM waves as a function of time can always be expressed as superposition of waves at different frequencies through a Fourier transform. It is therefore sufficient to investigate the characteristics of a wave at a single frequency in most of the cases. The wave at a single frequency is called a *time-harmonic* or *monochromatic wave*. For example, vector electric field $\vec{E}(\vec{r},t)$, a real function of position $\vec{r}$ and time *t*, is given by

$$\vec{E}(\vec{r},t)=\mathrm{Re}\left\{\vec{E}(\vec{r})e^{-i\omega t}\right\}\qquad \text{[V/m]} \tag{A.21}$$

where $\vec{E}(\vec{r})$ is a phasor field (peak value), in general complex. In this case, Maxwell's equations can be re-written by just replacing $\partial/\partial t\rightarrow -i\omega$. This is also true for

$$\vec{E}(\vec{r},t)=\mathrm{Re}\left\{\vec{E}(\vec{r})e^{+i\omega t}\right\}\qquad \text{[V/m]} \tag{A.22}$$

with $\partial/\partial t\rightarrow +i\omega$. For example, the first two Maxwell's equations in (A.2) and (A.3) become

$$\nabla\times\vec{E}=i\omega\mu\vec{H} \tag{A.23}$$

$$\nabla\times\vec{H}=\vec{J}_v+\sigma\vec{E}-i\omega\varepsilon\vec{E}=-i\omega\varepsilon_0\left(\varepsilon_r-i\frac{\sigma}{\omega\varepsilon_0}\right)\vec{E} \tag{A.24}$$

under exp(−iωt) time dependence. Here,

$$\tilde{\varepsilon}_r = \left(\varepsilon_r - i\frac{\sigma}{\omega\varepsilon_0} \right) \tag{A.25}$$

is the complex relative permittivity.

A.6 Scattered Field Formalism

Electric and magnetic fields can be separated into two parts, namely, incident plus scattered fields,

$$\vec{E}(\vec{r},t) = \vec{E}_{inc}(\vec{r},t) + \vec{E}_{scat}(\vec{r},t) \qquad \text{[V/m]} \tag{A.26}$$

and it can be very useful, especially in problems such as RCS simulations, GPR modeling, etc. In these numerical simulation problems, incident fields can be represented analytically throughout the discrete problem space, therefore only the scattered field components are calculated numerically, which reduces numerical burden as well as error. Scattered fields arise on and within the interacting object and boundary conditions should be changed appropriately-for example, on and within a PEC object

$$\vec{E}_{scat}(\vec{r},t) = -\vec{E}_{inc}(\vec{r},t) \qquad \text{[V/m]} \tag{A.27}$$

For anything other than a PEC object, the scattered fields depend on the constitutive parameters of the material.

The two rotational Maxwell's equations in free space, for example, can be given as

$$\nabla\times\vec{E}_{\text{scat}}(\vec{r},t) = -\mu_0\frac{\partial}{\partial t}\vec{H}_{\text{scat}}(\vec{r},t) - \left[(\mu-\mu_0)\frac{\partial}{\partial t}\vec{H}_{\text{inc}}(\vec{r},t)\right] \tag{A.28}$$

$$\begin{aligned}\nabla\times\vec{H}_{\text{scat}}(\vec{r},t) &= \varepsilon\frac{\partial}{\partial t}\vec{E}_{\text{scat}}(\vec{r},t) + \sigma\vec{E}_{\text{scat}}(\vec{r},t)\\ &\quad + \left[(\varepsilon-\varepsilon_0)\frac{\partial}{\partial t}\vec{E}_{\text{inc}}(\vec{r},t) + \sigma\vec{E}_{\text{inc}}(\vec{r},t)\right]\end{aligned} \tag{A.29}$$

with ε, μ, and σ inside the object and ε_0, μ_0, and $\sigma = 0$ outside in free space.

A.7 Wave Polarization

The polarization of EM waves is defined by the geometrical locus of the electric field vector, because it varies during one wave period at a fixed point in space. In the most general case, EM waves are *elliptically polarized.* There are two limiting cases:

- *Linear polarization*: If the tip of the electric field vector moves along a straight line, then the polarization is said to be linear. The two cases of linear polarization are the vertical and horizontal polarizations. This is best understood when propagation over the earth's surface is the case. The polarization is said to be vertical (horizontal) when the electric field vector is perpendicular (parallel) to the Earth's surface.

- *Circular polarization*: When the locus of the tip of the electric field vector is a circle, then the polarization is said to be circular. Or, when vertically and horizontally polarized waves are combined with a 90° phase difference, the resultant wave will be circularly polarized. Circularly polarized waves are mostly used in satellite communication, since they can be generated and received using antennas which are oriented in any direction.

A.8 EM Sources

There are two basic sources of EM excitation: electric and magnetic. The simplest forms of these two sources are the electric dipole and the magnetic dipole.

- *Electric dipole*: It is a short wire carrying time-varying electric current. It has a strong electric and weak magnetic fields (i.e., high wave impedance) in the near vicinity.
- *Magnetic dipole*: It is a small wire loop with time-varying electric current. It has a strong magnetic and weak electric fields (i.e., low wave impedance) in the near vicinity.

Within a wavelength or two (with respect to the source), both dipoles have similar wave characteristics which is called a plane wave.

A.9 Plane-Wave Assumption

A point source radiates EM fields isotropically and its power radiates spherically. It has a spherical wavefront. In most of EM problems, such as antenna and RCS analyzes, far fields are of interest and EM wavefront may be assumed locally plane. This is true when the distance between EM source and the observer goes to infinity.

A plane wave is a non-physical wave where electric field, magnetic field and direction of propagation are perpendicular to each other. In free space, it has a *wave impedance* of

$$Z_0 = \frac{|\vec{E}|}{|\vec{H}|} = 120\pi = 377 \qquad [\Omega] \tag{A.30}$$

which is also called the *characteristic impedance* of the free space.

A.10 EM Power and Force

Maxwell's equations satisfy energy conservation. This can be illustrated by the Poynting's theorem, which can be written in free space as

$$\nabla \bullet (\vec{E} \times \vec{H}) = -\frac{\partial}{\partial t}\left[\frac{1}{2}\mu_0 \vec{H}.\vec{H} + \frac{1}{2}\varepsilon_0 \vec{E}.\vec{E}\right] - \vec{E}.\vec{J} \qquad \text{[joule]} \tag{A.31}$$

where

$$W_e = \frac{1}{2}\varepsilon_0 \vec{E}.\vec{E} \qquad \text{[joule]} \tag{A.32}$$

and

$$W_m = \frac{1}{2}\mu_0 \vec{H}.\vec{H} \qquad \text{[joule]} \tag{A.33}$$

are the stored electric and magnetic energies in the medium, respectively. Poynting's theorem states that a divergence of EM power density from a small volume is equal to the negative time rate change of the stored electric and magnetic densities inside the volume plus the power $-\vec{E}.\vec{J}$ supplied by the electric current source $\vec{J}$. Applying the divergence theorem to the Poynting's theorem yields

$$\oiint_S (\vec{E} \times \vec{H}) d\vec{S} = -\frac{\partial}{\partial t} \iiint_V \left[\frac{1}{2} \mu_0 H^2 + \frac{1}{2} \varepsilon_0 E^2 \right] - \iiint_V \vec{E}.\vec{J}\, d\vec{V} \qquad \text{(A.34)}$$

The left-hand side represents power flow out to the surface enclosing the volume *V*. The first term on the right-hand side represents the depletion of the electric energy and magnetic energy inside the volume in order to supply the outflow of the Poynting power. The last term is the power generated by the source $\vec{J}$ inside the volume.

Line integral of electric field is defined as the electromotive force (EMF)

$$EMF = \oint_C \vec{E}\, d\vec{l} = -\frac{\partial}{\partial t} \iint_A \vec{B}\, d\vec{s} = -\frac{\partial}{\partial t} \psi \qquad [V] \qquad \text{(A.35)}$$

where ψ is the magnetic flux linking a loop with area *A* bounded by a closed contour C. Equation (A.35) states that the EMF is equal to the negative time derivative of the magnetic flux linking the loop. This means that if ψ increases the EMF decreases the flux, and vice versa, which is known as Len's law. The EMF has unit of voltage and the voltage drop around the loop *V* is equal to the negative of the induced EMF

$$V = -EMF = \frac{\partial}{\partial t} \psi \qquad [V] \qquad \text{(A.36)}$$

A.11 Equivalence Principle

In most EM problems, the space is required to be divided into two parts: "a finite region of interest" and "all other regions of no interest." The effects of "all other regions of no interest" can be replaced with equivalent EM sources. This is called the equivalence principle.

A volume with EM sources and/or EM scatterers may be isolated from the outside. In this case EM activity outside this volume can be determined one-to-one by using the tangential components of electric and magnetic fields (caused by the sources inside) at the surface of the volume. This is known as the equivalence principle and can be given via the Stratton-Chu equations as

$$\begin{aligned} \vec{E}_{\text{scat}}(\vec{r},t) = & -\frac{\mu}{4\pi r} \int_S \vec{e}_r \times \left(\frac{\partial}{\partial t} \vec{J}(t-\tau) \times \vec{e}_r \right) dS' \\ & -\frac{1}{4\pi r c} \int_S \left(\frac{\partial}{\partial t} \vec{M}(t-\tau) \times \vec{e}_r \right) dS' \end{aligned} \qquad \text{(A.37)}$$

$$\vec{H}_{\text{scat}}(\vec{r},t) = \frac{1}{4\pi rc}\int_S\left(\frac{\partial}{\partial t}\vec{J}(t-\tau)\times\vec{e}_r\right)dS' - \frac{\varepsilon}{4\pi r}\int_S \vec{e}_r\times\left(\frac{\partial}{\partial t}\vec{M}(t-\tau)\times\vec{e}_r\right)dS' \quad \text{(A.38)}$$

where,

$$\vec{J}_S = \vec{n}\times\vec{H} \quad \text{(A.39)}$$

$$\vec{M}_S = -\vec{n}\times\vec{E} \quad \text{(A.40)}$$

are the equivalent surface electric and magnetic currents, respectively. In (A.37) and (A.38) r is the radial distance to the point of equivalent currents at the surface *S* that encloses the volume *V* (*c* is the speed of light in free space).

Stratton - Chu equations in TD in (A.37) and (A.38) are used for extrapolating far fields from the simulated near fields inside a finite discrete volume, such as in FDTD and TLM.

A.12 Strategies for Wave Field Solutions

The strategy in solving Maxwell equations under given boundary and initial conditions and sources is a problem dependent. It is quite different for an antenna radiation in open space than a radiating element inside a waveguide or for a radar cross-section problem. In most of EM engineering problems there are two approaches:

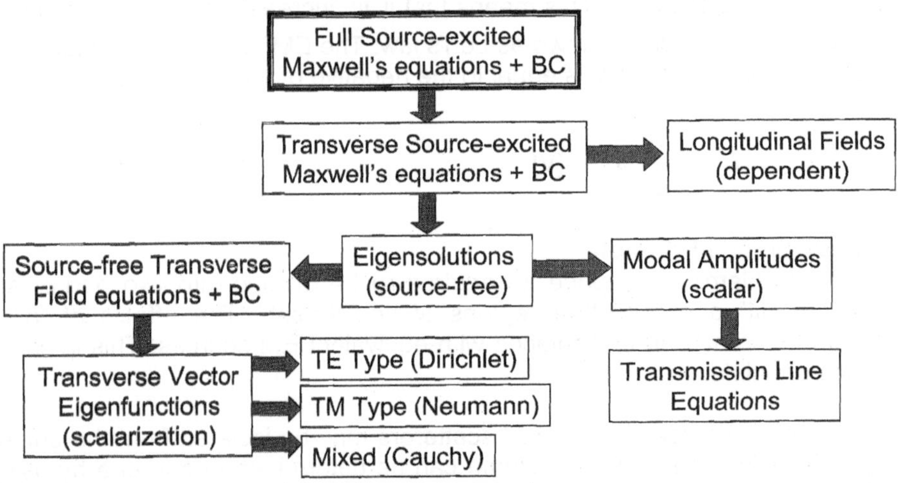

Figure A.1: *Transverse and longitudinal reduction of Maxwell's equations in waveguding environments, in which a symmetry occurs along one direction*

- Approach 1: Reduce and solve the problem in terms of transverse and longitudinal components for most of the waveguding environments (see Fig. A.1). Here, the problem not only reduced but also is solved.

- Approach 2: Simplify the full-wave equations at the beginning by using vector and scalar potentials. Use vector potential $\vec{A}$ and scalar potential U. Then scalarization of these auxiliary functions is the second step. In this case, one needs to use equations that relate vector/scalar potentials to the electric and magnetic fields.

For example, rectangular waveguides, resonators, and microstrip networks are better handled via Approach 1. On the other hand, Approach 2 is used in most of antenna radiation and RCS problems.

A.13 Sturm - Liouville Problem

Many of the boundary value problems in EM lead to the Sturm - Liouville equation. The Sturm - Liouville equation is of the form of

$$\left[\frac{d}{dz} p(z) \frac{d}{dz} + q(z) + \lambda w(z)\right] g(z, z'; \lambda) = -\delta(z - z') \tag{A.41}$$

and the boundary conditions at z_1 and z_2 ($z_1 < (z, z') < z_2$). Here, p,q,w are real (in general functions of z) and correspond to geometrical quantities of the problem, and λ is a spectral parameter (not the wavelength). The three common boundary conditions are

- $\psi = 0$ at $z = z_1$ and z_2 (DIRICHLET TYPE)
- $d\psi / dz = 0$ at $z = z_1$ and z_2 (NEUMANN TYPE)
- $\psi + \alpha d\psi / dz = 0$ at $z = z_1$ and z_2 (CAUCHY TYPE)

where α is a parameter.

Sturm - Liouville equation in (A.41) (see Fig. (A.1)) models:

- Transmission line system directly if λ is a given fixed parameter.
- An eigenvalue problem (when the equation is homogeneous, i.e., right-hand side is zero) where λ is a free parameter (that CANNOT be an eigenvalue). So, the condition for uniqueness for this case can easily be satisfied when λ is assumed complex, ie., Im $\{\lambda\} \neq 0$ (since eigenvalues are real when p, q, are w are real).

Once the boundary conditions are specified, the homogeneous Sturm - Liouville equation

$$\left[\frac{d}{dz} p(z) \frac{d}{dz} + q(z) + \lambda w(z)\right] \psi = 0 \tag{A.42}$$

has an infinite set of solutions $\psi = \psi_n$ that correspond to particular values of $\lambda = \lambda_n$. The functions ψ_n are called *eigenfunctions* and the values λ_n are called *eigenvalues*. In all cases, ψ_n form an orthogonal set of functions over the finite interval of $z_1 < z < z_2$ and orthogonality is given as

$$\int_{z_1}^{z_2} w(z) \psi_n \psi_m = 0 \tag{A.43}$$

i.e., ψ_n form an orthogonal set with respect to the weighing function $w(z)$. Each of the functions in the orthogonal set better be unit energy, so they are also normalized. Orthonormality condition is then given as

$$\int_{z_1}^{z_2} w(z)\psi_n\psi_m = \begin{cases} 1 & n = m \\ 0 & n \neq m \end{cases} \tag{A.44}$$

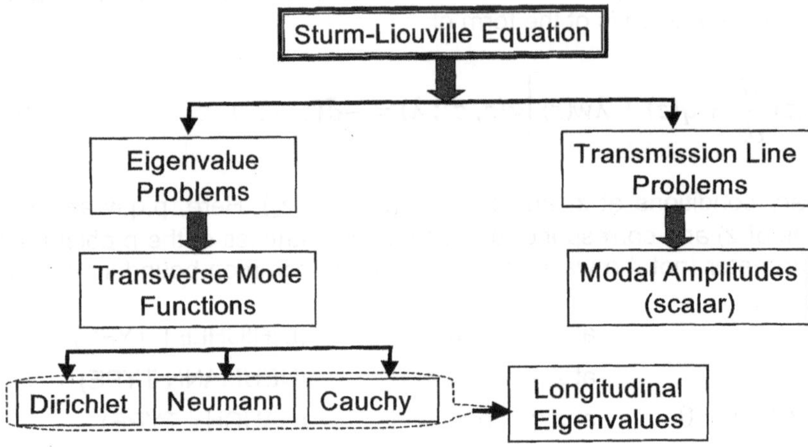

Figure A.2: *Sturm-Liouville equation and its representations*

Superposition of source-free solutions (eigenfunctions) must form a complete set. A set of functions defined over some interval and subject to stated boundary conditions at its ends is said to be complete, if any function in this interval can be expressed as a linear combination of them. Since the most peculiar function is the delta function, a complete set of functions is the set that can be represented in terms of delta functions:

$$\frac{\delta(z - z')}{w(z')} = \sum_n \psi_n(z)\ \psi_n{}^*(z') \tag{A.45}$$

where (*) denotes complex conjugate.

It should be noted that solutions which are in series forms should be convergent (also well convergent) so only a few terms are adequate to represent wave fields.

A.14 Green's Function Representation

The Green's function $g(z,z';\lambda)$ is the response of a linear system (solution of the Sturm-Liouville equation) to a point source of unit strength. A source of unit strength at the position $z = z'$ can be represented by the Dirac delta function as given in (A.45) above.

When boundary conditions are specified, the Green's function can be built via either

$$g(z,z';\lambda)=\sum_{n=1}^{\infty}\frac{\psi_n(z)\,\psi_n{}^*(z')}{\lambda-\lambda_n} \tag{A.46}$$

or

$$g(z,z')=-\frac{f_1(z_<)f_2(z_>)}{p(z')W(z')} \tag{A.47}$$

where $z_<$ is the lesser of z or z' and $z_>$ is the greater of z or z'. $W(z')$ is called Wronskian, is a global constant, and can be calculated from

$$W=\left(f_1(z)\frac{df_2(z)}{dz}-f_2(z)\frac{df_1(z)}{dz}\right)\text{ at a point between } z_1 \text{ and } z_2 \tag{A.48}$$

Constructing the Green's function is adequate to solve Sturm - Liouville function for any kind of excitation under specified boundary conditions, since the excitation can be represented as a superposition of point sources.

A.15 Wave Equation in 1D

Assume $u(x,t)$ is any component of electric or magnetic fields and depends only one spatial coordinate (x) and time (t). In lossless and source-free region

$$\frac{\partial^2}{\partial x^2}u(x,t)-\frac{1}{c^2}\frac{\partial^2}{\partial t^2}u(x,t)=0 \tag{A.49}$$

represent 1D wave equation. General solution of $u(x,t)$ is given as

$$u(x,t)=u_+(x-ct)+u_-(x+ct) \tag{A.50}$$

which is nothing but the superposition of waves traveling positive x direction (u_+) and negative x direction (u_-) with the speed of light (c). If the sources are far on the left (i.e., at infinity along negative x axis) then only one way traveling $u_+(x-ct)$ wave (along positive x direction) will be present, and vice versa.

A.16 Stoke's Theorem

Any surface integral can be reduced to a line integral around the rim of the surface. In terms of electric fields, this is stated as

$$\int_S\left(\nabla\times\vec{E}\right)\vec{n}dS=\oint_C\vec{E}\,d\vec{l} \tag{A.51}$$

where $\vec{n}$ is the unit vector pointing out of the surface S.

A.17 Divergence Theorem (Gauss Law)

The divergence theorem relates the divergence of a function within some region to the values of that function on the boundary of that region. If, for example, V is a volume of space and S is the surface enclosing this volume, then

$$\iiint_V \left(\nabla \bullet \vec{E}\right) dv = \oint\!\!\oint_S \left(\vec{n} \bullet \vec{E}\right) dS \tag{A.52}$$

where $\vec{n}$ is the unit vector pointing out of the surface S. This is known as the Gauss law and it states that the electric flux through any closed surface is proportional to the amount of electric charge contained within that surface.

A.18 Amprere's Law

The algebric sum of the currents passing through a surface enclosed by a closed line C is determined by the magnetic flux along that line:

$$\oint_C \vec{H} d\vec{l} = \Sigma I \tag{A.53}$$

A.19 Well-Posed Problem

A problem is said to be well-posed if the following conditions are met:

(a) *Existence*; implies that the conditions do not over determine the problem and the solution exists.
(b) *Uniqueness*; implies that the conditions do not under determine the solution and it assures that the solution, no matter how obtained, is the correct and the only one.
(c) *Stability*; implies that arbitrary small perturbations in data, e.g., sources, do not, for any physical quantity, let to infinity.

References

[1] L. B. Felsen, N. Marcuvitz, *Radiation and Scattering of Waves*, Prentice-Hall, New Jersey, 1973
[2] R. E. Collin, *Field Theory of Guided Waves*, McGraw Hill, New York, 1960
[3] A. Ishimaru, *Electromagnetic Wave Propagation, Radiation and Scattering*, Prentice-Hall, Englewood Cliffs, NJ, 1991

B

Tables of Computer Codes

Table B.1: GRWAVE package

CODE	GRWAVE
TECHNIQUE	• Based on Norton (ray) and Wait (mode) formulations
ASSUMPTIONS	• Smooth spherical earth • Wave equation in 3D spherical coordinates is reduced to 2D Parabolic form under azimuthal symmetry assumption • Homogeneous atmosphere above earth's surface • Heights and distances are much smaller than earth's radius • Vertically polarized EM waves
APPROXIMATIONS	• Ray approximations in interference region (within the LOS) • Transmitter and receiver heights are included as height gain functions in diffraction region
CAPABILITIES	• Can handle propagation over smooth earth's surface • Can handle propagation under Dirchlet-, Neumann-, and Cauchy (impedance)-type boundary conditions • Direct, ground-reflected and surface contributions can be obtained separately • Surface-wave contribution up to diffraction region when transmitter/receiver are located on the ground
VALIDITY RANGE	• Frequency: MF, HF, VHF, UHF, MW • Distance: within LOS and in diffraction (shadow) region • Height: within a few kilometers above the ground • Medium: for any permittivity and conductivity of the ground and ($\mu = \mu_0$)
CANONICAL TEST CASES	• Longitudinal interference effects of direct and ground-reflected waves at MW frequencies within LOS • Transverse interference effects of direct and ground-reflected waves at MW frequencies within LOS
COMPARISONS	• SSPE (split-step parabolic equation) technique • TDWP (time-domain wave propagator) technique
COMPUTER REQUIREMENTS	• None for today's computers
COMPLEXITY	• Easy to use • Can't be extended to other kind of propagation problems by the user • Requires analytical knowledge
PACKAGE	• Executable package is included

Table B.2: HFMIX package

CODE	HFMIX
TECHNIQUE	• Based on Norton (ray) and Wait (mode) formulations
ASSUMPTIONS	• Smooth spherical earth • Wave equation in 3D spherical coordinates is reduced to 2D Parabolic form under azimuthal symmetry assumption • Homogeneous atmosphere above earth's surface • Heights and distances are much smaller than earth's radius • Vertically polarized EM waves
APPROXIMATIONS	• Ray approximations in interference region (within the LOS) • Transmitter and receiver heights are included as height gain functions in diffraction region
CAPABILITIES	• Can handle propagation over smooth earth's surface • Can handle propagation under Dirchlet-, Neumann-, and Cauchy (impedance)-type boundary conditions • Surface-wave contribution beyond diffraction region when transmitter/receiver are located on the ground • Mixed-path calculations for surface-wave path losses
VALIDITY RANGE	• Frequency: MF, HF, VHF, UHF, MW • Distance: within LOS and in diffraction (shadow) region • Height: within a few kilometers above the ground • Medium: for any permittivity and conductivity of the ground and ($\mu = \mu_0$) • Multi-mixed paths can be chosen only with two different ground parameter sets (e.g., sea/land/sea/land, or, lake/sea/lake, etc.)
CANONICAL TEST CASES	• MF and HF path-loss calculations over land and/or sea (compare against CCIR curves) • HF mixed-path calculations (Millington effect), sea/land transitional effects
COMPARISONS	• SSPE (split-step parabolic equation) technique • TDWP (time-domain wave propagator) technique
COMPUTER REQUIREMENTS	• None for today's computers
COMPLEXITY	• Easy to use • Can't be extended to other kind of propagation problems by the user • Requires analytical knowledge
PACKAGE	• Executable package is included

Table B.3: Split-step parabolic equation (SSPE) package

CODE	SSPE
TECHNIQUE	• Based on a marching FFT solution of 2D parabolic wave equation in rectangular coordinates
ASSUMPTIONS	• Flat earth (i.e., 2D rectangular coordinates) • Wave equation in 3D spherical coordinates is reduced to 2D parabolic form under azimuthal symmetry assumption • Earth's curvature may be included as refractivity variation
APPROXIMATIONS	• One-way (forward) propagation • Propagation within paraxial regions
CAPABILITIES	• Can handle propagation over smooth earth's surface • Can handle propagation under Dirchlet-and Neumann-type boundary conditions • Cauchy-type boundary conditions can also be included • PEC, non-flat terrain can be included and forward propagation effects can be calculated • Can do calculations at all ranges and altitudes while marching so wave contour maps can be plotted • Atmospheric losses may also be modeled via complex reflection coefficient
VALIDITY RANGE	• Frequency: MF, HF, VHF, UHF, MW • Distance: within LOS and a small distance in diffraction (shadow) region • Height: No physical limitation (only limited by computer memory requirements) • Medium: for any permittivity and conductivity of the ground ($\mu = \mu_0$), and any refractivity profile above ($\sigma = 0$, $\mu = \mu_0$ for air)
CANONICAL TEST CASES	• MF and HF path loss calculations over land and/or sea (compare against CCIR curves in Rec. 386) at short ranges • HF mixed-path calculations (Millington effect), sea/land transitional effects at short ranges • Multi knife-edge effects along propagation • Urban wireless scenarios (propagation within buildings)
COMPARISONS	• WAVEPROB (hybrid ray-mode formulations) • TDWP (time-domain wave propagator) technique • KNIFE (four-ray knife-edge code)
COMPUTER REQUIREMENTS	• None for today's computers for typical calculations • Large memory and high speed may be required if high altitude, high-frequency simulations are of interest
COMPLEXITY	• Easy to use • Can be improved to handle backward propagation problems • Requires analytical knowledge
PACKAGE	• Executable package and Fortran source code are included • Matlab m.file is included for off-line illustrations

Table B.4: Time-domain wave propagator (TDWP) package

CODE	TDWP
TECHNIQUE	• Based on a sliding window FDTD technique in 2D • PML-type boundaries at left, right, and top (ground at bottom)
ASSUMPTIONS	• Flat earth (i.e., 2D rectangular coordinates) • Earth's curvature may be included as refractivity variation • TE- or TM-type propagation problems
APPROXIMATIONS	• One-way (forward) propagation • Backscatter effects are handled locally
CAPABILITIES	• Can handle propagation over smooth earth's surface • Can handle propagation under Dirchlet-, Neumann- and Cauchy-type boundary conditions • PEC, non-flat terrain can be included and forward (and locally backward) propagation effects can be calculated • Can do calculations at all ranges and altitudes while marching so wave contour maps can be plotted • Parallel processing techniques may extend its capabilities • Atmospheric losses may also be modeled via complex reflection coefficient
VALIDITY RANGE	• Frequency: MF, HF, VHF, UHF, MW • Distance: Short ranges • Height: No physical limitation (only limited by computer memory requirements) • Medium: Dirichlet- or Neumann-type boundaries on the ground and any refractivity profile above ($\sigma = 0$, $\mu = \mu_0$ for air)
CANONICAL TEST CASES	• Path-loss calculations over ground at short ranges • HF mixed-path calculations (Millington effect), sea/land transitional effects • Short-pulse diffraction scenarios • Urban wireless propagation and diffraction scenarios
COMPARISONS	• WAVEPROB (hybrid ray-mode formulations) • SSPE (split-step parabolic equation) technique • KNIFE (four-ray knife-edge code)
COMPUTER REQUIREMENTS	• High for realistic calculations • Large memory and high speed may be required if high-altitude, short-pulse simulations are of interest
COMPLEXITY	• Difficult to use, one needs to trace the propagating pulse within the sliding window • A promising technique and open to improvements • Requires FDTD knowledge
PACKACE	• Executable package is included • Matlab m.file is included for off-line graphical illustrations

Table B.5: FDTD-based antenna Simulator (FDANT) package

CODE	FDANT
TECHNIQUE	• Based on 3D FDTD technique • MUR-type free-space simulators • Near fields are simulated with total field formulation for short-pulse excitation • Far fields are extrapolated by using Stratton/Chu equations directly in TD along all directions on a desired plane
ASSUMPTIONS	• The antenna under investigation is located at the origin of rectangular coordinate system
APPROXIMATIONS	• No approximations in formulations • Antennas are discretized with the staircase approach
CAPABILITIES	• Can calculate TD fields both in the near and far fields • FD variations can be obtained via off-line DFT analysis • Can visualize EM scattering directly in TD • *E*- or *H*-plane radiation patterns can be plotted via off-line • Beam steering is possible • Broad band responses are obtained via single FDTD run
VALIDITY RANGE	• Frequency: MF, HF, VHF, UHF, MW (limited by numerical dispersion) • Dimension: Antenna arrays up to$10\lambda \times 10\lambda \times 10\lambda$ dimensions • Medium: any ε, σ and μ (here $\mu = \mu_0$ is used)
CANONICAL TEST CASES	• Radiation pattern of a resonant wire antenna • 1D, 2D, and 3D antenna arrays • Mobile phone antenna (wire on a PEC box) • Mobile antenna near a human head
COMPARISONS	• Method of moments (MoM) technique • Transmission line matrix (TLM) technique • Finite element method (FEM)
COMPUTER REQUIREMENTS	• Very high for realistic calculations • Large memory and high speed may be required if short pulse simulations are of interest
COMPLEXITY	• Easy to use • Requires FDTD knowledge • Can easily be improved to handle RCS problems
PACKAGE	• Executable package is included • AN-DFT code is included for off-line DFT analysis

Table B.6: FDTD-based RCS Simulator (F-SNRCS) package

CODE	F-SNRCS
TECHNIQUE	• Based on 3D FDTD technique • MUR-type free-space simulators • Scattered field-based Maxwell's equations are used • Incident fields are injected analytically, and scattered near fields are simulated • Far fields are extrapolated by using Stratton/Chu equations directly in TD only along a desired direction
ASSUMPTIONS	• Object under investigation is located at the origin of the rectangular coordinate system
APPROXIMATIONS	• No approximations in formulations • Objects are discretized with the staircase approach
CAPABILITIES	• Mono or bi-static RCS calculations • Can calculate TD scattered fields along a given direction • FD variations can be obtained via off-line DFT analysis • Rayleigh and resonance region RCS calculations • Can visualize EM scattering directly in TD • Broadband responses are obtained via single FDTD run
VALIDITY RANGE	• Frequency: MF, HF, VHF, UHF, MW (limited by numerical dispersion) • Dimension: 3D objects with $10\lambda \times 10\lambda \times 10\lambda$ typical dimensions • Object parameters: any ε, σ and μ (here $\mu = \mu_0$ is used)
CANONICAL TEST CASES	• RCS of a PEC plate, a PEC dihedral and a PEC trihedral • RCS of resonant wires • RCS of a PEC rectangular prism
COMPARISONS	• Method of moments (MoM) technique • Fast multipole method (FMM) • Transmission line matrix (TLM) technique • Finite element method (FEM)
COMPUTER REQUIREMENTS	• Very high for realistic calculations • Large memory and high speed may be required if short-pulse simulations are of interest
COMPLEXITY	• Easy to use • Requires FDTD knowledge • Can easily be improved to handle antenna radiation problems
PACKAGE	• Executable package and Fortran source codes are included • F-SNDFT code is included for off-line DFT analysis

Table B.7: TLM-based RCS Simulator (T-SNRCS) package

CODE	T-SNRCS
TECHNIQUE	• Based on 3D TLM technique • Reflection-type free-space simulators • Incident and reflected voltages are updated in TD • Incident fields are injected analytically and scattered near fields are simulated • Far fields are extrapolated by using Stratton/Chu equations directly in TD only along a desired direction
ASSUMPTIONS	• Object under investigation is located at the origin of the rectangular coordinate system
APPROXIMATIONS	• No approximations in formulations • Objects are discretized with the staircase approach
CAPABILITIES	• Mono or bi-static RCS calculations • Can calculate TD scattered fields along a given direction • FD variations can be obtained via off-line DFT analysis • Rayleigh and resonance region RCS calculations • Can visualize EM scattering directly in TD • Broadband responses are obtained via single TLM run
VALIDITY RANGE	• Frequency: MF, HF, VHF, UHF, MW (limited by numerical dispersion) • Dimension 3D objects with $10\lambda \times 10\lambda \times 10\lambda$ typical dimensions • Object parameters: any ε, σ and μ (here $\mu = \mu_0$ is used)
CANONICAL TEST CASES	• RCS of a PEC plate, a PEC dihedral and a PEC trihedral • RCS of resonant wires • RCS of a PEC rectangular prism
COMPARISONS	• Method of moments (MoM) technique • Fast multipole method (FMM) • Finite difference time domain (FDTD) technique • Finite element method (FEM)
COMPUTER REQUIREMENTS	• Very high for realistic calculations • Large memory and high speed may be required if short pulse simulations are of interest
COMPLEXITY	• Easy to use • Requires TLM knowledge • Can easily be improved to handle antenna radiation problems
PACKAGE	• Executable package and Fortran source codes are included • T-SNDFT code is included for off-line DFT analysis

Table B.8: FDTD-based RCS Simulator (F-BIRCS) package

CODE	F-BIRCS
TECHNIQUE	• Based on 3D FDTD technique • MUR-type free-space simulators • Scattered-field based Maxwell's equations are used • Incident fields are injected analytically, and scattered near fields are simulated • Far fields are extrapolated by using Stratton/Chu equations directly in TD along all directions on a desired plane
ASSUMPTIONS	• Object under investigation is located at the origin of the rectangular coordinate system
APPROXIMATIONS	• No approximations in formulations • Objects are discretized with the staircase approach
CAPABILITIES	• Bi-static RCS calculations • Can calculate TD scattered fields along all directions • FD variations can be obtained via off-line DFT analysis • Rayleigh and resonance region RCS calculations • Can visualize EM scattering directly in TD • *E*- or *H*-plane bi-static RCS patterns can be obtained • Broadband responses are obtained via single FDTD run
VALIDITY RANGE	• Frequency: MF, HF, VHF, UHF, MW (limited by numerical dispersion) • Dimension: 3D objects with $10\lambda \times 10\lambda \times 10\lambda$ typical dimensions • Object parameters: any ε, σ and μ (here $\mu = \mu_0$ is used)
CANONICAL TEST CASES	• RCS of a PEC plate, a PEC dihedral, and a PEC trihedral • RCS of resonant wires • RCS of a PEC rectangular prism
COMPARISONS	• Method of moments (MoM) technique • Fast multipole method (FMM) • Transmission line matrix (TLM) technique • Finite element method (FEM)
COMPUTER REQUIREMENTS	• Very high for realistic calculations • Large memory and high speed may be required if short pulse simulations are of interest
COMPLEXITY	• Easy to use • Requires FDTD knowledge • Can easily be improved to handle antenna radiation problems
PACKAGE	• Executable package and Fortran source codes are included • F-BIDFT code is included for off-line DFT analysis

Table B.9: FDTD-based microstrip network simulator (F-MSTRIP) package

CODE	F_MSTRIP
TECHNIQUE	• Based on 3D FDTD technique • PML-type boundary simulators • Total near fields are simulated for short-pulse excitation
ASSUMPTIONS	• The microstrip under investigation is located longitudinally along the *z* direction
APPROXIMATIONS	• No approximations in formulations • Microstrips are discretized with the staircase approach
CAPABILITIES	• Can calculate TD fields along the strip • FD variations can be obtained via off-line DFT analysis • Can visualize EM scattering directly in TD • Incident and reflected fields can be obtained • Voltages and currents at input and output ports can be calculated in TD • *S* parameters vs. frequency can be obtained • Broadband responses are obtained via single FDTD run
VALIDITY RANGE	• Frequency: MF, HF, VHF, UHF, MW (limited by numerical dispersion) • Dimension: 3D microstrip networks with $10\lambda \times 10\lambda \times 10\lambda$ typical dimensions • Medium: any ε, σ and μ (here $\mu = \mu_0$ is used)
CANONICAL TEST CASES	• Quarter-wavelength impedance transformer • Effective dielectric constant calculation of a simple strip line • Low-pass, high-pass, band-pass filters • Input impedance of radial stubs, butterfly
COMPARISONS	• Transmission line matrix (TLM) technique • Finite element method (FEM) • Analytical approaches
COMPUTER REQUIREMENTS	• High for realistic calculations • Large memory and high speed may be required if short pulse simulations are of interest
COMPLEXITY	• Easy to use • Requires FDTD knowledge
PACKAGE	• Executable package and Fortran source codes are included • SPAR code is included for off-line DFT analysis

Table B.10: TLM-based microstrip network simulator (TLMMS) package

CODE	TLMMS
TECHNIQUE	• Based on 3D TLM technique • PML-type boundary simulators • Total near fields are simulated for short-pulse excitation
ASSUMPTIONS	• The microstrip under investigation is located longitudinally along the *z* direction
APPROXIMATIONS	• No approximations in formulations • Microstrips are discretized with the staircase approach
CAPABILITIES	• Can calculate TD fields along the strip • FD variations can be obtained via off-line DFT analysis • Can visualize EM scattering directly in TD • Incident and reflected fields can be obtained • Voltages and currents at input and output ports can be calculated in TD • *S* parameters vs. frequency can be obtained • Broadband responses are obtained via single TLM run
VALIDITY RANGE	• Frequency: MF, HF, VHF, UHF, MW (limited by numerical dispersion) • Dimension: 3D microstrip networks with $10\lambda \times 10\lambda \times 10\lambda$ typical dimensions • Medium: any ε, σ and μ (here $\mu = \mu_0$ is used)
CANONICAL TEST CASES	• Quarter-wavelength impedance transformer • Effective dielectric constant calculation of a simple strip line • Low-pass, high-pass, band-pass filters • Input impedance of radial stubs, butterfly
COMPARISONS	• Finite difference time domain (FDTD) technique • Finite element method (FEM) • Analytical approaches
COMPUTER REQUIREMENTS	• High for realistic calculations • Large memory and high speed may be required if short pulse simulations are of interest
COMPLEXITY	• Easy to use • Requires FDTD knowledge
PACKAGE	• Executable package and Fortran source codes are included • SPART code is included for off-line DFT analysis

Table B.11: FDTD-based EMC analyzer (FDSE)

CODE	FDSE
TECHNIQUE	• Based on 3D FDTD technique • PML-type boundary simulators • Total near fields are simulated for short-pulse excitation
ASSUMPTIONS	• The screen under investigation is located at the origin
APPROXIMATIONS	• No approximations in formulations • Objects are discretized with the staircase approach
CAPABILITIES	• Can calculate TD fields inside and outside of the screen • FD variations can be obtained via off-line DFT analysis • Can visualize EM scattering directly in TD • Shielding effectiveness (SE) calculations are possible • Broadband responses are obtained via single FDTD run
VALIDITY RANGE	• Frequency: MF, HF, VHF, UHF, MW (limited by numerical dispersion) • Dimension: 3D microstrip networks with $10\lambda \times 10\lambda \times 10\lambda$ typical dimensions • Medium: any ε, σ and μ (here $\mu = \mu_0$ is used)
CANONICAL TEST CASES	• A grounding screen • A PEC box with holes • A PEC box with apertures • Uniform metal box without discontinuities
COMPARISONS	• Transmission line matrix (TLM) technique • Finite element method (FEM) • Methods of moments (MoM)
COMPUTER REQUIREMENTS	• High for realistic calculations • Large memory and high speed may be required if short pulse simulations are of interest
COMPLEXITY	• Easy to use • Requires FDTD knowledge
PACKAGE	• Executable package and Fortran source codes are included • SEDFT code is included for off-line DFT analysis

Table B.12: TLM-based EMC analyzer (TLMSE)

CODE	TLMSE
TECHNIQUE	• Based on 3D TLM technique • PML-type boundary simulators • Total near fields are simulated for short-pulse excitation
ASSUMPTIONS	• The screen under investigation is located at the origin
APPROXIMATIONS	• No approximations in formulations • Objects are discretized with the staircase approach • A center-fed wire antenna is used
CAPABILITIES	• Can calculate TD fields inside and outside of the screen • FD variations can be obtained via off-line DFT analysis • Can visualize EM scattering directly in TD • Shielding effectiveness (SE) calculations are possible • Broadband responses are obtained via single FDTD run
VALIDITY RANGE	• Frequency: MF, HF, VHF, UHF, MW (limited by numerical dispersion) • Dimension: 3D microstrip networks with $10\lambda \times 10\lambda \times 10\lambda$ typical dimensions • Medium: any ε, σ and μ (here $\mu = \mu_0$ is used)
CANONICAL TEST CASES	• A grounding screen • A PEC box with holes • A PEC box with apertures • Uniform metal box without discontinuities
COMPARISONS	• Finite difference time domain (FDTD) technique • Finite element method (FEM) • Method of moments (MoM)
COMPUTER REQUIREMENTS	• High for realistic calculations • Large memory and high speed may be required if short pulse simulations are of interest
COMPLEXITY	• Easy to use • Requires TLM knowledge
PACKAGE	• Executable package is included • SEDFT code is included for off-line DFT analysis

Table B.13: FDTD-based SAR simulator (FDSAR) package

CODE	FDSAR
TECHNIQUE	• Based on 3D FDTD technique • PML-type boundary simulators • Total near fields are simulated for a GMS CW excitation (900 MHz or 1800 MHz) • Radiated and absorbed powers are calculated in TD
ASSUMPTIONS	• The discrete lossy materials under investigation are located at the origin near the antenna • Near fields are simulated inside FDTD volume • A center-fed wire antenna is used
APPROXIMATIONS	• No approximations in formulations • Objects are discretized with the staircase approach
CAPABILITIES	• Can calculate TD fields inside and outside of the screen • FD variations can be obtained via off-line DFT analysis • Can visualize EM scattering directly in TD • SAR (SAR) calculations are possible
VALIDITY RANGE	• Frequency: MF, HF, VHF, UHF, MW (limited by numerical dispersion) • Dimension: 3D objects with $10\lambda \times 10\lambda \times 10\lambda$ typical dimensions • Medium: any ε, σ and μ (here $\mu = \mu_0$ is used)
CANONICAL TEST CASES	• Power consumption (absorbed + radiated powers = input power) • Lossy rectangular materials • A discrete human head model near an antenna
COMPARISONS	• Transmission line matrix (TLM) technique • Finite element method (FEM) • Method of moments (MoM)
COMPUTER REQUIREMENTS	• High for realistic calculations • Large memory and high speed may be required if short pulse simulations are of interest
COMPLEXITY	• Easy to use • Requires FDTD knowledge • Requires care in building the discrete phantoms
PACKAGE	• Executable package and Fortran source codes are included

Table B.14: TLM-based SAR simulator (TLMSAR) package

CODE	TLMSAR
TECHNIQUE	• Based on 3D TLM technique • PML-type boundary simulators • Total near fields are simulated for a GMS CW excitation (900 MHz or 1800 MHz) • Radiated and absorbed powers are calculated in TD
ASSUMPTIONS	• The discrete lossy materials under investigation are located at the origin near the antenna • Near fields are simulated inside TLM volume
APPROXIMATIONS	• No approximations in formulations • Objects are discretized with the staircase approach
CAPABILITIES	• Can calculate TD fields inside and outside of the screen • FD variations can be obtained via off-line DFT analysis • Can visualize EM scattering directly in TD • SAR (SAR) calculations are possible
VALIDITY RANGE	• Frequency: MF, HF, VHF, UHF, MW (limited by numerical dispersion) • Dimension: 3D objects with $10\lambda \times 10\lambda \times 10\lambda$ typical dimensions • Medium: any ε, σ and μ (here $\mu = \mu_0$ is used)
CANONICAL TEST CASES	• Power consumption (absorbed + radiated powers = input power) • Lossy rectangular materials • A discrete human head model near an antenna
COMPARISONS	• Finite difference time domain (FDTD) technique • Finite element method (FEM) • Method of moments (MoM)
COMPUTER REQUIREMENTS	• High for realistic calculations • Large memory and high speed may be required if short pulse simulations are of interest
COMPLEXITY	• Easy to use • Requires FDTD knowledge • Requires care in building the discrete phantoms
PACKAGE	• Executable package is included

Index

ABOUT THE AUTHOR

Levent Sevgi was born in Akhisar, Turkey on January 1, 1958 and received his B.S.E.E., M.S.E.E. and Ph.D. degrees in Electronic Engineering from the Technical University of Istanbul (ITU) in 1982, 1984 and 1990, respectively. In 1982, he joined the faculty as a research assistant. In 1987, while working on his Ph.D. Dr. Sevgi was awarded a two-year fellowship to work with Professor L. B. Felsen at Weber Research Institute/Polytechnic University in New York. His work at Polytechnic concerned the propagation phenomena in non-homogeneous open and closed waveguides. In 1990, he returned to the Electrical and Electronics Engineering Faculty of ITU as an assistant professor and became an associate professor in 1995. He was with the Center for Defense Studies, ITUV-SAM between 1993–1997, while working on the Long Horizon (UZUN UFUK) Project studies for the Turkish Navy. After an invitation from the Scientific Research Group of Raytheon Systems Canada, Dr. Sevgi spent almost a year working on Integrated Maritime Surveillance System trials, where surface wave HF Radars are of interest. He has also worked with the TUBITAK-MRC, Information Technologies Research Institute as the chair of Electronic Systems Department during 1999 and 2000. He was also with the Center for Defense Studies, ITUV-SAM, in 2000–2002, for Vessel Traffic System installation for Turkish Straits. Since 2002, he has been with the Electronics-Communication Department of the Engineering Faculty of Doğuş University in Kadlköy, Istanbul. His research study has focused on propagation in complex environments, antenna design, radar cross section prediction, numerical methods in electromagnetics, and radar systems and recent interests include EMC/EMI modeling and measurement, bio-electromagnetics, surface-wave HF, and VHF radars. Dr. Sevgi is the author or co-author of nearly 30 journal and 50 international conference papers.

Printed and bound by CPI Group (UK) Ltd, Croydon, CR0 4YY
16/09/2026
14963447-0005